T0319392

WAVELET ANALYSIS AND TRANSIENT SIGNAL PROCESSING APPLICATIONS FOR POWER SYSTEMS

WAVELET ANALYSIS AND TRANSIENT SIGNAL PROCESSING APPLICATIONS FOR POWER SYSTEMS

Zhengyou He
Southwest Jiaotong University, China

This edition first published 2016

Published by John Wiley & Sons Singapore Pte. Ltd., 1 Fusionopolis Walk, #07-01 Solaris South Tower, Singapore 138628, under exclusive license granted by China Electric Power Press for all media and languages excluding Simplified and Traditional Chinese and throughout the world excluding Mainland China, and with non-exclusive license for electronic versions in Mainland China.

For details of our global editorial offices, for customer services and for information about how to apply for permission to reuse the copyright material in this book please see our website at www.wiley.com.

Library of Congress Cataloging-in-Publication Data

Names: He, Zhengyou, author.
Title: Wavelet analysis and transient signal processing applications for power systems / Zhengyou He.
Description: Singapore; Hoboken, NJ: John Wiley & Sons, 2016. | Includes bibliographical references and index.
Identifiers: LCCN 2016003740 | ISBN 9781118977002 (cloth) | ISBN 9781118977033 (epub)
Subjects: LCSH: Wavelets (Mathematics) | Signal processing.
Classification: LCC QA403.3 .H4 2016 | DDC 621.3101/515723–dc23
LC record available at http://lccn.loc.gov/2016003740

Cover Image: saicle/Getty

Set in 10.5/12.5pt Times by SPi Global, Pondicherry, India
Printed and bound in Singapore by Markono Print Media Pte Ltd

1 2016

Contents

Preface

The fault-generated voltage and current contain abundant fault information, such as time of fault occurrence, fault location, fault direction, and so on. This information varies according to different fault conditions. It is important to analyze the fault transient signal and extract the fault features for fast protection, fault type identification, and fault location.

As a new branch of mathematics, wavelet analysis has made many achievements in seismic exploration, atmospheric and ocean wave analysis, speech synthesis, image processing, computer vision, and data compression, among others. With the fast development of computers and the application of large-scale scientific computing, wavelet analysis has been applied to power systems, especially in transient signal analysis.

This book provides the research results from recent years and the author's many years of teaching experience in wavelet theory for engineering applications. A primary study of wavelet analysis theory applied to analyzing transient signals in power systems was carried out. This book is organized into 12 chapters: Chapter 1 briefly introduces the evolution from Fourier transform to time–frequency analysis and wavelet transform and gives a review about wavelet transform application in transient signal processing of power systems. Chapter 2 summarizes the fundamental theory of wavelet transform, including the author's many years of teaching experience. Chapter 3 introduces the wavelet singularity detection theory and noise elimination capacity of wavelet transform. Chapter 4 presents the sampling techniques in wavelet analysis of transient signals, wavelet sampling in direct wavelet transform, and pre-sampling in indirect wavelet transform. Chapter 5 provides the method for selecting wavelet bases for transient signal analysis of power systems based on large simulations and validations. The guidance principle is provided, aimed at selecting the right wavelet basis for different conditions, such as detection of a high-order singular signal, detection of weak transients of low-frequency carriers, detection of transients in narrow-band interference, data compression, de-noising of transients, and location of transients. Chapter 6 introduces the construction method of practical wavelets in power system transient signal analysis. The construction and application of a class of M-band wavelets, recursive wavelets, and optimal wavelets are presented. Chapter 7 describes the wavelet post-analysis methods put forward by the author in detail and presents several typical post-analysis methods. Chapters 8 to 11 introduce the application of wavelet analysis in different power system fields. Chapter 12 introduces the definitions and physical significances of six wavelet entropies on the basis of traditional wavelet entropy. Meanwhile, the applications of six wavelet entropies in detection and identification of power systems transient signals have been presented.

The book is written by Prof. Zhengyou He, with the guidance of the author's doctoral supervisor, Qingquan Qian. Dr Jing Zhao, Dr Xiaopeng Li, Master Haishen Zhang, Master Wen He, and Master Shu Zhang also contributed to the compilation and arrangement of this book. Some research of doctoral candidates and master's degree candidates instructed or aided by the author (e.g., Dr Zhigang Liu, Dr Linyong Wu, Dr Ling Fu, Dr Ruikun Mai, Dr Sheng Lin, Dr Jianwei Yang, and PhD candidate Yong Jia) are included in this book. The author hereby thanks the teachers and students above for their contribution.

The book is supported by the National Natural Science Foundation of China (No. 50407009, "Wavelet Entropy Theory and Its Application in Power System Fault Detection and Classification"; and No. 50877068, "Research on Multi-source Power System Fault Diagnosis Method and System Based on Information Theory"), New Century Excellent Talents in University of Ministry of Education of China (No. NCET-06-0799, "Theory and Application of Power System Fault Diagnosis System Based on Information Theory"), Sichuan Province Youth Fund Projects (No. 06ZQ026-012, "Generalization of Information Entropy and Its Application on Power Grid Fault Diagnosis"), and Research Fund for the Doctoral Program of Higher Education of China (No. 200806130004, "A Novel Method for Transmission Line Fault Location Based on Single Ended Traveling Wave Natural Frequency"). This work is also supported by the Electrical Engineering School of Southwest Jiaotong University and colleagues of the National Rail Transit Electrification and Automation Engineering Technique Research Center. The author greatly acknowledges their help.

Sincere thanks also go to the researchers and experts whose research is referred to or cited in this book.

1

Introduction

1.1 From Fourier transform to wavelet transform

1.1.1 Fourier transform [1]

Information in the time domain and frequency domain is the basic characteristic of the description of a signal $x(t)$. Information in the time domain is easy to observe, whereas information in the frequency domain is not observable unless the signal transforms. Fourier first proposed the method to get frequency domain information when he was researching an equation of heat conduction. Moreover, he suggested transforming the equation of heat conduction from the time domain to the frequency domain, which is the famous Fourier transform concept. The definition of Fourier transform of a continuous signal $x(t)$ is

$$X(\omega)=\int_{\mathbb{R}} x(t)\mathrm{e}^{-\mathrm{i}\omega t}\mathrm{d}t, \mathrm{i}=\sqrt{-1} \tag{1.1}$$

Fourier transform established the relation between the time domain and frequency domain of a signal.

With the development of computer technology, all the computing problems in science and engineering now relate to computers inextricably. A typical feature of computer computing is discretization. Fourier transform, defined in Equation (1.1), is essentially integral computation, which reflects continuous characteristics. Meanwhile, discrete sampling in applications obtains signals. Equation (1.1) needs to be discretized with high efficiency and high accuracy in order to sample information through discretization and compute Fourier transform effectively using computers. Thus, one must derive the definition of discrete Fourier transform (DFT).

Wavelet Analysis and Transient Signal Processing Applications for Power Systems, First Edition. Zhengyou He.

Let $x(t)$ be a limited signal over the interval $[-\pi,\pi]$, so Fourier transform of $x(t)$ can be simplified as

$$X(\omega)=\int_{-\pi}^{\pi} x(t)\mathrm{e}^{-i\omega t}\mathrm{d}t \tag{1.2}$$

Moreover, let the signal be equidistant sampled. The sampling number is N, the input signal in the time domain is x_k, and the output signal required in the frequency domain is X_k. To get more accurate output X_k of Equation (1.1) using sampling points of input x_k, DFT is a polynomial of best approximation $S(t)$ fitting from $x(t)$, according to x_k, and having $S(t)$ instead of $x(t)$ in Equation (1.1) to get X_k. The following paragraph briefly discusses the solution of $S(t)$ and X_k.

Using a given group of orthogonal basis ($\Phi_k=\left\{1, \mathrm{e}^{\frac{2k\pi}{N}}, \mathrm{e}^{\frac{2k\cdot 2\pi}{N}}, \ldots, \mathrm{e}^{\frac{2k\cdot(N-1)\pi}{N}\mathrm{i}}\right\}, k=0,1,2,\cdots, N-1$) to verify the inner product relations of the vectors: $<\Phi_k,\Phi_l>=N\delta_{k,l}\mathbf{I}_N$, in which $\mathbf{I}_N$ is an identity matrix for N dimensions, and $\delta_{k,l}=\begin{cases}1, k=l\\ 0, k\neq l\end{cases}$.

Let $S(t)=\frac{1}{N}\sum_{k=0}^{N-1}c_k\mathrm{e}^{ikt}$, and use orthogonal basis $\{\mathbf{\Phi_k}\}$, to solve the least-square problem:

$$\min_{c_k\in\mathbb{R},0\le k\le N-1} X(c_0,c_1,\ldots,c_k,\ldots,c_{N-1})=\min_{c_k\in\mathbb{R},0\le k\le N-1}\sum_{n=0}^{N-1}\left[x_n-S\left(\frac{2n\pi}{N}\right)\right]^2 \tag{1.3}$$

Solution of Equation (1.3):

$$c_k=\sum_{n=0}^{N-1}x_nW_N^{nk},\ k=0,1,2,\cdots,N-1;\ W_N=\mathrm{e}^{-\frac{2\pi}{N}\mathrm{i}} \tag{1.4}$$

Use definition of $S(t)$ and coefficient c_k from Equation (1.4) to approach X_k.

Its DFT is a translation by coefficient c_k of the polynomial $S(t)$ of $x(t)$:

$$X_l=\int_{-\pi}^{\pi}S(t)\mathrm{e}^{-ilt}\mathrm{d}t=\frac{1}{N\pi}\sum_{k=0}^{N-1}c_k\int_{-\pi}^{\pi}\mathrm{e}^{i(k-l)t}\mathrm{d}t=\frac{2\pi}{N}\sum_{n=0}^{N-1}x_nW_N^{nl} \tag{1.5}$$

Except for 2π, DFT is defined by Equation (1.5), where the input x_n and output X_l are the time domain information and frequency domain information of the signal, respectively.

1.1.2 *Short-time Fourier transform [1, 2]*

Although Fourier transform and DFT have been applied to signal processing, especially time–frequency analysis, extensively, the Fourier integral cannot be localized in both the time and frequency neighborhoods at one time. For example, according to Equation (1.1),

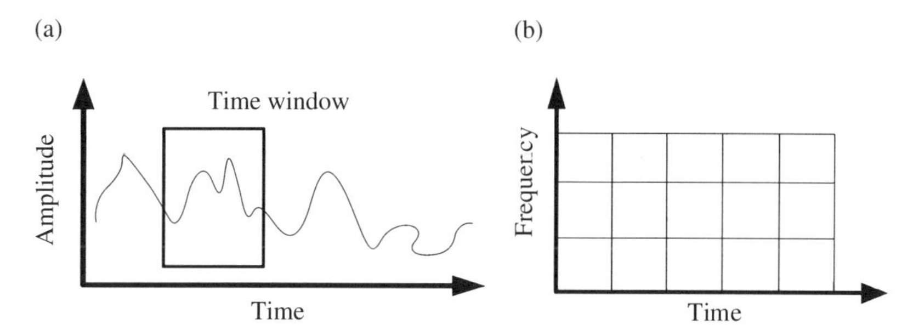

Figure 1.1 Diagram of short-time Fourier transform: (a) window in time domain, (b) time–frequency plane division

Fourier transform cannot analyze the time evolution of such spectral components and the signal cycle. That is to say, Fourier transform cannot localize in the frequency neighborhoods. Actually, the signal is transformed from the time domain to frequency domain by Fourier transform, where $x(t)\mathrm{e}^{-\mathrm{i}t\varpi}$ was added at all points in time and $\mathrm{e}^{-it\varpi}$ was the limitation of frequency. Thus, Fourier transform is not a survey of frequency domain information of a signal in a period. In contrast, for signal processing, especially nonstationary signal processing (voice, seismic signal, etc.), local frequency of signal and its period should be realized. Because standard Fourier transform has the ability of local analysis in the frequency domain rather than time domain, Dennis Gabor proposed short-time Fourier transform (STFT) in 1946. The basic idea behind STFT is dividing the signal into many time intervals to analyze each of them and determine their frequencies. Figure 1.1 is a sketch of signal analysis using Fourier transform.

We assume that we are only interested in the frequency of $x(t)$ neighboring $t = \tau$, which is the value of Equation (1.1) in a certain period I_τ. We have

$$X(\omega,\tau)=\frac{1}{|I_\tau|}\int_{I_\tau} x(t)\mathrm{e}^{-\mathrm{i}t\omega}\mathrm{d}t \tag{1.6}$$

With $|I_\tau|$ representing the length of the period I_τ, we definite the square wave function $g_\tau(t)$ as

$$g_\tau(t)=\begin{cases}\dfrac{1}{|I_\tau|}, t\in I_\tau\\ 0, others\end{cases} \tag{1.7}$$

So Equation (1.6) can be written as

$$X(\omega,\tau)=\int_{\mathbb{R}} x(t)g_\tau(t)\mathrm{e}^{-\mathrm{i}t\omega}\mathrm{d}t \tag{1.8}$$

with i representing the whole real axis. According to Equations (1.1), (1.7), and (1.8), to analyze the local frequency domain information at time τ, Equation (1.6) is the windowed function $g_\tau(t)$ of function $x(t)$. Obviously, the smaller the length $|I_\tau|$ of the window is, the more easily the function could reflect the local frequency domain information of signals.

The definition of STFT is given in Reference [2].

For a given signal $x(t) \in L^2(\mathbb{R})$, STFT can be written as

$$\begin{aligned} STFT_x(t,\omega) &= \int x(\tau) g_{t,\omega}^*(\tau) d\tau \\ &= \int x(\tau) g^*(\tau - t) e^{-i\omega\tau} d\tau = \left\langle x(\tau), g(\tau - t) e^{i\omega\tau} \right\rangle \end{aligned} \tag{1.9}$$

with

$$g_{t,\omega}(\tau) = g(\tau - t) e^{i\omega\tau} \tag{1.10}$$

and

$$\|g(\tau)\| = 1, \ \|g_{t,\omega}(\tau)\| = 1$$

The windowed function $g(\tau)$ should be symmetrical. The meaning of STFT is as follows. Add a window function $g(\tau)$ to $x(\tau)$. The time variable of $x(t)$ and $g(t)$ changes from t to τ. The windowed signal transforms to get the Fourier transform of the signal at time t. A moving time t, which means a moving central position of the windowed function, can derive the Fourier transform at different times. The set of these Fourier transforms is $STFT_x(t,\omega)$, as shown in Figure 1.2. $STFT_x(t,\omega)$ is a two-dimensional function of variables (t,ω).

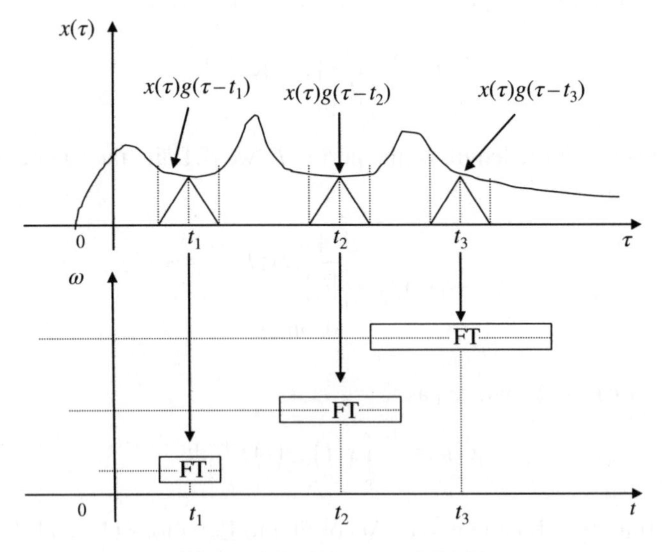

Figure 1.2 Diagram of STFT

The windowed function $g(\tau)$ is finite supported in the time domain, and $e^{i\omega\tau}$ is a line spectrum in the frequency domain. Thus, the basis function of STFT is finite supported in both the time and frequency domains. In this way, the inner product of Equation (1.9) can realize the function of time–frequency locating for $x(t)$. Fourier transform on both sides of Equation (1.10):

$$\begin{aligned} G_{t,\omega}(\upsilon) &= \int g(\tau \quad t)e^{i\omega\tau}e^{-i\upsilon\tau}d\tau \\ &= e^{-i(\upsilon-\omega)t}\int g(t')e^{-i(\upsilon-\omega)t'}dt' \\ &= G(\upsilon-\omega)e^{-i(\upsilon-\omega)t} \end{aligned} \tag{1.11}$$

in which υ is the equivalent frequency variable the same as ω.

$$\langle x(t), g_{t,\omega}(\tau)\rangle = \frac{1}{2\pi}\langle X(\upsilon), G_{t,\omega}(\upsilon)\rangle = \frac{1}{2\pi}\int_{-\infty}^{\infty} X(\upsilon)G^*(\upsilon-\omega)e^{i(\upsilon-\omega)t}d\upsilon \tag{1.12}$$

Thus,

$$STFT_x(t,\omega) = e^{-i\omega t}\frac{1}{2\pi}\int_{-\infty}^{\infty} X(\upsilon)G^*(\upsilon-\omega)e^{i\upsilon t}d\upsilon \tag{1.13}$$

The equation indicates that windowed $x(\tau)$ in the time domain (which is $g(\tau-t)$) is equal to windowed $X(\upsilon)$ in the frequency domain (which is $G(\upsilon-\omega)$).

STFT solved the problem that standard Fourier transform has the ability of local analysis only in the frequency domain rather than the time domain. However, STFT is itself flawed because a fixed windowed function whose form and shape don't change determine the resolution of STFT. To change the resolution is to choose a different windowed function. STFT is effective when you use it to analyze segment-wise stationary signals or approximate-stationary signals. But nonstationary signals, which change dramatically, demand high time resolution. When the signal is relatively flat, such as a low-frequency signal, it demands a windowed function with high-frequency resolution. STFT cannot balance the demand of both frequency resolution and time resolution.

1.1.3 Time–frequency analysis and wavelet transform

In view of the analysis in Section 1.1.2, Fourier transform can reveal a signal's frequency domain feature and the energy feature of a stationary signal. That is to say, Fourier transform is an overall transformation of signal, which is completely in the time domain or frequency domain. Therefore, it cannot reveal the law of the time-varying signal spectrum. For a nonstationary signal, because of large variations in a spectrum, analysis method is demanded to reflect local time-varying spectrum features of signal instead of simply those in the time domain or frequency domain. To make up the shortcoming of Fourier transform, a novel method that can realize time–frequency localization for signals is essential. Consequently, the time–frequency analysis was proposed, which can represent signals in the time and frequency domain at the same time.

During the development of Fourier transform theory, people realized its shortcomings (mentioned in this chapter). Therefore, in 1946, Gabor proposed representing one signal with both a time axis and frequency axis. The Gabor expansion of signal $x(t)$ is [3–5]:

$$x(t)=\sum_{m}\sum_{n}C_{m,n}g_{m,n}(t)=\sum_{m=-\infty}^{\infty}\sum_{n=-\infty}^{\infty}C_{m,n}g(t-mT)e^{in\omega t} \tag{1.14}$$

The windowed function is called a Gabor basis function or Gabor atom; $C_{m,n}$ is the expansion coefficient; and m and n, respectively, are the time domain coefficient and frequency domain coefficient.

In 1932, Wigner proposed the concept of Wigner distribution during his study of quantum mechanics. When Ville introduced the concept to the field of signal processing in 1948, the famous theory of Wigner–Ville distribution (WVD) was put forward; the equation is

$$W_x(t,\omega)=\int x\left(t+\frac{\tau}{2}\right)x^*\left(t-\frac{\tau}{2}\right)e^{-i\omega\tau}d\tau \tag{1.15}$$

Because $x(t)$ appeared twice in integration, it is also called *bilinear time–frequency distribution*. The result $W_x(t,\omega)$ is a two-dimension functional with variables t,ω, which has a number of properties and is the most widely used signal time–frequency analysis method.

In 1966, Cohen proposed the following time–frequency distribution:

$$C_x(t,\omega:g)=\frac{1}{2\pi}\iiint x\left(u+\frac{\tau}{2}\right)x^*\left(u-\frac{\tau}{2}\right)g(\theta,\tau)e^{-i(\theta t+\omega\tau-u\theta)}dud\tau d\theta \tag{1.16}$$

in which $g(\theta,\tau)$ is a weight function on two dimensions (θ,τ). If $g(\theta,\tau)=1$, Cohen distribution would turn into WVD. For different given weight functions, we would get different time–frequency distributions. Time–frequency distributions proposed around the 1980s have more than 10 types, which were called *Cohen class time–frequency distributions* or *Cohen class* for short.

Wavelet transform theory, developed in the late 1980s and early 1990s, has become a great tool for signal analysis and signal processing. Actually, wavelet transform is also another form of time–frequency analysis.

For a given signal $x(t)$, we hope to find a basic function, and we regard its dilation coefficient and shifting coefficient as a class of function:

$$\psi_{a,b}(t)=\frac{1}{\sqrt{a}}\psi\left(\frac{t-b}{a}\right) \tag{1.17}$$

$x(t)$ and inner products of the class of functions are defined as a wavelet transform of $x(t)$.

$$WT_x(a,b) = \int x(t)\psi^*_{a,b}(t)dt = \langle x(t), \psi_{a,b}(t) \rangle \tag{1.18}$$

where a is the scaling constant, b is the shift factor, and $\psi(t)$ is called basic the wavelet or mother wavelet.

According to the properties of Fourier transform, if the Fourier transform of a signal $\psi(t)$ is $\Psi(\omega)$, then signal $\psi\left(\frac{t}{a}\right)$'s Fourier transform is $a\Psi(a\omega)$. When $a > 1$, $\psi\left(\frac{t}{a}\right)$ is $\psi(t)$ stretched on a time axis. And when $a < 1$, $\psi\left(\frac{t}{a}\right)$ is $\psi(t)$ compressed on a time axis. The changes that a brought to $\Psi(\omega)$ is opposite to what a brought to $\psi(t)$. If $\psi(t)$ is regarded as a one-window function, the length of $\psi\left(\frac{t}{a}\right)$ on a time axis varies with a, which influences its form in the frequency domain ($\Psi(a\omega)$). So we can get different time domain resolution and frequency domain resolution. From the discussions that follows in this chapter, a fairly small a corresponds to the high-frequency analysis of a signal, whereas a fairly large a corresponds to the low-frequency analysis of a signal. Parameter b reflects shifting along the time axis. The result is the scaling-shifting analysis that is also a kind of time–frequency distribution. Wavelet theory has substantial content, and its theoretical basis will be discussed in detail in Chapter 2.

1.2 Application of time–frequency analysis in transient signal processing

Transient signals in power system are typical stationary signals. In traditional signal processing, Fourier transform builds up the connection between the time domain and frequency domain of signal. Therefore, Fourier transform has become the most common and most direct tool to analyze and process signals. However, for nonstationary signals, Fourier transform has numerous deficiencies, such as lack of time orientation function, fixed resolution, and so on, which limited its application for transient signal processing in power systems. For this reason, many algorithms were presented to improve traditional Fourier transform, which contributed to a higher time resolution. These kinds of transforms are called *time–frequency analysis*, which includes STFT, quadric time–frequency distribution (bilinear time–frequency distribution), wavelet transform, and so on. The window length of STFT is fixed, so that it cannot guarantee that signals within a windowed function are locally stationary for signals with multifrequency components or discontinuous transient process, and it cannot depict local spectrum characteristics of all times [6]. Therefore, numerous studies focused on window selection, and many improved algorithms, were presented, such as various modified forms of STFT.

In power systems, STFT has been applied widely. In 2000, Yu Hua and H.J. Bollen wrote a paper, “Time-frequency and time-scale domain analysis of voltage disturbances,” in which they tried to detect the spectrum feature of each frequency component when voltage sags. The method has excellent properties [7]. To select and locate the STFT window, one method combined wavelet transform with STFT for measuring transient frequency [8]; it overcame the blindness of window selection and detected the main high-order harmonic component of a transient fault signal. In Reference [9], the author presented the method for detecting the real-time amplitude of signals using STFT fundamental frequency amplitude curves when voltage sags, and the location method for when voltage sags using a high-frequency signal that was generated when the voltage started and finished sagging. This method can effectively distinguish voltage sags caused by a short-circuit fault from voltage sags caused by an induction motor's starting process. Generally, STFT reflects signals' local frequency spectrum feature well and is easily understandable. Moreover, it has an intuitive performance in the time–frequency domain. However, because of the difficulties of its window selection and quantified results extraction, the application for STFT in power system transient signal analysis needs in-depth study.

Except for STFT, bilinearity time–frequency distribution is also a typical form of time–frequency distribution. WVD connected signals in the time domain with those in the frequency domain through a correlation function, which is a real time–frequency wedding distribution. However, cross terms exist in WVD. To conquer the influence of the cross term, improved methods such as smoothing pseudo-WVD, Rihazek distribution, and Page distribution were presented, which can be integrated as the same form (called *Cohen class time–frequency distribution*). However, different signals adapt to different kernel functions, so fixed kernel distribution cannot guarantee analysis precision. Thus, one can use adaptive kernel time–frequency distribution (AOK), whose kernel function changes over time and frequency. AOK with the design criterion of an adaptive optimized kernel function was proposed by G.B. Richard and L.J. Douglas in 1993. According to the design criterion, some typical adaptive kernel distributions were presented later, such as radial Gaussian kernel function distribution, adaptive Butterworth kernel distribution, and so on. In industrial applications, Cohen class time–frequency distribution and adaptive kernel time–frequency distribution have many research results. In power system signal analysis, harmonic wave and voltage changes were detected by smoothing pseudo-WVD in Reference [10] through energy changes, which realized feature detection of power system disturbance signals. In Reference [11], a feature vector was extracted in fuzzy fields of time–frequency domain distribution, and a neural network recognized disturbance of signals. Simulation results indicated that the method accurately classifies different kinds of single disturbances. Moreover, the possibility for this method to detect cross disturbance is discussed in Reference [11]. Analysis of cross power quality disturbance using smoothing pseudo-WVD appears in Reference [12], and simulation certified that the method could reflect different frequency components of disturbance and its time of duration effectively. Adaptive time–frequency kernel function was used to analyze power quality disturbance signals, and then extract the base frequency ridge information and disturbance ridge information of the results [13]. Analysis results indicated that differences exist among kinds of disturbance ridge information, which can recognize different kinds of signals.

1.3 Wavelet transform application in transient signal processing of power systems

Wavelet theory and its engineering application have received much attention from mathematicians and engineers since the 1990s. Wavelet analysis is an important breakthrough for Fourier analysis. Wavelet transform provides an adjustable time–frequency window compared with STFT. The window automatically becomes narrower when observing a high-frequency signal, whereas it becomes wider when researching a low-frequency signal, functioning in the same way that the variable focal length does. The ability to characterize a signal's singularity is another feature of wavelet transform. The maximum modulus or Lipschitz exponent of a signal's wavelet transform under different scales can reflect the sudden change of the signal. The application of wavelet transform on a power system has been well developed in recent years, and proved its advantage and wide application prospects in the field of analyzing and processing transient signals. Its main applications include noise reduction of electric signals, data compression, power equipment fault diagnosis, analysis of disturbance signals of power quality, relay protection, and fault location.

1.3.1 Signal de-noising

Under ideal conditions, an electric signal contains only a signal of power frequency (50 Hz); however, the actual signal is a combination of the fundamental component of power frequency, the harmonic component of different orders, the transient component (when fault occurs), and the noise signal. The purpose of noise reduction is to extract the fundamental component, harmonic component, and high-frequency transient component, which could be useful when noise exists, and to remove all possible influence of noise. Wintkin was the first to propose the idea of using the spatial correlation of signals of different scales in wavelet transform to reduce noise. Then, Mallat put forward the wavelet maximum modulus method to de-noise, according to the fact that the maximum modulus of wavelet transform fully reflects the singularity of the signal, and the white noise has distinct properties from the signal under wavelet transform. The maximum modulus of wavelet transform of noise decreases with the increase of scale, whereas that of other signals grows larger with the increase of scale (for step signals, the maximum modulus remains unchanged).

In Reference [14], neutral point current data during transformer impulse tests are analyzed by a soft-threshold shrinkage de-noising method based on wavelet frame, and the conflicts between de-noising and reserving the signal's partial features were well solved. In Reference [15], the authors applied wavelet de-noising to transformer online monitoring, and the method performed well in reducing the influence of white noise. Reference [16] proposed an improved soft-threshold de-noising method based on the 3σ principle of normal distribution, and it simulated typical power quality disturbance signals. Simulation results showed that the method improved the detection accuracy of the signal. In Reference [17], performance analysis of a wavelet-based de-noising system for power quality disturbances (including sag, swell, flicker, harmonic, transient, and notch) is presented, and the db6 wavelet can be applied effectively in the temporary voltage rise or drop signals and flicker signals. Researchers proposed a wavelet packet transform method that can eliminate the influence of noise components and implement transient power quality disturbance detection and localization [18], thus providing good foundations for transient power quality disturbance monitoring under a noise environment. In general, a de-noising method based on wavelet transform can effectively reduce white noise in real time without committing any damage.

1.3.2 Power quality signal analysis

In the field of evaluating power quality, Santoso first proposed that wavelet transform is an effective method to evaluate power quality. In 1996, he applied biorthogonal wavelet to detect, locate, and recognize power quality problems resulting from various reasons. In 2000, he conducted feature extraction to power quality disturbance signals by applying Fourier transform and wavelet transform. After that, he proposed the complete neural network classifier implementation method based on wavelet transform.

The disturbance that causes power quality degradation is essentially a kind of transient disturbance signal containing a singularity point. Reference [19] describes a systematic summary of detecting and dealing with the singularity of signal by means of wavelet transform, and it discusses the inner relationship between the character of signal singularity and the modulus maximum of wavelet transform. It shows that when wavelet function is the first derivative of smooth function, the corresponding point with the modulus maximum of wavelet coefficient is the signal singularity, also called the *catastrophe point*. Based on this conclusion, many researchers have carried out further studies. In References [20, 21], a dyadic wavelet transform approach for the detection and location of the power quality disturbance is presented, and it revealed that wavelet transform is sensitive to singularities. Reference [22] combines wavelet transform and the root mean square (rms) value of the disturbance signal to extract the characteristics of disturbance signals that successfully achieved the classification of transient signals. Its idea of selecting characteristics is of great value. Reference [23] analyzed power quality disturbance signals with wavelet packet transform. The energy and entropy of terminal nodes through wavelet packet decomposition were selected as feature vectors that were put into a Fisher linear classifier to carry out classification simulation. Results showed that wavelet packet transform can effectively reflect the power quality disturbance signal, and it helps to conduct classification. Reference [24] analyzed a power quality problem caused by capacitor switching. The wavelet transform coefficient of the special layer is adopted to enlarge useful signals and mitigate disturbing ones. Then, the entropy eigenvalue extracted from coefficients can be utilized to identify the capacitor switching disturbance at low voltage.

At present, with the development of distributed generation and the application of power electronic devices and nonlinear load, harmonic pollution in power systems is becoming more and more serious. Harmonics have a big influence on the safe and economical operation of power systems and have become a severe problem. However, harmonic detection methods applied in power systems are mostly based on fast Fourier transform or its modified algorithms proposed by Cooly and Tukey, which cannot satisfy the requirement of application in obtaining the time when harmonics occurs or solving time-varying harmonic estimation problems. In 1994, Ribeiro first proposed that wavelet transform is the new tool for analyzing nonstationary harmonic distortion in power systems [25]. After that, more researchers applied wavelet transform in harmonic detection. Because frequency domain information is handled by means of frequency band and full-period sampling is not required in wavelet transform, compared with Fourier transform, wavelet transform under certain circumstances can perform a relatively stable harmonic detection result. Reference [26] describes the principles and main achievements of harmonic detection methods based on wavelet transform; the advantages and disadvantages of those methods were analyzed, and Reference [26] discussed the situation that wavelet transform cannot adequately deal with harmonic detection problems of asymmetrical systems. The frequency-mixing phenomenon that exists in almost all the harmonic detection methods based on the wavelet function family caused low accuracy, bad robustness, and poor resolution in the detection of harmonics. Reference [27] considers the aliasing phenomenon and applies Morlet continuous wavelet transformation to detect harmonics. Reference [28] proposed an aliasing compensation method to reduce aliasing in discrete wavelet transform for power system harmonic detection, which can effectively eliminate aliasing and precisely extract harmonic information. Reference [29] used wavelet packet transform for harmonic detection, and simulation results proved that the detection method for nonintegral harmonics and interharmonics based on wavelet transform performed much better compared with the method recommended by the International Electrotechnical Commission (IEC).

1.3.3 Relay protection

The principle of traditional relay protection is based on the calculation of power frequency signals and steady components, and the high-frequency components are filtered as interference signals. Frequently used methods include Fourier transform, Kalman filtering, least-square filtering, and finite impulse response (FIR) filtering, all of which are practical methods capable of dealing with stationary signals. For the new-generation relay protection that deals with the extraction and recognition of complex nonstationary transient information in power systems, if only power frequency or steady components are used to achieve protection, problems will come up, for example, in the conflict of quick action and protection reliability when applying traveling wave protection and ultra-high-speed protection. Besides, the inrush current in differential protection of transform cannot be distinguished accurately. Wavelet transform provides effective methods for the transient protection developed during recent years.

For relay protection, Reference [30] proposes the multiresolution analysis technique based on wavelet transform. Researchers applied dyadic wavelet transform to fault recognition, designed microprocessor circuits, and proved that the application of wavelet transform in relay protection was of good accuracy and anti-interference. Meanwhile, it effectively improved the sensitivity and selectivity. Reference [31] applied a Morlet complex wavelet amplitude algorithm in digital relay protection. Researchers deduced the most appropriate parameters for the measurement of fundamental frequency signals through analyzing the influence of parameters m, c in a Morlet complex wavelet.

This method has a great frequency characteristic. It can filter decaying direct-current (DC) components and applies well in digital relay protection combined with the powerful computational and signal-processing capabilities of digital signal processing (DSP). Research in References [32–34] pointed out that a traveling wave signal would undergo a sudden change when it reached the measurement point. If the signal is analyzed with wavelet transform and calculates the maximum modulus, the point where sudden change occurs to the traveling wave signal will have the characteristics of the maximum modulus. Thus, the maximum modulus of wavelet transform is connected with the sharp variation point of the traveling wave, and analysis of the traveling wave can be conducted though analysis of the maximum modulus of wavelet transform. According to the values of the maximum modulus under different scales, the startup condition of traveling wave protection can be determined. According to the polarity of the modulus maximum, the traveling wave current comparison method can be applied to accurately locate a fault area.

The application of wavelet transform in the inrush current analysis of transformers can effectively utilize a wavelet's advantage of a detection singular signal to make the protection block during an inrush current. Because a no-load switching differential current of transform is discontinuous and the current waveform changes continuously when an internal fault occurs, the wavelet coefficient of differential current of inrush current has different characteristics from when an internal fault occurs. Reference [35] used a db5 wavelet as the mother wavelet for inrush current and current waveform of internal fault. On the basis of the wavelet decomposition result at scale 1, researchers conducted analysis by wavelet packet transform and found significant differences in singularities between the inrush waveform and various fault currents. The transform result of the inrush waveform showed significant singularity, whereas the transform of the fault waveform goes gently, which can work as a criterion to distinguish an inrush current. Reference [36] conducted simulation analysis because the wavelet transform under different scales will undergo a corresponding abrupt change when a sudden change to the current occurs. It proved that the wavelet transform results of scale $S = 2^1$ and $S = 2^4$ (especially the high-frequency part of $S = 2^1$) under an inrush current will show abrupt change, whereas no such change occurs during internal or external fault. Reference [37], aimed at the waveform of fault current and symmetrical inrush current, applied the third-scale wavelet transform. The transform results indicate that the maximum modulus of fault current wavelet transform appeared to be positive and negative, and evenly spaced, whereas the adjacent maximum modulus of the inrush current has the same sign and the two maximum moduli correspond with the dead angle of the inrush current waveform. As for the identification of fault current and symmetrical inrush current, the author utilized the difference method and wavelet transform, obtained the asymmetric inrush current, and then applied wavelet transform, and the adjacent maximum modulus corresponding with the dead angles of wavelet transform is of the same symbol. The adjacent maximum modulus under the third-level wavelet transform is of a different symbol, which makes a difference with the fault state. In this method, it is not necessary to calculate the dead angle, and the inrush current can be identified qualitatively. The fuzzy set method can be applied to conduct identification. Theoretical analysis and experiment results proved the method is of great robustness to current transformer saturation.

1.3.4 Fault location

1.3.4.1 Fault line selection

In a power system's fault information analysis and maintenance system, faulty line selection is achieved mainly based on a negative sequence phase-comparison algorithm. This method is based on the fact that, only toward the positive direction of the fault line, the negative-sequence voltage phasor of the first period after the fault occurs lags behind the negative-sequence current phasor 90~120° [38]. The fault line selection problem also can be solved by comparing the maximum modulus of wavelet transform of a circuit traveling signal. In view of the amplitude, the current traveling wave signal of a nonfault line is only the transmission component of the traveling wave of a fault line. Thus, the maximum modulus of its wavelet transform is small, whereas the maximum modulus of a fault line is large. According to this principle, Reference [39] presents a fault line selection method that employs the theory of wavelet transform maximum moduli and only uses the current traveling wave. This method did not use discrete sample points, and it proved to be reliable. Based on a current transient traveling wave, Reference [40] proposed a grounding line selector and had passed field tests of a single-phase-grounding line selection in a nonsolidly grounding system.

1.3.4.2 Fault phase selection

In traditional protection methods, a phase-current fault phase selector, low-voltage fault phase selector, and impedance fault phase selector are usually applied. In digital microcomputer relay protection, software is often applied as a fault phase selector, which makes it flexible to design selectors. At present, a combination of break variables of phase current differences and sequence components is mainly used in the phase selector. Many domestic researchers have carried out research on fault phase selectors with various kinds of algorithms; and, in sum, the methods can be classified into two categories: the conventional method and modern method.

Conventional methods include break variable phase selection, steady fault component phase selection, transient fault component phase selection [41], and so on. Modern methods include the neural network method [42], phase selection based on fuzzy theory [43], and the method based on wavelet transform [44–52]. As for the phase selection method based on wavelet transform, in Reference [44], the author combined sharp transitions generated on the faulted phase with wavelet transform to select the fault phase, and the single-phase earth fault is mainly identified in this paper for the purpose of automatic reclosing. In Reference [45], fault detection and fault location indices are derived by using two-terminal synchronized measurements incorporated with a distributed line model and modal transformation theory, but the specific fault type of two phase earth faults cannot be determined. Reference [41] applied wavelet transform to reflect the transient energy of a current fault component, and by comparing the energies of three phases, the fault type and fault phase can be identified. In Reference [51], after intensive study of the energy characteristics of a current traveling wave after fault occurs, researchers extracted the aerial mode and zero mode of the current traveling wave by using wavelet transform, and proposed a novel fault phase selection method based on mode current energy of the traveling wave. According to the method, the energy of the first one-eighth period of the current traveling wave is utilized to select the fault phase. Reference [52] introduced a method that synthetically uses the magnitude and polarity characteristics of fault-induced current traveling waves to identify fault type, and designed the speedy wavelet-based algorithm of the fault type identification. The algorithm improved the reliability in identifying single- or two-phase-to-ground faults; previously proposed phase selection schemes have been based on traveling waves. In summary, researchers have carried out many studies on the application of wavelet transform on fault phase selection based on transient signals, but in practical application, these methods still need to be further analyzed and demonstrated.

1.3.4.3 Fault location

Fault location requires the time and accurate location of the fault line so that the fault can be handled and the power supply restored as soon as possible. There are various fault location methods in practical application. The size and polarity of the maximum modulus of wavelet transform can well reflect the characteristics of traveling wave signals when fault occurs (catastrophe point), so wavelet transform is applied in the traveling wave fault location [33, 53–55]. The main principle is to decompose fault signals of singularity and instantaneity, thus obtaining fault information reflected by the maximum modulus of wavelet transform under different scales. Wavelet singularity detection is applied to determine the fault time and the time interval for two traveling wave heads reaching the detection point, and in this way, the fault location can be calculated. In Reference [56], the maximum modulus is solved by dyadic wavelet transform; by use of points corresponding to the maximum modulus at both sides, the moments, when traveling waves arrive at the buses, can be determined. By means of comparing the polarity, the reflected waves coming from the faulty point and adjacent bus can be well recognized. Thus, one finds the accurate fault location. Reference [57] proposed a method to identify the fault-reflecting wave by relation of the polarity of the zero module and line module. In general, the method using the maximum modulus alone to determine fault location is immature. The key problem of selecting the proper wavelet function and decomposition level is still unsolved. Practical applications often use the maximum modulus of the decomposition level 1, which cannot properly deal with the influence of noise signals.

1.3.5 Data compression

To record the complete distortion information in a power system, the sampling frequency is required to range from 1 to 4 MHz; this results in a large amount of data. Uploading those data to a dispatch center is time-consuming and causes communication channel congestion, making it a big challenge to improve the efficiency of data compression. The data reduction techniques used in the disturbance-monitoring apparatus in a power system often utilize the overlapping method, in which existing data are constantly overlapped, or the quantization technique, in which a series of similar waveforms are taken as a whole. Both techniques save storage space but are not data compression in a general sense. The Fourier transform is an efficient data-compressing method to the stable sinusoidal signal, but it could cause data loss when failure occurs.

Wavelet transform provides a generic way for data compression. First, after wavelet decomposition, adjust the transformation coefficients whose absolute value is lower than threshold to zeros. Second, record only the positions and the values of each nonzero coefficient. Third, reconstruct the signal. Reference [58] shows implementation of the method. For signals in a real power system, the length of the compressed signal is between one-sixth and one-third of the signal, indicating that the corresponding compression ratio reaches 3–6 times and the normalized mean squared error (NMSE) of the reconstructed signal is less than 10^{-6}~10^{-5}. Transient data with disturbance were appropriately compressed in Reference [59] using a spline wavelet. And the data compression method based on the optimum wavelet packet base for power quality transient data was presented in Reference [60], which has a smaller relative error, a larger energy restitution coefficient, and minimum distortion compared to the method based on a wavelet with a similar compression ratio.

1.3.6 Fault diagnosis of power equipment

The fault diagnosis of power equipment (such as motors) aims to analyze the electromagnetic, mechanical, and sound signals in operation, and evaluate its operational state in real time. Power equipment generates steady signals (such as its vibration signal) when it operates in good condition, but once fault occurs, fault transient signals can be detected. By conducting multiresolution analysis on fault transient signals, we can obtain the mutation amplitude and mutation time of the fault signal, thus achieving fault diagnosis. Wavelet analysis in fault diagnosis of power equipment is now mainly focused on diagnosis of the generator [61–63], motor [64–66], transformer [67], and current transformer [68,69]. In addition, it is also used in fault diagnosis of circuit breakers and various types of partial discharge detection.

Reference [61] studies application of the continuous wavelet transformation in turbine generator internal fault diagnosis, including the mother function choice, the fault detection results, and their comparisons with the discrete Fourier transformation methods. Laboratory tests prove the feasibility and priority of the designed wavelet transform–based method. In Reference [62], incipient bearing fault characteristic signals obscured by the noise background are extracted by using the wavelet decomposition method. The results of theoretical and experimental research show that the spectra of intrinsic mode functions obtained by the above method clearly reveal the characteristic information in bearings, and this can be used to detect incipient faults in bearings. Reference [63] uses the multiwavelet's characteristics of multiple wavelet functions and scale functions in the fault diagnosis of a generator. It indicates that the CL4 multiwavelet based on proper pre-processing methods excels at detecting the generator's fault. Based on discussing the fast-searching algorithm of the best wavelet packet basis (BWPB) adopting Shannon entropy, Reference [64] presents a new method based on BWPB to de-noise and detect the faulted motor signal. Reference [64] demonstrates, by analysis of an actual signal example, that adopting BWPB produces a better signal de-noising effect compared with signal de-noising adopting wavelet analysis or ordinary wavelet packet analysis. Reference [65], aiming at a motor's startup procedure during which the characteristic frequency component (CFC) of a faulted rotor approaches the line frequency gradually, adopts the wavelet ridge–based method to analyze this transient procedure, and the CFC is extracted. The influence of line frequency can be effectively eliminated and the accuracy of detection can be greatly improved by using the method presented in this chapter. Reference [66] introduces a novel approach for the detection of rotor faults in asynchronous machines based on wavelet analysis of the stator phase current, and filters the measured stator phase current through a complex wavelet. Reference [67] proposes a new scheme to distinguish inrush and internal fault that has occurred in the transformer, based on a wavelet transform algorithm. The algorithm can accurately identify the typical fault type of transformers. Accurate detection of the core saturation of the current transformer (TA) helps to inhibit its bad influence on secondary equipment. Based on the lag characteristic of saturation of the current transformer when short-circuit occurs, Reference [68] proposed a method for analyzing the transient voltage of the fault bus by the wavelet transform. The saturation of TA can be distinguished based on the time difference between the time of the fault and the time of differential current. Simulation results showed the effectiveness of the proposed method. Reference [70] published a wavelet-based insulation diagnosis method and equipment dealing with nondestructive testing of insulation aging conditions of large generators. By knocking on the surface of the major insulation stator bar, sound sensors receive the reflected sound wave and the sound signals are dealt with via wavelet transform. The maximum modulus of wavelet transform under scale 1 can be obtained; thus, the aging condition of the generator's major insulation can be determined. The equipment consists of an impulse source, a signal detection circuit, a sound sensor, data-processing components, and other numerical control plate and controlling software.

Note that in the condition detection and fault diagnosis of equipment, the opinions of experts in corresponding fields must be considered properly to determine what wavelet to apply and what characteristic to extract, so that the fault characteristic can be accurately described.

1.3.7 Remaining unsolved problems in wavelet analysis of transient signals in power systems

Successful application of wavelet transform in transient analysis of power systems (such as transient signal detection, fault location, and protection of traveling waves) shows its huge application potential in related fields. Although wavelet analysis of transient signals is a promising research direction, there are still some remaining problems in this field:

1. *The mathematical foundation of wavelet transform and the mechanism of signal processing*. We need to realize that there are still some kinds of flaws in wavelets. The fundamental theory, especially the theory of complex wavelets, is still incomplete. Further study of the mechanism of signal processing based on wavelets is crucial to applying wavelets in power systems.

2. *The criteria for choosing a wavelet basis.* Wavelet basis is the core of wavelet transform. When we analyze transient signals of power systems, we need to study the time–frequency characteristics and phase characteristics of the wavelet base (including a complex wavelet base). Furthermore, we need to use the various features of transient signals in the power system to find or create a wavelet base with those similar features; thus, we can try to make the energy of the transform domain of wavelet transform more convergent, and then improve the precision when extracting the characteristics of transient signals. The study on the criteria for choosing wavelet base is of high value both theoretically and practically.
3. *The algorithm of feature extraction.* Potential ways for developing a proper algorithm contain continuous wavelet transform, discrete wavelet transform, multiresolution analysis, wavelet packet transform, and singularity detection. An effective algorithm can detect the features of transient information, harmonics, and disturbance in a power system, offering reliable pre-processing algorithms in fault diagnosis, relay protection, and harmonic analysis.
4. *Identify the operation mode, and diagnose the fault type.* With the rapid development of a power system, the operation mode and fault types turn out to be more and more complex. The potential of applying wavelet transform in transient signal analysis of a power system is to achieve complex self-adaptive relay protection by conducting an automatic identification and diagnosis process of fault types and system operation modes. Such a system requires a combination of study of wavelet transform, nonlinear theory, and the modern intellectual technology, like neural networks, fuzzy identification, and pattern recognition.
5. *The study of real time.* Applying wavelet transform in relay protection requires a high sampling rate and an enormous calculated amount; besides, it is difficult to build different startup criteria for relay protection, and this is why there is a requirement for fast wavelet transform algorithms in engineering application. Moreover, developing the special processing chips with the development of DSP is also a direction for the study of applying wavelet transform in relay protection of power systems.
6. *Other applications in power systems.* The signal analysis based on wavelet transform is of great potential in other fields of power systems. For example, the multiscale concept of wavelet transform can be applied well in the prediction of transient stability of power systems, dynamic safety analysis, and load prediction. The characteristic of "variable focus" can identify and trace the slight changes of variables, and improve the instantaneity and accuracy of the dynamic safety analysis, the load prediction, and the prediction of transient stability of power system.

The time–frequency localization is a unique feature of wavelet transform. Applying wavelet transform to power systems, especially in the direction of analysis and processing of transient signals, is crucially need because of its complicated development in power systems. This is because it will be able to offer a technical foundation to achieve the transient protection and new impending relay protection, and exploit research ideas by intelligently diagnosing faults of power system and equipment.

References

[1] Cheng Li-zhi, *The wavelet and the discrete transformation theory and its engineering practice. Tsinghua university press*, 2004. (in Chinese)

[2] Hu Guang-shu, *Modern signal processing tutorial. Tsinghua university press*, pp. 417, 2004. (in Chinese)

[3] Gabor D., Theory of communication. Part 1: The analysis of information. *Electrical Engineers – Part III: Radio and Communication Engineering, Journal of the Institution*, vol. 93, no. 26, pp. 429–441, 1946.

[4] Gabor D., Theory of communication. Part 2: The analysis of hearing. *Electrical Engineers – Part III: Radio and Communication Engineering, Journal of the Institution*, vol. 93, no. 26, pp. 442–445, 1946.

[5] Gabor D., Theory of communication. Part 3: Frequency compression and expansion. *Electrical Engineers – Part III: Radio and Communication Engineering, Journal of the Institution*, vol. 93, no. 26, pp. 445–457, 1946.

[6] Leon Cohen, Time-frequency analysis. *Simon & Schuster, Inc.*, pp. 267, 1998.

[7] Gu Y. H., Bollen M. H. J., Time-frequency and time-scale domain analysis of voltage disturbances. *IEEE Transactions on Power Delivery*, vol. 15, no. 4, pp. 1279–1284, 2000.

[8] Zhao Chengyong, He Mingfeng, Short time fourier analysis based on the special frequency component. *Automation of electric Power Systems*, vol. 28, no. 14, pp. 41–44, 2004. (in Chinese)

[9] Zhao Feng-zhan, Yang Ren-gang, Voltage sag disturbance detection based on short time fourier transform. *Proceedings of the CSEE*, vol. 27, no. 10, pp. 28–34, 2007. (in Chinese)

[10] Le Ye-qing, Xu Zheng, Application of smoothed pseudo Wigner-Ville distribution in detecting harmonics and short duration voltage variations. *RELAY*, vol. 34, no. 16, pp. 39–43, 2006. (in Chinese)

[11] Min W., Ochenkowski P., Mamishev A., Classification of power quality disturbances using time-frequency ambiguity plane and neural networks, *Power Engineering Society Summer Meeting*, 2001.

[12] Le Ye-qing, Xu Zheng, Crossed power quality disturbances detection based on the time-frequency distribution. *Proceedings of the CSU-EPSA*, vol. 19, no. 6, pp. 114–117, 2007. (in Chinese)

[13] Li Ling, Jin Guo-bin, Huang Shao-ping, Power quality disturbances detection based on ridges of time-frequency plain. *High Voltage Engineering*, vol. 34, no. 4, pp. 772–776, 2008. (in Chinese)

[14] Fu Chen-zhao, Ji Sheng-chang, Li Yan-ming, et al., Application of soft-thresholding wavelet de-noising method in the diagnosis of transformer during impulse test. *Proceedings of the CSEE*, vol. 21, no. 7, pp. 103–106, 2001. (in Chinese)

[15] Yang Ji, Li Jian, Wang You-yuan, et al., Application of wavelet denoising in partial discharge online monitoring of transformer. *Journal of Chongqing University*, vol. 27, no. 10, pp. 67–70, 2004. (in Chinese)

[16] Wang Ji-dong, Wang Cheng-shan, Power quality disturbance signals de-noising based on improved soft-threshold method. *Advanced Technology of Electrical Engineering and Energy*, vol. 25, no. 2, pp. 34–38, 2006. (in Chinese)

[17] Tan R. H. G., Ramachandaramurthy V. K., Performance analysis of wavelet based denoise system for power quality disturbances, *Proceeding of the ICIP*, 2009.

[18] Xue Hui, Yang Ren-gang, Power quality disturbance detection method using wavelet package transform based de-noising scheme. *Proceedings of the CSEE*, vol. 24, no. 3, pp. 85–90, 2004. (in Chinese)

[19] Mallat S., Hwang W. L., Singularity detection and processing with wavelets. *IEEE Transactions on Information Theory*, vol. 38, no. 2, pp. 617–643, 1992.

[20] Hu Ming, Chen Heng, Detection and location of power quality disturbances using wavelet transform modulus maxima. *Power System Technology*, vol. 25, no. 3, pp. 12–16, 2001. (in Chinese)

[21] Shang Jie, Chen Hongwei, Li Yan, Wavelet transform in the application of the transient signal detection of power quality. *RELAY*, vol. 31, no. 2, pp. 27–30, 2003. (in Chinese)

[22] Vega V., Duarte C., Ordoez G., et al., *Selecting the best wavelet function for power quality disturbances identification patterns. Proceeding of the IEEEAPPIEEC*, 2008.

[23] Wang Cheng-shan, Wang Ji-dong, Classification method of power quality disturbance based on wavelet packet decomposition. *Power System Technology*, vol. 28, no. 15, pp. 78–82, 2004. (in Chinese)

[24] Zhang Wentao, Wang Chengshan, Recognition and locating of wavelet entropy based cpacitor switching disturbances, *Automation of Electric Power Systems*, vol. 31, no. 7, pp. 71–74, 2007. (in Chinese)

[25] Ribeiro P. F., *Wavelet transform: An advanced tool for analyzing non-stationary harmonic distortions in power systems*. Bologna, Italy: *Proceeding of the IEEEICHPS*, 1994.

[26] Xia Xue, Zhou Lin, Wan Yun-jie, et al., Wavelet transform for harmonic measurement. *Electrical Measurement & Instrumentation*, vol. 43, no. 3, pp. 13–19, 2006. (in Chinese)
[27] Xue Hui, Yang Ren-gang, Morlet wavelet based detection of noninteger harmonics. *Power System Technology*, vol. 26, no. 12, pp. 41–44, 2002. (in Chinese)
[28] Du Tian-jun, Chen Guang-ju, Lei Yong, A novel method for power system harmonic detection based on wavelet transform with aliasing compensation. *Proceedings of the CSEE*, vol. 25, no. 3, pp. 54–59, 2005. (in Chinese)
[29] Barros J., Diego R. I., Application of the wavelet-packet transform to the estimation of harmonic groups in current and voltage waveforms. *IEEE Transactions on Power Delivery*, vol. 21, no. 1, pp. 533–535, 2006.
[30] Hua D Y, Ming L C, Yan S. Wavelet-based fault diagnosis scheme for power system relay protection, in *IEEE International Conference on Machine Learning and Cybernetics*, 2008, pp. 2187–2191.
[31] Zhongwei L, Weiming T. Study on application of morlet complex wavelet in digital protective relays, in *IEEE International Conference on Sustainable Power Generation and Supply*, 2009, pp. 1–4.
[32] Dong Xinzhou, He Jiali, Ge Yaozhou, Xu Bingyin, Research of fault phase selection with transient current travelling waves and wavelet transform part two: Result of simulation and test. *Automation of Electric Power Systems*, vol. 23, no. 4, pp. 20–22, 1999. (in Chinese)
[33] Ge Yaozhong, Dong Xinzhou, Dong Xingli, Travelling wave-based distance protection with fault location part one: Theory and technology. *Automation of Electric Power Systems,* vol. 26, no. 6, pp. 34–40, 2002. (in Chinese)
[34] Kong Rui-zhou, Dong Xin-zhou, BI Jian-guang, Test of fault line selector based on current travelling wave. *Automation of Electric Power Systems*, vol. 30, no. 5, pp. 63–67, 2006. (in Chinese)
[35] Lin Xiangning, Liu Pei, Cheng Shijie, A wavelet packet based new algorithm used to identify the inrush. *Proceedings of the CSEE*, vol. 19, no. 8, pp. 15–19, 1999. (in Chinese)
[36] Yang Zhonghao, Dong Xinzhou, Method for exciting inrush detection in transformer using wavelet transform. *Tsinghua Science and Technology*, vol. 42, no. 9, pp. 1184–1187, 2002. (in Chinese)
[37] Jiao Shaohua, Liu Wanshun, Liu Jianfei, Zhang Zhenhua, Yang Qixun, a new principle of discrimination between inrush current and fault current of transformer based on wavelet. *Proceedings of the CSEE*, vol. 19, no. 7, pp. 1–5, 1999. (in Chinese)
[38] Ren Jian-wen, Zhou Ming, Li Geng-yin, Research on comprehensive analysis and management system for power system fault information, *Power System Technology*, vol. 26, no. 4, pp. 38–41, 2002. (in Chinese)
[39] Shu Hong-chun, Dong Jun, Si Da-jun, Ge Yao-zhong, and Chen Xue-yun, Research on detecting fault traveling wave and selecting fault line and fault phase based on wavelet theory. *Yunnan Water Power,* vol. 18, no. 2, pp. 6–9, 2002. (in Chinese)
[40] Wang Kan, Shi Shenxing, Yang Jianming, Dong Xinzhou, Transient-traveling-wave-based grounding line selector and its field tests. *Electric Power Automation Equipment*, vol. 28, no. 6, pp. 118–121, 2008. (in Chinese)
[41] Duan Jian-dong, Zhang Bao-hui, Zhou Yi, Luo Si-bei, Ren Jin-feng, etc., Transient-based faulty phase selection in EHV transmission lines. *Proceedings of the CSEE*, vol. 26, no. 3, pp. 1–6, 2006. (in Chinese)
[42] Wang Xiaoru, Wu Sitao, Qian Qingquan, A neural network fault classifier for HV transmission line. *Automation of Electric Power Systems*, vol. 22, no. 11, pp. 28–31, 1998. (in Chinese)
[43] Zheng Tao, Liu Wanshun, Yang Qixun, Yin Zhiliang, A new phase selector based on fault sequence components and fuzzy logic theory. *Automation of Electric Power Systems*, vol. 27, no. 12, pp. 41–44, 2003. (in Chinese)
[44] Youssef O. A. S., New algorithm to phase selection based on wavelet transforms. *Power Engineering Review*, vol. 22, no. 6, pp. 60–61, 2002.
[45] Jiang J. A., Ching-Shan C, Chih-Wen L., A new protection scheme for fault detection, direction discrimination, classification, and location in transmission lines. *IEEE Transactions on Power Delivery*, vol. 18, no. 1, pp. 34–42, 2003.
[46] Zhang Bao-hui, Ha Heng-xu, Lu Zhi-lai, Study of non-unit transient-based whole-line high speed protection for EHV transmission line part 3: Fault phase selection and problems to be studied. *Electric Power Automation Equipment*, vol. 21, no. 8, pp. 1–4, 2001. (in Chinese)
[47] Duan J, Zhang B H, Ha H X, *A novel approach to faulted-phase selection using current traveling waves and wavelet analysis*, in *IEEE International Conference on Power System Technology*, 2002, pp. 1146–1150. (in Chinese)

[48] Wei Ren-yong, Liu Chun-fang, Approach of fault location and fault phase selection in UHV power transmission line based on the wavelet theory. *Proceedings of the CSEE*, vol. 20, no. 5, pp. 85–88, 2000. (in Chinese)
[49] Qin Jian, Chen Xiangxun, Zheng Jianchao, Study on dispersion of travelling wave in transmission line. *Proceedings of the CSEE*, vol. 19, no. 9, pp. 27–30, 1999. (in Chinese)
[50] Yang Wei, Hu Jun, Wu Yi-ang, Wu Jun-ji, Improved criteria of grounding fault selection based on wavelet transform. *Electric Power Automation Equipment*, vol. 25, no. 10, pp. 33–35, 2005. (in Chinese)
[51] Mai Rui-kun, He Zheng-you, Fu Ling, and Qian Qing-quan. Study on faulty phase selection based on energy of current traveling wave and wavelet transform. *Power System Technology*, vol. 31, no. 3, pp. 38–43, 2007. (in Chinese)
[52] Duan Jian-dong, Zhang Bao-hui, and Zhou Yi. Study of fault-type identification using current traveling-waves in extra-high-voltage transmission lines. *Proceedings of the CSEE*, vol. 25, no. 7, pp. 58–63, 2005. (in Chinese)
[53] Li Ze-wen, Yao Jian-gang, Zeng Xiang-jun, Chu Xiang-hui, and Deng Feng. Fault location based on traveling wave time difference in power grid. *Proceedings of the CSEE*, vol. 29, no. 4, pp. 60–64, 2009. (in Chinese)
[54] Xie Min. Networking operation practice of traveling wave based fault locating system for 220kV grid. *Electric Power Automation Equipment*, vol. 30, no. 5, pp. 136–138, 141, 2010. (in Chinese)
[55] Xu Wei-zong, and Tang Kun-ming. An improving derivation algorithm to recognize wave heads of fault generated traveling waves. *Power System Technology*, vol. 34, no. 1, pp. 198–202, 2010. (in Chinese)
[56] Zheng Zhou, Lu Yan-ping, Wang Jie, and Wu Fan. A new two-terminal traveling wave fault location method based on wavelet transform. *Power System Technology*, vol. 34, no. 1, pp. 203–207, 2010. (in Chinese)
[57] Shi Shen-xing, Dong Xin-zhou, and Zhou Shuang-xi. *Automation of Electric Power System*, vol. 30, no. 1, pp. 41–44, 59, 2006. (in Chinese)
[58] Santoso S, Powers E J, Grady W M. Power quality disturbance data compression using wavelet transform methods. *IEEE Transactions on Power Delivery*, vol. 12, no. 3, pp. 1250–1257, 59, 1997.
[59] P. K. Dash, B. K. Panigrahi, D. K. Sahoo, and G. Panda. Power quality disturbance data compression, detection, and classification using integrated spline wavelet and S-transform. *IEEE Transactions on Power Delivery*, 2003, 18(2): 595–600.
[60] Pan Wen-xia, Li Chun-lin, and Shi Lin-jun. Power quality transient disturbance compression based on optimum wavelet packet. *Proceedings of the CSU-EPSA*, vol. 17, no. 3, pp. 50–54, 2003. (in Chinese)
[61] Guan Lin, Wu Guo-pei, and Huang Wen-ying, Ren Zhen, and Zhou Hong. Study on the application of wavelet transform in fault diagnosis of electric devices. *Proceedings of the CSEE*, vol. 20, no. 10, pp. 46–49, 54, 2000. (in Chinese)
[62] Luo Zhong-hui, Xue Xiao-ning, Wang Xiao-zhen, Wu Bai-hai, and He Zhen. Study on the method of incipient motor bearing fault diagnosis based on wavelet transform and EMD. *Proceedings of the CSEE*, vol. 25, no. 14, pp. 125–129, 2005. (in Chinese)
[63] Tang Hui-ling. Pre-processing methods of CL4 multiwavelet and their applications in generator fault detection. *Relay*, vol. 35, no. 20, pp. 43–46, 59, 2007. (in Chinese)
[64] Ren Zhen, Zhang Zheng-ping, Huang Wen-ying, Guan Lin, Yang Chu-ming, and Hu Guo-sheng. *Proceedings of the CSEE*, vol. 22, no. 8, pp. 53–57, 2002. (in Chinese)
[65] Zhang Zheng-ping, Ren Zhen, Huang Wen-ying, Guan Lin, Yang Chu-ming, and He Jian-jun. *Proceedings of the CSEE*, vol. 23, no. 1, pp. 97–101, 2003. (in Chinese)
[66] Ioannis P. Tsoumas, George Georgoulas, Epaminondas D. Mitronikas, and Athanasios N. Safacas. Asynchronous Machine Rotor Fault Diagnosis Technique Using Complex Wavelets. *IEEE Transactions on Energy Conversion*, vol. 23, no. 2, pp. 444–459, 2008.
[67] Lin Xiangning, Liu Pei, and Cheng Shijie. A wavelet packet based new algorithm used to identify the inrush. *Proceedings of the CSEE*, vol. 19, no. 8, pp. 15–19, 38, 1999. (in Chinese)
[68] Li Guicun, Liu Wanshun, Jia Qingquan, CAO Feng, Li Ying, and Deng Huiqiong. A new method for detecting saturation of current transformer based on wavelet transform. *Automation of Electric Power System*, vol. 25, no. 5, pp. 36–39, 44, 2001. (in Chinese)
[69] Hong Y Y, Chang-Chian P C. Detection and correction of distorted current transformer current using wavelet transform and artificial intelligence. *Generation, Transmission & Distribution, IET*, vol. 2, no. 4, pp. 566–575, 2008.
[70] Xie Heng-kun, Ma Xiao-qin, and Xiang Tian-chun. Insulation diagnosis method and equipment for generator based on wavelet transform: China, 02139425.3. 2002-09-13.

2

The Fundamental Theory of Wavelet Transform

2.1 Continuous wavelet transform [1–4]

2.1.1 The definition of continuous wavelet transform

If $\psi \in L^2(\mathbb{R})$ satisfies the admissibility condition $\int_R \psi(t)dt = 0$, then ψ is called a basic wavelet, or the mother wavelet. The continuous wavelet transform (CWT) of signal $x(t)$ is thus defined as

$$WT_x(a,b) = a^{-1/2} \int_{-\infty}^{\infty} x(t)\psi^*\left(\frac{t-b}{a}\right)dt = \langle x(t), \psi_{a,b}(t) \rangle \tag{2.1}$$

where $a,b \in \mathbb{R}$; $a > 0$ is the scale parameter corresponding to frequency; and b is the position parameter corresponding to time. By translating, compressing, or stretching the given mother wavelet, we get a wavelet family of the same analytic function: $\psi_{a,b}(t) = a^{-1/2}\psi\left[(t-b)/a\right]$, also known as the wavelet basis.

If signal $x(t)$ and the mother wavelet $\psi(t)$ are both real functions, then $WT_x(a,b)$ is also a real function; otherwise, the mother wavelet $\psi(t)$ and $WT_x(a,b)$ are complex functions.

In wavelet basis $\psi_{a,b}(t)$, the scale parameter a plays the role of stretching or compressing the mother wavelet; the position parameter b determines the analyzed position of time, or the time domain center of the analysis of $x(t)$. Considering the scale parameter a alone, $\psi(t)$ becomes $\psi(t/a)$. When $a > 1$, if a increases, the time domain width of $\psi(t/a)$ will become wider compared with that of $\psi(t)$; conversely, when $a < 1$, if a decreases, the time domain width of $\psi(t/a)$ will become narrower. Therefore, the time domain center and the wavelet width could be determined by a and b, as depicted in Figure 2.1.

The CWT as described in Equation (2.1) could be interpreted as a kind of analysis of signal $x(t)$ that uses a set of mother wavelets with different analysis widths, which adjust themselves to change the analysis resolutions according to the needs of different frequency scopes.

Wavelet Analysis and Transient Signal Processing Applications for Power Systems, First Edition. Zhengyou He.

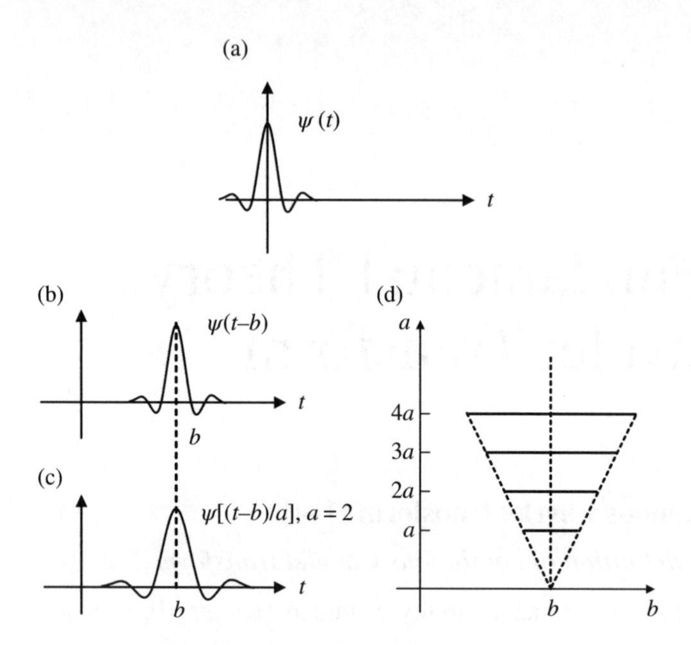

Figure 2.1 The stretched or compressed versions of the mother wavelet and the variation of the wavelet widths corresponding to different a and b parameters: (a) the mother wavelet, (b) when $b > 0, a = 1$, (c) when $a = 2$, while b is unchanged, (d) the wavelet widths with different combinations of a and b

Letting $X(\omega)$ be the Fourier transform of $x(t)$, and letting $\Psi(\omega)$ be the Fourier transform of $\psi(t)$, then the Fourier transform of $\psi_{a,b}(t)$ can be expressed by

$$\Psi_{a,b}(\omega) = a^{1/2}\Psi(a\omega)e^{-j\omega b} \tag{2.2}$$

According to Parsevals theorem, the expression of Equation (2.1) becomes

$$\begin{aligned} WT_x(a,b) &= \frac{1}{2\pi}\langle X(\omega), \Psi_{a,b}(\omega)\rangle \\ &= \frac{\sqrt{a}}{2\pi}\int_{-\infty}^{+\infty} X(\omega)\Psi^*(a\omega)e^{j\omega b}d\omega \end{aligned} \tag{2.3}$$

Here, Equation (2.3) gives the expression of wavelet transform in the frequency domain.

2.1.2 Characteristics of wavelet transform

If the time domain center and width of $\psi(t)$ are set to be t_0 and Δ_t, and the frequency domain center and band width of $\Psi(\omega)$ are ω_0 and Δ_ω, then the time domain center of $\psi\left(\frac{t}{a}\right)$ will still be t_0, while the width will turn to be $a\Delta_t$. Besides, the central frequency of $a\Psi(a\omega)$, the frequency spectrum of $\psi\left(\frac{t}{a}\right)$, will become ω_0/a, and the band width will become Δ_ω/a. Thus, the product of the time domain width and the band width of $\psi\left(\frac{t}{a}\right)$ is still $\Delta_t\Delta_\omega$, and does not changc with a. This rcflccts that the time frequency relation of wavelet transform is subject to the uncertainty principle; and, more importantly, it reveals the constant Q feature of wavelet transform defined as follows:

$$Q = \Delta_\omega / \omega_0 = \text{the band width/the center frequency} \tag{2.4}$$

where Q is the quality factor of the mother wavelet $\psi(t)$. For $\psi\left(\frac{t}{a}\right)$, the band width/the center frequency $= \dfrac{\Delta_\omega / a}{\omega_0 / a} = \Delta_\omega / \omega_0 = Q$.

Hence, no matter how $a\,(a > 0)$ changes, $\psi\left(\frac{t}{a}\right)$ will have the same quality factor as that of $\psi(t)$. Constant Q is an important feature of wavelet transform that distinguishes wavelet transform from other transforms. Figure 2.2 shows the changing rules of the bandwidth and the central frequency of $\Psi(\omega)$ and $\Psi(a\omega)$ against a.

As seen from Figure 2.2, when a decreases, the time domain observation width of $x(t)$ narrows, but the frequency domain observation width of $X(\omega)$ widens, and the central observation frequency shifts to a higher frequency (Figure 2.2c). In contrast, if a increases, the time domain observation width of $x(t)$ widens, the frequency domain observation width narrows, and the center frequency shifts to a lower frequency (Figure 2.2b). Considering the above two situations on the time–frequency plane, we can picture the relations between the time domain width, the bandwidth, the central frequency, and the time center of wavelet transform as in Figure 2.3.

In Figure 2.3, because of the constant Q feature, when a changes, the three analysis domains (the three shaded rectangles) still possess identical surface areas. From this, wavelet transform provides us with an adjustable analysis window on the time–frequency plane. The analysis window at the top has a coarse frequency resolution (long frequency range) but a fine time resolution (short time range) in the high-frequency range ($2\omega_0$); conversely, it has fine frequency resolution but coarse time resolution in the low-frequency range ($\omega_0/2$). The area of the analysis window remains unchanged with different a (see Figure 2.3). That is to say, we can adjust the time–frequency resolution according to different analysis targets. Without further explanation, we know that the high-frequency

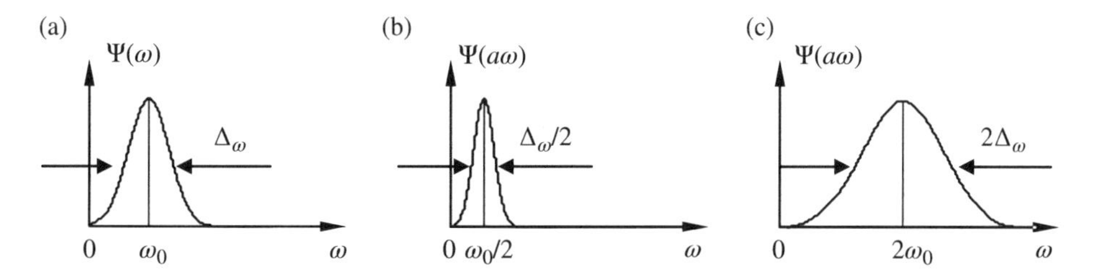

Figure 2.2 Demonstrations of $\Psi(a\omega)$ versus a. (a) $a = 1$; (b) $a = 2$; (c) $a = 1/2$

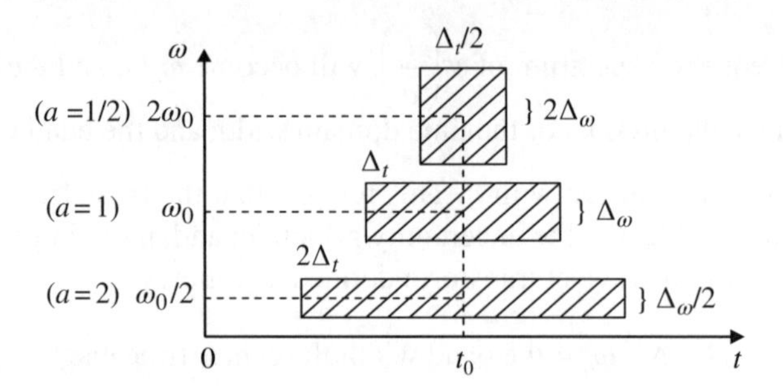

Figure 2.3 The time-frequency domains of wavelet transform with different a

components usually correspond to the fast-changing components of signals such as steep leading edges, lagging edges, and spike pulses. When analyzing these kinds of signals, the time resolution is expected to be fine enough for the short-time intervals of fast-changing signals, whereas the frequency resolution is not equally restrictive, but the analysis window shall still be positioned in the high-frequency range. Contrary to this, the low-frequency signal usually corresponds to the slow-changing signals. Consequently, when analyzing these kinds of signals, the frequency resolution shall be fine enough and the time resolution is not highly restricted. At the same time, the central analysis frequency shall shift to lower frequencies. Obviously, the features of wavelet transform indicate the automatic satisfaction of these practical requirements.

2.2 The discretization of continuous wavelet transform [1, 5]

If wavelet $\psi \in L^2(\mathbb{R})$ satisfies the perfect reconstruction condition $C_\psi = \int_R \frac{|\Psi(\omega)|^2}{\omega} d\omega < +\infty$, then the signal can be recovered by reconstructing its wavelet transform:

$$x(t) = C_\psi^{-1} \iint_{RR} \psi_{a,b}(t) WT_x(a,b) db \frac{da}{a^2} \tag{2.5}$$

In fact, the scale parameter and the position parameter must be discretized because the information of the two-dimensional $WT_x(a,b)$ is redundant after performing CWT on the one-dimensional signal $x(t)$. Besides, the fact that the transform needs to be realized by computers also strengthens the need for discretization. Finally, we can get the discrete wavelet transform by discretizing the scale parameter a and the position parameter b.

2.2.1 *The discretization of the scale parameter and the dyadic wavelet*

When performing CWT, the scale parameter ($a = 2^j, j \in \mathbb{Z}$) is only discretized by dyadic discretization, whereas the position parameter remains continuous; hence, we get the semidiscrete wavelet transform:

$$WT_x\left(2^j,b\right)=2^{-j/2}\int_{-\infty}^{\infty}x(t)\overline{\psi}\left[2^{-j}(t-b)\right]dt \tag{2.6}$$

This wavelet transform is called a dyadic wavelet transform, and the corresponding wavelet is named the dyadic wavelet, which is a kind of admissible wavelet. The dyadic wavelet shall satisfy the following stability condition: for $\psi \in L^2(\mathbb{R})$, there exist constant A and B, $0 < A \le B < \infty$, so that

$$A \le \sum_{j=-\infty}^{j=\infty}\left|\Psi\left(2^{-j}\omega\right)\right|^2 \le B \tag{2.7}$$

The stability condition guarantees the existence of the Fourier transform of the dual wavelet. In other words, we can always find a stable dual wavelet that could recover $x(t)$.

2.2.2 *The discretization of the position parameter and the frame wavelet*

Letting $a = a_0^j, j \in \mathbb{Z}$, we can realize the discretization of a. If $j=0$, then $\psi_{j,b}(t)=\psi(t-b)$. The simplest way to discretize b is to sample b uniformly; for example, let $b = kb_0$, and the value of b_0 shall guarantee the recovery of $x(t)$ from $WT_x(j,k)$. When $j \ne 0$, if a changes from a_0^{j-1} to a_0^j (a is expanded a_0 times), then the sampling interval of b will be a_0 times expanded correspondingly. This reflects that, when the scale parameter a is sequentially set to be $a_0, a_0^1, a_0^2, \cdots$, the sampling of b will be $a_0b_0, a_0^1b_0, a_0^2b_0, \cdots$, respectively. In this way, after the discretization of a and b, the result will be

$$\begin{aligned}\psi_{j,k}(t) &= a_0^{-j/2}\psi\left[a_0^{-j}\left(t-ka_0^jb_0\right)\right]\\ &= a_0^{-j/2}\psi\left(a_0^{-j}t-kb_0\right) \qquad j,k\in\mathbb{Z}\end{aligned} \tag{2.8}$$

For a given signal $x(t)$, the CWT of Equation (2.1) can be modified into the form of a wavelet transform computed on a discrete grid, and it is given as

$$WT_x(j,k)=\int x(t)\psi^*_{j,k}(t)dt \tag{2.9}$$

Equation (2.9) is defined as a discrete wavelet transform (DWT).

If the CWT of Equation (2.1) is replaced by the DWT of Equation (2.9), the following questions will emerge:

1. Is it completely feasible to represent $x(t)$ with $WT_x(j,k)$? In other words, can the values of $WT_x(j,k)$ facilitate robust reconstruction of signal $x(t)$?
2. Is it possible for an arbitrary signal $x(t)$ to be expressed as the weighted sum of the basic units of $\psi_{j,k}(t)$?

The answers to the two questions are consistent, and they are founded on the mathematical basis of frame theory. The definition of a frame wavelet is given as follows:

The set $\{\psi_{b_0,j,k}(t) \mid j \in \mathbb{Z}, k \in \mathbb{Z}\}$ is said to be a wavelet frame (the corresponding wavelet is known as the frame wavelet) if the wavelet family $\psi_{j,k}(t) = 2^{-j/2}\psi\left[2^{-j}(t - kb_0)\right]$ of scaled and positioned mother wavelets $\psi(t)$ satisfies

$$A\|x\|^2 \leq \sum_{j,k \in Z} \left|\left\langle x, \psi_{b_0,j,k} \right\rangle\right|^2 \leq B\|x\|^2, 0 < A \leq B < +\infty \tag{2.10}$$

where A, B are the frame bonds of wavelet frames. If $A = B$, then the frame is called a compact frame. Under the frame condition, we can define the frame wavelet transform and its inversion formula (reconstruction formula). The wavelet frame is the generalization of orthonormal wavelet bases, and it is the simplest and commonest form of a discretized version of CWT, which guarantees the existence of its inversion formula. A wavelet frame is not necessarily a linearly independent family.

2.2.3 Riesz wavelet

For $\psi \in L^2(\mathbb{R})$, if its linear span $\{\psi_{b_0,j,k}, j,k \in \mathbb{Z}\}$ is dense in $L^2(\mathbb{R})$, and there exist constants $A, B, 0 < A \leq B < +\infty$, so that

$$A\sum_j \sum_k \left|c_{j,k}\right|^2 \leq \| \sum_j \sum_k c_{j,k}\psi_{b_0,j,k}(t) \|^2 \leq B\sum_j \sum_k \left|c_{j,k}\right|^2 \tag{2.11}$$

is valid for all $\{c_{j,k}\} \in l^2(\mathbb{Z}^2)$, then ψ is defined as a Riesz basis for $\{\psi_{b_0,j,k}, j,k \in \mathbb{Z}\}$. A, B are called the upper bound and the lower bound of Riesz bases, respectively.

Because a generic wavelet frame is not a linearly independent family, when the signal is expanded according to the frame, there is a high redundancy in wavelet coefficients. To reduce the redundancy, we shall find a wavelet form that could generate Riesz bases in $L^2(\mathbb{R})$. However, it is not guaranteed that for all Riesz functions, we can find dual Riesz bases of wavelets that are dilated or positioned from one single function. (A dual Riesz basis is also called a Riesz wavelet.) We can prove that, if ψ generates the wavelet frame $\{\psi_l\}$ in $L^2(\mathbb{R})$, then ψ is a dyadic wavelet. Moreover, if $\{\psi_l\}$ is linearly independent, then ψ is a Riesz basis in $L^2(\mathbb{R})$. We can classify wavelets into three groups according to the definition of Riesz basis (Figure 2.4):

1. *Orthogonal wavelet*: Wavelets that satisfy $\left\langle \psi_{j,k}, \psi_{l,m} \right\rangle = \delta_{jl}\delta_{km}$.
2. *Semiorthogonal wavelet*: Wavelets that satisfy $\left\langle \psi_{j,k}, \psi_{l,m} \right\rangle = 0$ when $j \neq l$.
3. *Biorthogonal wavelet*: Wavelets that satisfy $\left\langle \psi_{j,k}, \bar{\psi}_{l,m} \right\rangle = \delta_{j,k}\delta_{l,m}$ for $\psi_{j,k}$ and its dual function $\bar{\psi}_{l,m}$.

Apparently, for the three wavelet groups, (i) the dual function of an orthogonal wavelet is itself; (ii) the dual function of a semiorthogonal wavelet can be obtained with the

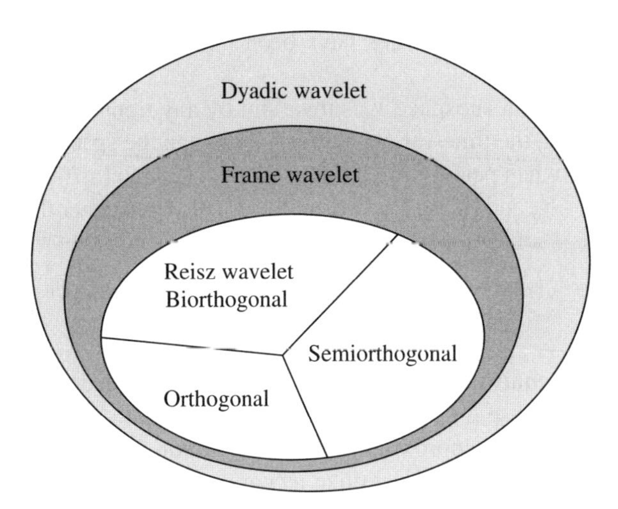

Figure 2.4 The classification of wavelets

orthogonalization method; and (iii) it is comparatively complicated to obtain the dual function of a biorthogonal wavelet (the method of which is omitted here for brevity and can be seen in related literature).

2.3 Multiresolution analysis and the Mallat algorithm

2.3.1 Multiresolution analysis [2]

The square-integrable function could be seen as the gradual approach of a function, that is to say, it is the smoothing result of $x(t)$ with a low-pass smoothing function $\phi(t)$, which gradually changes its scales, so that the signal being analyzed could be approximated in different resolutions. This is also the original intention of multiresolution analysis.

Assuming that $\{V_j\}, j \in \mathbb{Z}$ is a collection of closed subspaces in $L^2(\mathbb{R})$, if the subspaces comply with the six properties that follow, then $\{V_j\}, j \in \mathbb{Z}$ is said to be a multiresolution approximation. The six properties are:

1. *Translation invariance*: $\forall (j,k) \in \mathbb{Z}^2$, if $x(t) \in V_j$, then $x(t - 2^j k) \in V_j$;
2. *Monotonicity*: $\forall j \in \mathbb{Z}, V_j \supset V_{j+1}$, or $\cdots V_0 \supset V_1 \supset V_2 \cdots V_j \supset V_{j+1} \cdots$;
3. *Scaling regularity*: $\forall j \in \mathbb{Z}$, if $x(t) \in V_j$, then $x\left(\frac{t}{2}\right) \in V_{j+1}$;
4. *Approximability*: $\lim_{j \to \infty} V_j = \bigcap_{j=-\infty}^{\infty} V_j = \{0\}$;
5. *Approximability*: $\lim_{j \to \infty} V_j = Closure\left(\bigcup_{j=-\infty}^{\infty} V_j\right) = L^2(\mathbb{R})$;
6. *Existence of Riesz basis*: There exists a basic function $\theta(t)$ so that $\{\theta(t-k)\}; k \in \mathbb{Z}$ is a Riesz basis in V_0.

Explanations to these six properties have been given in Reference [2], and they are restated here.

Property 1 indicates that subspace V_j is invariant by any translation proportional to the scale 2^j; in other words, the time variance does not change the space the function belongs to. As we said before when performing dyadic discretization to (a, b), if we let $a = 2^j$, then b shall satisfy $b = 2^j kb_0$, and b_0 shall be normalized to 1. Consequently, we have

$$\psi_{a,b}(t) = \frac{1}{\sqrt{a}}\psi\left(\frac{t-b}{a}\right) = a^{-j/2}\psi\left(2^{-j}t - k\right) = \psi_{j,k}(t) \tag{2.12}$$

Thus, property 1 is equivalent to the following statement: for $\forall j \in \mathbb{Z}$, if $x(t) \in V_j$, then $x(t-k) \in V_j$. This is because for $\forall j \in \mathbb{Z}$, there always exists $2^j \in \mathbb{Z}$.

Property 2 implies that an approximation at a resolution 2^{-j} contains all the necessary information to compute an approximation at a coarser resolution 2^{-j-1}.

Property 3 is a direct derivation of Property 2. In V_{j+1}, the function is twice stretched, and the resolution is degraded to 2^{-j-1}; then, $x\left(\frac{t}{2}\right)$ shall belong to V_{j+1}.

Property 4 indicates that when $j \to \infty$, $2^{-j} \to 0$, all the information about $x(t)$ will be lost.

$$\lim_{j\to\infty} P_j x(t) = 0 \tag{2.13}$$

From the perspective of space, the intersection of $V_j (j = -\infty \sim +\infty)$ would be a null space.

Property 5 is the contrary side of Property 4, when $j \to -\infty$ and $2^{-j} \to \infty$. Then, the signal approximation converges to the original signal:

$$\lim_{j\to-\infty} \left|P_j x(t) - x(t)\right| = 0 \tag{2.14}$$

From the perspective of space, the union of $V_j (j = -\infty \sim +\infty)$ would converge to the whole $L^2(\mathbb{R})$ space.

Property 6 describes the existence of the Riesz basis in V_0. The definition of the Riesz basis is as follows. Assume V_0 is a Hilbert space (note: a Hilbert space is a space with limited energy), where $\{\theta_k = \theta(t-k)\}, k \in \mathbb{Z}$ is a vector in V_0. The number of $\{\theta_k = \theta(t-k)\}, k \in \mathbb{Z}$ is consistent with the dimension of V_0. Obviously, an arbitrary element x in V_0 could be expressed as the linear combination of θ_k:

$$x(t) = \sum_{k=-\infty}^{\infty} c_k \theta(t-k) \tag{2.15}$$

If elements in $\{\theta_k = \theta(t-k)\}, k \in \mathbb{Z}$ are linearly independent, and there exists a constant number $0 < A \le B < \infty$, so that

$$A\|x\|^2 \leq \sum_{k=-\infty}^{\infty} |c_k|^2 \leq B\|x\|^2 \tag{2.16}$$

then $\theta(t-k), k \in \mathbb{Z}$ is a Riesz basis in V_0.

2.3.2 The Mallat fast algorithm based on multiresolution analysis [2, 4]

Given that $\phi_{j,k}(t)$ is the orthonormal basis in V_j, and $\psi_{j,k}(t)$ is the orthonormal basis in W_j, at the same time, $V_j \perp W_j$, $V_{j-1} = V_j \oplus W_j$. This relation implies that there must be some connections between the scales' functions, and between the scale function and the wavelet function under adjacent scales (e.g., j and $j-1$).

Because $\phi_{j,0}(t) = 2^{-j/2}\phi(2^{-j}t) \in V_j$, in which V_j is included in V_{j-1}, we can assume that $\phi_{j,0}(t)$ is one element of V_{j-1}. Thus, $\phi_{j,0}(t)$ can be expressed as the linear combination of an orthogonal basis in V_{j-1}, given as

$$\phi_{j,0}(t) = \sum_{k=-\infty}^{\infty} h_0(k)\phi_{j-1,k}(t) \tag{2.17}$$

where $h_0(k)$ is the weighted coefficient, which is a discrete sequence. Further expansion of the equation above will result in the following expressions:

$$2^{-j/2}\phi(2^{-j}t) = 2^{-(j-1)/2}\sum_{k=-\infty}^{\infty} h_0(k)\phi(2^{-(j-1)}t-k) \tag{2.18}$$

which is equivalent to

$$\phi\left(\frac{t}{2^j}\right) = \sqrt{2}\sum_{k=-\infty}^{\infty} h_0(k)\phi\left(\frac{t}{2^{j-1}}-k\right) \tag{2.19}$$

In the same way, W_j is also included in V_{j-1}; thus, $\psi_{j,0}(t)$ in W_j could also be expressed as a linear combination of orthogonal basis $\varphi_{j-1,k}(t)$ in V_{j-1}, which is

$$\psi\left(\frac{t}{2^j}\right) = \sqrt{2}\sum_{k=-\infty}^{\infty} h_1(k)\phi\left(\frac{t}{2^{j-1}}-k\right) \tag{2.20}$$

where $h_1(k)$ is, again, the weighted coefficient. Equations (2.19) and (2.20) are called *two-scale difference equations*, which reveals the mutual relationship between the scale function and the wavelet function in multiresolution analysis. This relationship exists between any two adjacent scales. For example, for $j=1$, there are

$$\phi\left(\frac{t}{2}\right) = \sqrt{2}\sum_{k=-\infty}^{\infty} h_0(k)\phi(t-k) \tag{2.21}$$

$$\psi\left(\frac{t}{2}\right) = \sqrt{2}\sum_{k=-\infty}^{\infty} h_1(k)\phi(t-k) \tag{2.22}$$

Equations (2.21) and (2.22) are further equivalent to

$$\phi(t)=\sqrt{2}\sum_{k=-\infty}^{\infty}h_0(k)\phi(2t-k) \tag{2.23}$$

$$\psi(t)=\sqrt{2}\sum_{k=-\infty}^{\infty}h_1(k)\phi(2t-k) \tag{2.24}$$

Thus, the two-scale difference equation is an important property of wavelet function and scale function in multiresolution analysis.

Because of the orthogonality of $\phi_{j,k}$ and $\psi_{j,k}$, $h_0(k)$ and $h_1(k)$ could be obtained from Equation 2.25.

$$h_0(k)=\langle\phi_{j,0}(t),\phi_{j-1,k}(t)\rangle=\frac{1}{\sqrt{2^j 2^{j-1}}}\int\phi\left(\frac{t}{2^j}\right)\phi^*\left(\frac{t}{2^{j-1}}-k\right)dt \tag{2.25}$$

Letting $\frac{t}{2^{j-1}}=t'$, there is

$$h_0(k)=\frac{1}{\sqrt{2}}\int\phi\left(\frac{t'}{2}\right)\phi^*(t'-k)dt' \tag{2.26}$$

or

$$h_0(k)=\langle\phi_{1,0}(t),\phi_{0,k}(t)\rangle \tag{2.27}$$

In the same way, there is

$$h_1(k)=\langle\psi_{1,0}(t),\phi_{0,k}(t)\rangle \tag{2.28}$$

Equations (2.27) and (2.28) reflect an important feature: that $h_0(k)$ and $h_1(k)$ are irrelevant to scale j. This property is applicable between any two adjacent scales. That is to say, $h_0(k)$ and $h_1(k)$, obtained from the two-scale difference equations when $j=0$ and $j=1$, could be applicable to any two-scale difference equations with any integer j. Hence, we may find out that $h_0(k)$ and $h_1(k)$ are similar to a two-channel filter bank: $h_0(k)$ corresponds to the low-pass filter $H_0(z)$, and $h_1(k)$ correspond to the high-pass filter $H_1(z)$, where at each scale $H_0(z)$ and $H_1(z)$ remain unchanged. If the assumption is valid, we can connect the wavelet transform with filter banks. In fact, the assumption does hold true.

Assuming $\phi\in L^2(\mathbb{R})$, $\psi\in L^2(\mathbb{R})$ are the scale function and the wavelet function in multiresolution analysis, respectively, and assuming $h_0(k)$, $h_1(k)$ are coefficients of the filter bank that satisfy two-scale difference equations (Equations (2.21) and (2.22)), then

$$|H_0(\omega)|^2+|H_0(\omega+\pi)|^2=2 \tag{2.29}$$

$$|H_1(\omega)|^2+|H_1(\omega+\pi)|^2=2 \tag{2.30}$$

$$H_0(\omega)H_1^*(\omega)+H_0(\omega+\pi)H_1^*(\omega+\pi)=0 \tag{2.31}$$

On the basis of the multiresolution analysis above, the wavelet transform and its inverse transform of signals for filter banks can be described here: letting $a_j(k)$, $d_j(k)$ be the approximate coefficients of multiresolution analysis, and letting $h_0(k)$, $h_1(k)$ be two filter banks that satisfy the two-scale difference equations (Equations (2.21) and (2.22)), then the recursive relations in terms of $a_j(k)$, $d_j(k)$ are

$$a_{j+1}(k) = \sum_{n=-\infty}^{\infty} a_j(n) h_0(n-2k) = a_j(k) * \bar{h}_0(2k) \tag{2.32}$$

$$d_{j+1}(k) = \sum_{n=-\infty}^{\infty} a_j(n) h_1(n-2k) = a_j(k) * \bar{h}_1(2k) \tag{2.33}$$

where $\bar{h}(k) = h(-k)$.

Now, let us think of the mathematical meaning of Equations (2.32) and (2.33): assume $j=0$, $a_0(k)$ is the coefficient of the decomposition of $x(t)$ in V_0 with orthogonal basis $\phi(t-k)$. It is the discrete smooth approximation of $x(t)$ in V_0. Let $a_0(k)$ pass through a filter, and get the discrete smooth approximation of $x(t)$ in V_1. The filter obtained through $\bar{h}_0(2k)$ in Equation (2.32) represents the downsampling by 2, as shown in Figure 2.5a. The input and output relations of Equation (2.33) are shown in Figure 2.5b. Assuming we start the decomposition from $j=0$, then the combination of Figure 2.5a and Figure 2.5b results in Figure 2.5c.

Letting j gradually increase from 0, the multiresolution analysis could be gradually achieved, as shown in Figure 2.6. This block diagram exhibits the procedure of the Mallat algorithm, or the fast algorithm of wavelet transform.

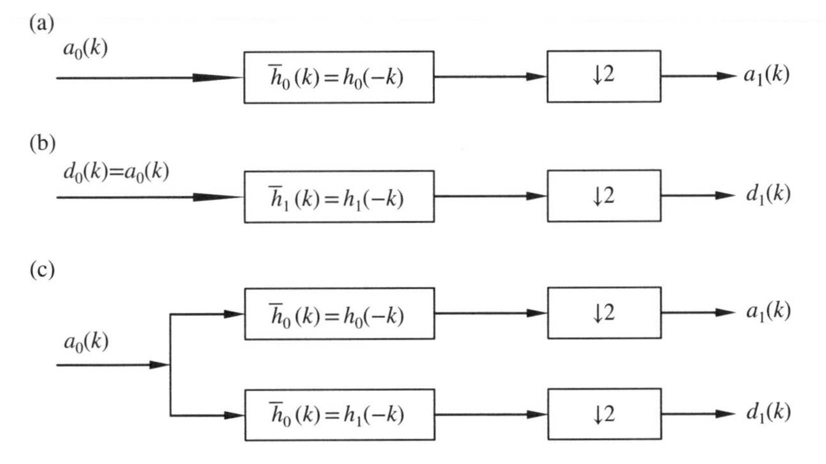

Figure 2.5 The network structure of Equations (2.32) and (2.33): (a) low-pass decomposition, (b) high-pass decomposition, (c) the combination of the above

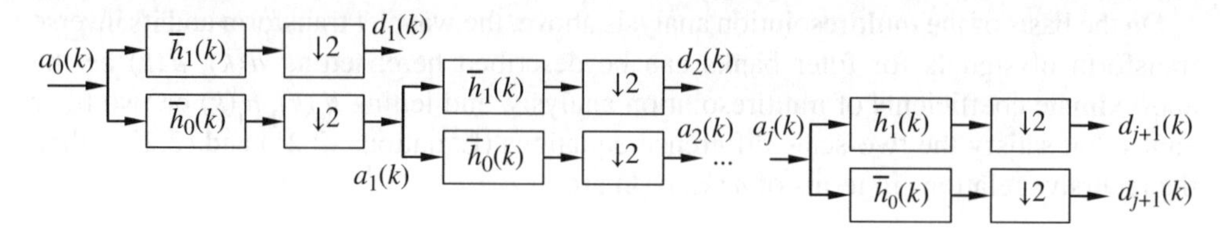

Figure 2.6 The implementation of filter banks based on multiresolution

From Figure 2.6, we can find out the basic idea of Mallat multiresolution analysis.

1. From the perspective of filter banks, if the frequency band of $h_0(-k)$ ranges in $0 \sim \pi/2$, and the frequency band of $h_1(-k)$ ranges in $\pi/2 \sim \pi$, then the frequency band of $a_0(k)$ ranges in $0 \sim \pi$, the frequency band of $a_1(k)$ ranges in $0 \sim \pi/2$, and the frequency band of $d_1(k)$ ranges in $\pi/2 \sim \pi$. Therefore, the respective division of frequency band is achieved. This kind of division not only guarantees the constant Q property but also guarantees the invariance property of $H_0(z)$ and $H_1(z)$ in each scale.
2. If we note the frequency band of $a_0(k)$ lies in space V_0, $a_1(k)$ is included in V_1 space, and $d_1(k)$ is included in W_1, we have $V_0 = V_1 \oplus W_1$, $V_1 \perp W_1$, $V_{j-1} = V_j \oplus W_j$, $V_j \perp W_j$, $j = 1 \sim +\infty$. At the same time, $V_0 = W_1 \oplus W_2 \oplus \cdots \oplus W_j \oplus V_j, j \in \mathbb{Z}$.

   When $j \to \infty$, the space (or frequency band) that $a_j(k)$, $d_j(k)$ occupy tend to infinitesimal; thus, there must be $\{V_j\}_{j\to\infty} = (0)$. At this time, the resolution is the coarsest, and $P_j x(t)\big|_{j\to\infty} = 0$. This is the main idea of multiresolution analysis.
3. Using the structure in Figure 2.6 to implement multiresolution analysis is equivalent to the use of multistep linear convolution in wavelet transform of discrete signals. If the coefficients of $h_0(n)$, $h_1(n)$ are not too long, the convolution can be implemented directly in the time domain; otherwise, it can be implemented with discrete Fourier transform or fast Fourier transform.

If $a_{j+1}(k)$, $d_{j+1}(k)$ are obtained from Equations (2.32) and (2.33), then we can reconstruct $a_j(k)$ with Equation 2.34, which is called the inverse wavelet transform:

$$a_j(k) = \sum_{n=-\infty}^{\infty} a_{j+1}(k) h_0(k-2n) + \sum_{n=-\infty}^{\infty} d_{j+1}(k) h_1(k-2n) \tag{2.34}$$

The corresponding network structure of Equation (2.34) is depicted in Figure 2.7. If j decreases from j to 0, then the whole reconstruction process is described in Figure 2.7b, which is exactly the inverse process of that in Figure 2.6. However, in the decomposition process, h_0 and h_1 shall exchange their positions; in contrast, in the reconstruction process, h_0 and h_1 do not exchange positions. The decomposition process involves downsampling by 2, whereas the reconstruction process involves two-dimensional interpolation As seen in Figure 2.7, in the multiresolution decomposition and reconstruction of signals with orthogonal wavelets, the applied filters are identical for both decomposition and reconstruction processes, which are $H_0(z)$ and $H_1(z)$.

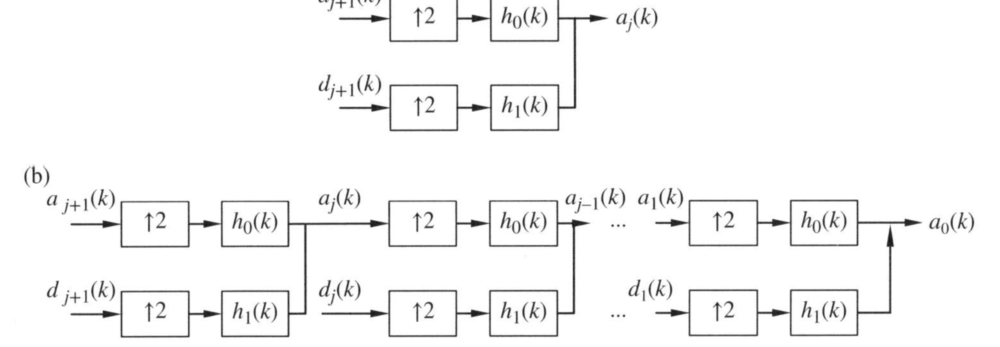

Figure 2.7 Inverse wavelet transform: (a) the jth step, (b) the process in terms of j decreasing from j to 0

2.3.3 The consistency of multiresolution analysis and dyadic orthogonal wavelet

As we know, in dyadic orthogonal wavelet transform, the family of the dyadic wavelet $\{\psi_{m,n}(t)\}$ composes an orthogonal basis in space $L^2(\mathbb{R})$. At the same time, multiresolution analysis is described as a subspace sequence of $L^2(\mathbb{R})$ that satisfies the above conditions. Here, you may wonder, is there any theoretical correspondence between multiresolution analysis and dyadic orthogonal wavelets? The answer is yes. This relation is one of the important factors that make wavelet transform theory so applicable and popular.

1. The standard orthogonal basis in $\{V_j, j \in \mathbb{Z}\}$ space

   The existence of the Riesz base: for an arbitrary scale function $\phi(t) \in V_0$, which makes $\{\phi(t-k), k \in \mathbb{Z}\}$ the Riesz base in V_0 (i.e., for arbitrary $\phi(t) \in V_0$), there exists the only sequence $\{a_k\} \in I^2$ so that $\phi(t) = \sum_k a_k \phi(t-k)$.

   From $\Phi(\omega) = \dfrac{\Phi(\omega)}{\left[\sum_{l \in Z} \left|\Phi(\omega + 2l\pi)\right|^2\right]^{1/2}}$, we can obtain $\{\phi(t-k), k \in \mathbb{Z}\}$, which thus forms the standard orthogonal basis in V_0.

   $\phi(k)$ is called the scale function; hence, we can get the standard orthogonal basis in V_j space. Given the above information, we can build an orthogonal basis from the Riesz base.
2. The standard orthogonal basis in the orthogonal complementary space of $\{V_j, j \in \mathbb{Z}\}$ space

   From the definition of multiresolution analysis, we know that $V_j \subset V_{j-1}$. In addition, from the orthogonal decomposition theory of functional analysis, we know that

$$V_j = W_{j+1} \oplus V_{j+1}, j \in \mathbb{Z} \tag{2.35}$$

   where W_j is the orthogonal complementary space of V_j.

If $\psi(x) \in W_0$, then $\{\psi(t-n), n \in \mathbb{Z}\}$ form the standard orthogonal basis in W_0 space. The orthogonal basis possesses the scalability of multiresolution analysis. Thus, we get the standard orthogonal basis in W_j space:

$$2^{-j/2}\psi\left(2^{-j}t-n\right), n \in \mathbb{Z} \tag{2.36}$$

3. The standard orthogonal basis in $L^2(\mathbb{R})$ space
   According to $L^2(\mathbb{R}) = \bigcup_{j=-\infty}^{\infty} V_j$ and $V_j = W_{j+1} \oplus V_{j+1}, j \in \mathbb{Z}$, we can get

$$L^2\left(\mathbb{R}\right) = \bigoplus_{j=-\infty}^{\infty} W_j \tag{2.37}$$

Equation 2.37 indicates that $L^2(\mathbb{R})$ is composed of the direct sum of infinite orthogonal complementary spaces. However, the orthogonal basis of $L^2(\mathbb{R})$ is the combination of the orthogonal basis of the subspaces of the direct sum. Thus, the standard orthogonal basis of $L^2(\mathbb{R})$ will be:

$$2^{-j/2}\psi\left(2^{-j}t-n\right), j \in \mathbb{Z}, n \in \mathbb{Z} \tag{2.38}$$

4. The corresponding relation of multiresolution analysis and dyadic orthogonal wavelet
   As is known from multiresolution analysis, the standard orthogonal basis of $L^2(\mathbb{R})$ is $2^{-j/2}\psi\left(2^{-j}t-n\right), j \in \mathbb{Z}, n \in \mathbb{Z}$.
   From the definition of wavelet transform, the dyadic orthogonal wavelet basis of $L^2(R)$ is $\psi_{m,n}(t) = 2^{-m/2}\psi(2^{-m}t-n), j \in \mathbb{Z}, n \in \mathbb{Z}$.
   Through careful examination, we can see that the two forms are consistent. If $\{\psi\left(t-n\right), n \in \mathbb{Z}\}$ is the orthonormal basis of the space W_0, then $\psi(t)$ is a binary wavelet, and $\psi_{m,n}(t)$ composed of $\psi(t)$ becomes the binary orthogonal wavelet in $L^2(\mathbb{R})$. Hence, the operator of multiresolution analysis space is equivalent to wavelet transform.

2.3.4 *Important properties of scale functions and wavelet functions*

First, let us review some of the computational methods of scale functions and wavelet functions, as shown in Figure 2.8.

The dyadic scale difference functions describe the inner relations among the basic functions $\phi_{j-1,k}(t)$, $\phi_{j,k}(t)$ of spaces $V_{j-1} \to V_j$ and the basic function $\psi_{j,k}(t)$ of space W_j. Some properties of $\phi(t)$, $\psi(t)$, $h(k)$, and $g(k)$ can be extended.

1. The sum of $h(k)$ and $g(k)$

$$\sum_k h\left(k\right) = \sqrt{2}, \sum_k g\left(k\right) = 0 \tag{2.39}$$

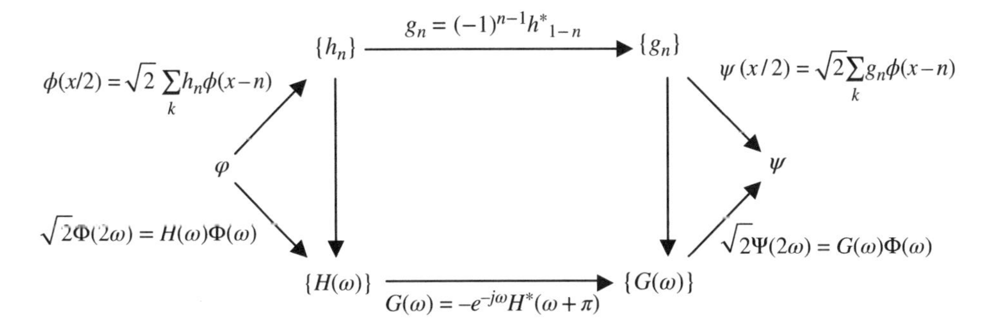

Figure 2.8 The computation of the scale function and the wavelet function

2. Frequency domain relations

$$\sqrt{2}\Phi(2\omega) = H(\omega)\Phi(\omega) \tag{2.40}$$

3. Frequency domain initial values

$$H(0) = \sqrt{2}, G(0) = 0 \tag{2.41}$$

4. Recurrence relations

$$\Phi(\omega) = \prod_{j=1}^{\infty}\left[H(2^{-j}\omega)/\sqrt{2}\right], \Psi(\omega) = (1/\sqrt{2})G(\omega/2)\prod_{j=1}^{\infty}\left[H(2^{-j}\omega)/\sqrt{2}\right] \tag{2.42}$$

5. The total energy of lossless energy spaces V_0 shall be equal to the sum of the energy of all $W_j(j = 1, \infty)$ spaces.

$$|\Phi(\omega)|^2 = \sum_{j=1}^{\infty}|\Psi(2^j\omega)|^2 \tag{2.43}$$

6. Orthogonal spaces shall comply with the following conditions:

$$|H(\omega)|^2 + |H(\omega+\pi)|^2 \longleftrightarrow \sum_k h(k)h(k+2m) = \sqrt{2}\delta_{0,m} \tag{2.44}$$

$$G(\omega) = e^{-j\omega}\bar{H}(\omega+\pi) \longleftrightarrow g(k) = (-1)^{1-k}\bar{h}(1-k) \tag{2.45}$$

2.4 The properties of basic wavelets [5]

Theoretically, there could be infinite numbers of wavelet functions, each with its own unique properties. When analyzing and processing signals in engineering, the selection of a wavelet basis shall consider not only its continuity, orthogonality, symmetry, and compact support properties, but also properties such as the center and area of its time–frequency window, vanishing moments, and attenuation. Sections 2.4 and 2.5 will introduce the description of wavelet properties and the properties of the most frequently used wavelets.

2.4.1 *The time–frequency window of mother wavelets*

Like in the short-time Fourier transform, the center and the radius of a time window could be defined as

$$t^* = \int_R t\left|\psi(t)\right|^2 dt \tag{2.46}$$

$$\Delta_\psi = \left[\int_R (t-t_0)^2 \left|\psi(t)\right|^2 dt\right]^{1/2} / \left\|\psi(t)\right\|_2^2 \tag{2.47}$$

If we only consider the positive part of the frequency domain, the corresponding center ω^* and radius Δ_Ψ of the frequency window could be

$$\omega^* = \int_0^\infty \omega\left|\Psi(\omega)\right|^2 d\omega \tag{2.48}$$

$$\Delta_\Psi = \left[\int_0^\infty (\omega-\omega_0)^2 \left|\Psi(\omega)\right|^2 d\omega\right]^{1/2} / \left\|\Psi(\omega)\right\|_2^2 \tag{2.49}$$

Then, the time–frequency window of the wavelet is

$$S = (2\Delta_\psi) \times (2\Delta_\Psi) = 4\Delta_\psi \Delta_\Psi \tag{2.50}$$

2.4.2 *Wavelet vanishing moments*

Define $L_r = \int_R t^r \psi(t) dt$ as the *r*th wavelet moment of the mother wavelet. If, for all $0 \le m \le M$, there are $L_m = 0$, then the mother wavelet $\psi(t)$ is defined to have *M*th vanishing moments. The higher value the vanishing moment is, the more capable the wavelet is to localize the signal, and the smoothness will be better.

2.4.3 *Linear phase and generalized linear phase*

If the Fourier transform of $\psi(t)$ satisfies $\Psi(\omega) = \left|\Psi(\omega)\right| e^{-ia\omega}$, where a is a real constant, and the sign is irrelevant to ω, then we say wavelet $\psi(t)$ has a linear phase, at the same time that the symmetrical real function also has a linear phase. If $\Psi(\omega) = \left|\Psi(\omega)\right| e^{-i(a\omega+b)}$, a, b are real constant values, then the function is called to have a generalized linear phase.

In engineering signal analysis, the scale function and wavelet functions can be used as the filter function. If the filter has a linear or generalized linear phase, distortion could be avoided in the decomposition and reconstruction of wavelets.

2.4.4 Compactly supported and attenuating properties

If wavelet function $\psi(t)$ is compactly supported, it is called a compactly supported wavelet. If, when $t \to +\infty$, it has a fast or exponential attenuation rate, the wavelet is called a sharp declining wavelet (the attenuation property is the property that, when $|t| \to +\infty$, the wavelet tends to be zero). The compactly supported property and the attenuation property are important properties of wavelets. If the compact support width is narrower or attenuating faster, the time domain localization of the wavelet is better; however, a function could not be compactly supported in the time domain and the frequency domain at the same time. The best situation for a function is that it is compactly supported in one domain and sharply declining in the other domain.

2.5 Frequently used properties of the mother wavelet in engineering signal analysis [5]

2.5.1 Several special mother wavelets

The Haar wavelet was first proposed in 1905; it is completely compactly supported in the time domain and completely loses its localization capabilities in the frequency domain. It is orthogonal, compactly supported, and symmetrical, but it has no vanishing moments. The Littlewood–Paley wavelet (Shannon wavelet) is another extreme that is contrary to the Haar wavelet. Littlewood and Paley proposed the orthogonal wavelet that is compactly supported in the frequency domain. It has a first-order vanishing moment, and it is continuous and is derivable of any order. The derived wavelet is orthogonal but not compactly supported, and it has the coarsest resolution in the time domain. Its time domain waveform is depicted in Figure 2.9. The Meyer wavelet was first constructed by the French mathematician Y. Meyer in 1985. The first orthonormal wavelet has smoothness in any order. There is no analytical expression for this wavelet. The Meyer wavelet is continuous in any order, symmetrical, and exponentially attenuating. Its frequency domain waveform can be seen in Figure 2.10. The Morlet wavelet is a single-frequency complex sinosoidal function under the Gaussian envelop. It is a Gaussian function in the frequency domain. The Morlet wavelet is a frequently used wavelet because of the simplicity of its analytical expression, and its good localization performance in both time and frequency domains (although it is not strictly compactly supported). Its frequency domain waveform can be seen in Figure 2.11. Additionally, the Morlet wavelet does not satisfy the admissible condition of wavelet function. In real applications, however, Morlet wavelets can essentially meet the admissible condition if we set $\omega_0 \geq 5$. Table 2.1 shows the properties of special wavelets, and Table 2.2 shows the time–frequency localization parameters of several frequently used wavelets.

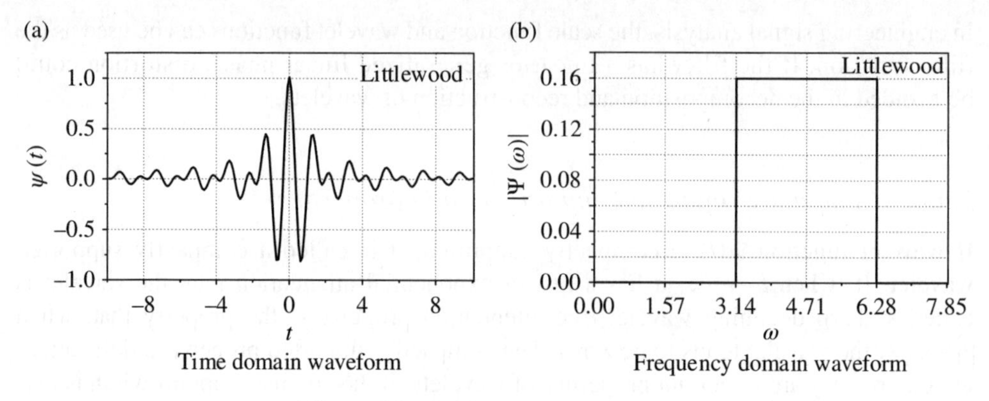

Figure 2.9 Littlewood–Paley wavelet

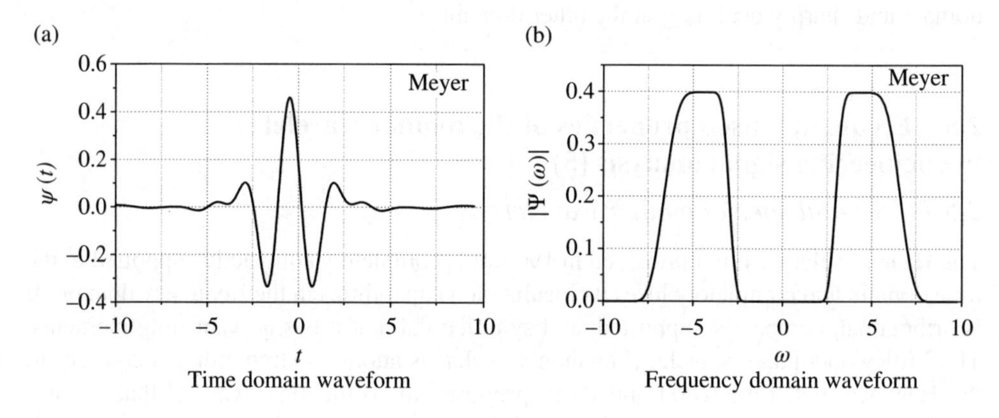

Figure 2.10 Meyer wavelet

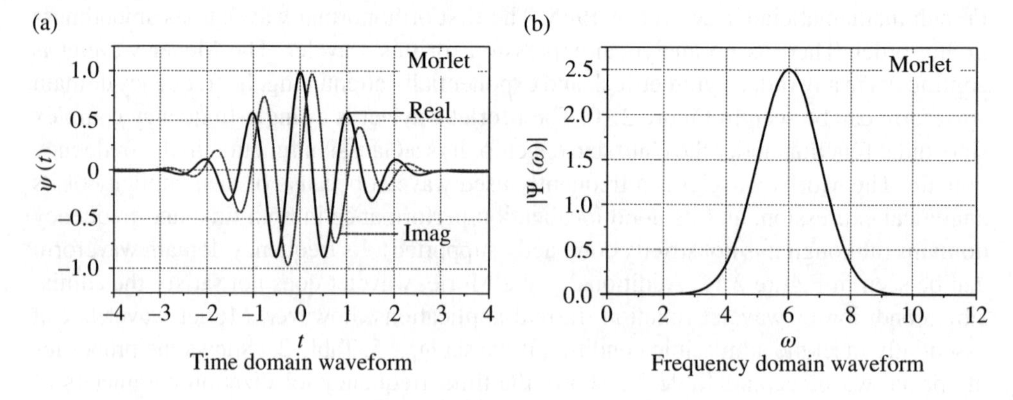

Figure 2.11 Morlet wavelet

Table 2.1 Properties of special wavelets

Wavelet \ Property	Orthogonality	Compactly support	Symmetry	Vanishing moments
Haar	Orthogonal	Compactly supported in the time domain	Symmetrical	None
Littlewood–Paley	Orthogonal	Compactly supported in the frequency domain	Symmetrical	First-order
Meyer	Orthogonal	Compactly supported in the frequency domain	Symmetrical	Any order
Morlet	Nonorthogonal	Not compactly supported	Symmetrical	Any order

Table 2.2 The time–frequency localization parameters of several frequently used wavelets

Wavelet \ Parameter	Center of the time window	Radius of the time window	Center of the frequency window	Radius of the frequency window	Area of the window
Mexican hat	0	1.0801	1.5045	0.4863	2.1012
Morlet	0	0.7071	5.0000	0.7071	2.0000
Meyer	−0.500	1.0880	4.7392	0.9810	4.2692
db4	0.4682	0.5005	5.0196	1.9821	3.9682

2.5.2 *Daubechies wavelet*

Band-limited compactly supported orthogonal wavelets play an important role in the decomposition process and data compression of wavelets. They feature fast computation and high resolution, and there is no need to manually truncate data during the process. In the aforementioned wavelets (except Haar wavelet), all symmetrical orthogonal wavelets are not compactly supported, only provided with an exponential or faster attenuation property. French researcher Daubechies conducted an in-depth exploration of the wavelet transform, with its scale function being the integer power of 2. She constructed the first kind of Daubechies wavelets, often called the *Daubechies wavelets*; by limiting the orthogonality and smoothness of scale functions, we can construct the second kind of Daubechies wavelets, also called the *Symlet wavelets*. The Symlet wavelet has the least symmetrical property. The third kind of Daubechies wavelets requires the scale function to have vanishing moments just like those of the wavelet functions, and they are called *Coiflets wavelets*. Daubechies wavelets are orthogonal and compactly supported, and they have N^{-1}-order vanishing moments, but they are not symmetrical, which is the drawback of Daubechies wavelets. Three kinds of Daubechies wavelets are shown in Figure 2.12, Figure 2.13, and Figure 2.14.

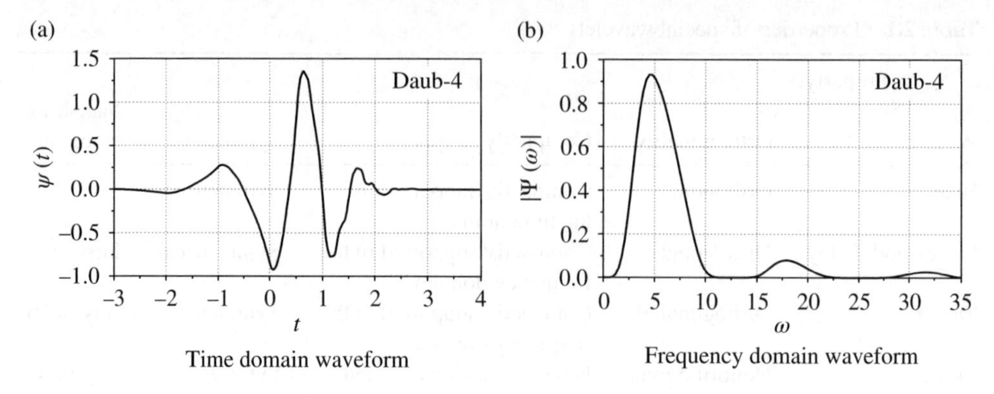

Figure 2.12 The first kind of Daubechies wavelets

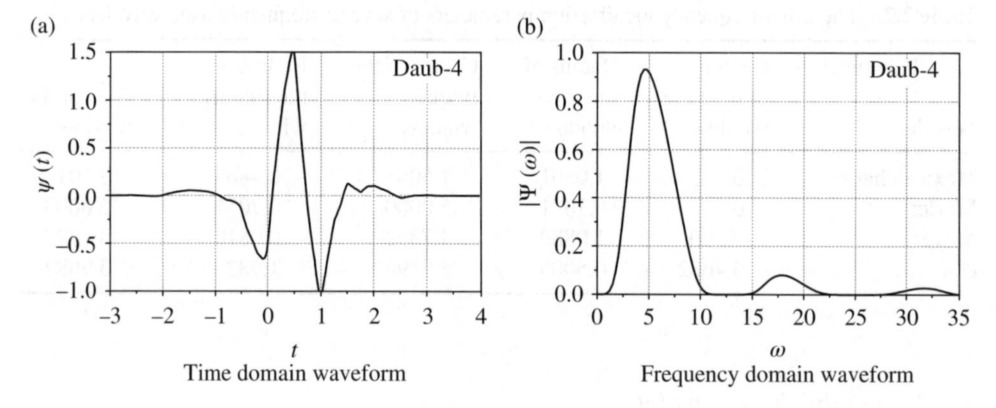

Figure 2.13 The second kind of Daubechies wavelets, also called Symlet wavelets

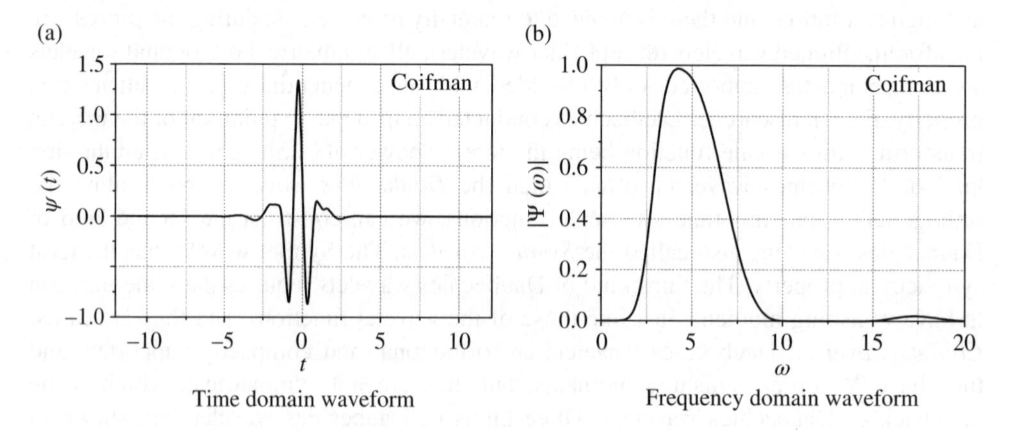

Figure 2.14 The third kind of Daubechies wavelets, also called Coiflets wavelets

2.5.3 Gaussian wavelets

Gaussian wavelets can be derived by the Gaussian function and its derivatives. Assuming the Gaussian function to be $g_0 = e^{-t^2/\sigma^2}$, we can normalize the n-order derivative and get

$$g_n(t) = (-1)^{[n/2]} \frac{\Gamma[(n+2)/2]}{\Gamma(n+1)} g_0(t) H_n(t) \tag{2.51}$$

where $H_n(t)$ is the n-order Hermite polynomial of t, and thus $g_n(t)$ is also called Gauss–Hermite wavelets, and $\Gamma(n)$ is the chi-square distribution function. Its frequency domain expression is

$$G_n(\omega) = \frac{(2\pi\omega)^n}{2} \frac{\Gamma(n/2)}{\Gamma(n)} G_0(\omega) \tag{2.52}$$

When $n = 2$, we can get the frequently used Bubble wavelet (seen in Equation (2.53)), the time–frequency waveform of which is shown in Figure 2.15.

$$\psi(t) = (2/\sqrt{3})\pi^{-1/4}\sigma^{-1/2}(1 - t^2/\sigma^2)e^{-\frac{t^2}{\sigma^2}} \tag{2.53}$$

When $\sigma = 1$, we can get the Mexican-hat wavelet. It is a special case of Mexican-hat wavelets.

The Gaussian wavelets are mostly used in the sharply transitioning parts of signals, which contain the richest information. If this wavelet is applied with a compact frame, it can also be reconstructed stably and precisely. These kinds of wavelets have fixed mother wavelets; thus, the time–frequency waveforms are identical in form. Another obvious drawback is that Gaussian wavelets do not have applicable fast algorithms. According to recent perspectives, wavelets shall have three major properties, one of which is the applicability of its fast algorithm. From this point of view, Gaussian wavelets are not perfect enough.

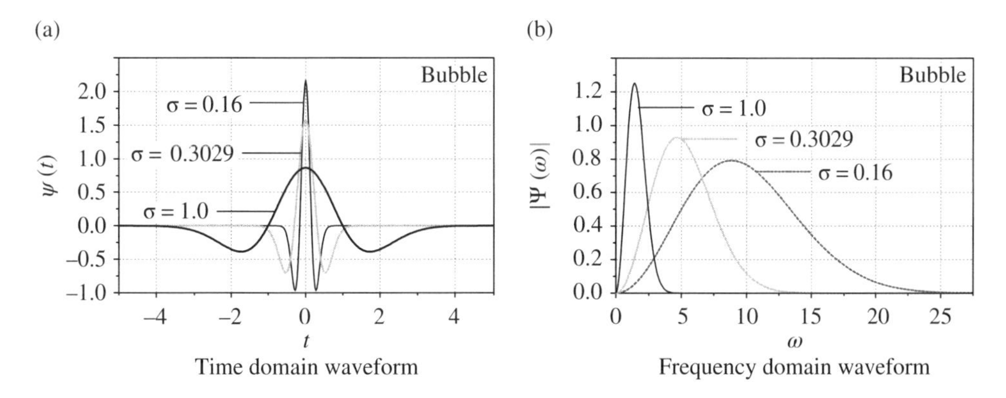

Figure 2.15 Bubble wavelets

2.5.4 Spline wavelets

A polynomial spline basis has explicit analytical expressions and good smoothness; B-spline bases for polynomial splines are compactly supported and can be obtained by convoluting rectangular pulses multiple times. They have many other advantageous properties. Cardinal spline functions are effective, simple functions for both software and hardware implementations. Hence, the wavelet and its fast transform based on spline functions have flexible time domain localization ability and easily applicable fast algorithms. It is a useful tool for real-time analysis of engineering signals, picture processing, and edge detection. Besides Haar wavelets, Stromberg first constructed the orthogonal spline wavelet with a polynomial spline basis; Mallat derived an orthogonal wavelet by orthogonalizing the nth-order B-spline basis; and Battle and Lemarie constructed an orthogonal spline wavelet using an exponentially attenuating symmetrical basis function. It is a special case of Mallat orthogonal wavelets; Cui Jing-tai and Wang Jian-zhong proposed a nonorthogonal wavelet constructed by the $(n+1)$th derivative of $(2n+1)$-order spline interpolation. In addition, researchers have proposed a method to construct compactly supported spline wavelets with the derivative of B-splines. The constructed B-spline derivative wavelets and the frequently used spline wavelets are listed here:

1. *B-spline wavelet* The cardinal spline space forms a multiresolution analysis space. On the basis of multiresolution analysis, the admissable wavelet function shall satisfy the two necessary properties: First, it must be included in the finite resolution scale space; and, second, the wavelet function ψ is orthogonal to the scale space, thus we can get the B-spline wavelet. There is an analytical expression for B-spline wavelets in specific conditions. It has symmetry or antisymmetry, and thus has a linear phase; it is a compactly supported wavelet with the smallest compact support. This is a group of semiorthogonal wavelets; when $n \geq 4$, the time–frequency window is closest to the optimal Heisenberg value.
2. *Orthogonal spline wavelet (O-spline)* Orthogonalizing the nth-order B-spline basis according to the orthogonality condition could result in the orthogonal spline wavelet, which, however, cannot be expressed analytically (except for lower order wavelets). It is symmetrical or antisymmetrical, and thus has a linear phase, and it is exponentially attenuating but not compactly supported.
3. *Cubic spline wavelet (C-spline)* According to the multiresolution analysis model, we can derive the cubic spline wavelet. The cubic spline wavelet has the same time domain waveform as that of the B-spline wavelet; it is exponentially attenuating and not compactly supported. This is a kind of biorthogonal wavelet, because the scale function satisfies the interpolation condition; thus, the corresponding cubic spline wavelet decomposition is a good approximation of the continuous signal. Figures 2.16, 2.17, and 2.18 show the time domain waveform and frequency domain waveform of three spline wavelets. Figure 2.19 depicts the curve of the changing principle of time–frequency localization parameters against the order.

Besides, the spline wavelet series also includes the dual wavelets, cubic spline derivative wavelets, B-spline derivative wavelets, and so on. See related articles for the details of the uniform multiresolution analysis model and its fast algorithms.

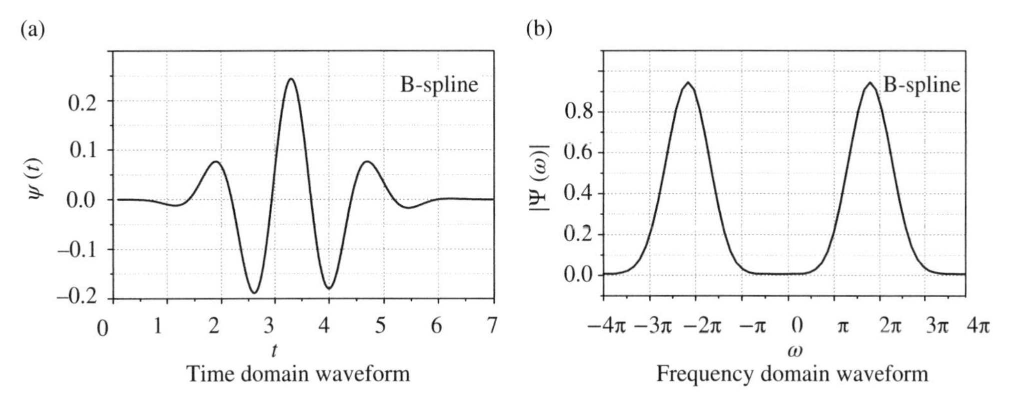

Figure 2.16 Fourth B-spline wavelet

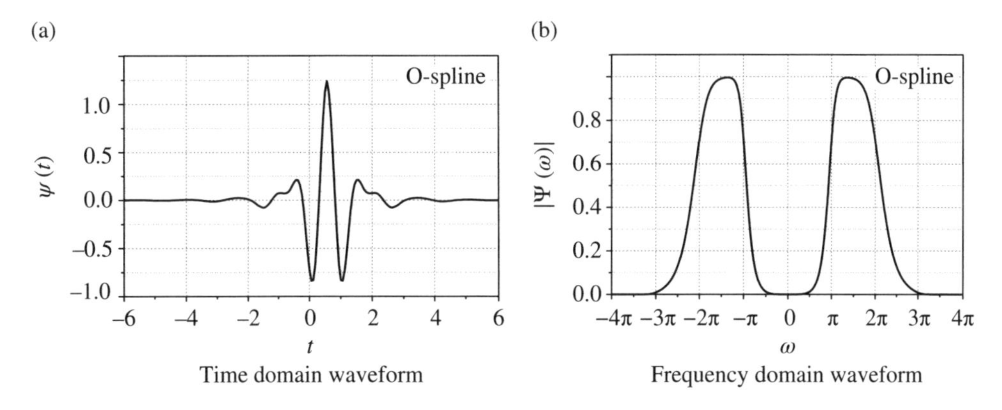

Figure 2.17 Fourth O-spline wavelet

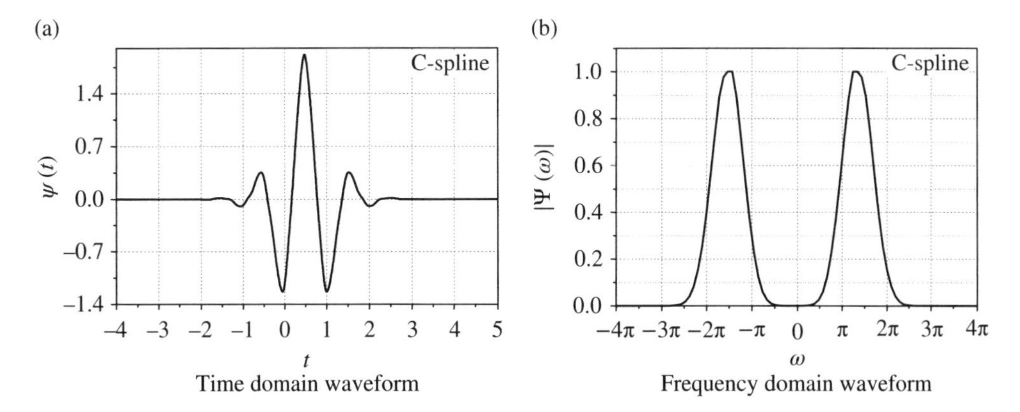

Figure 2.18 Fourth C-spline wavelet

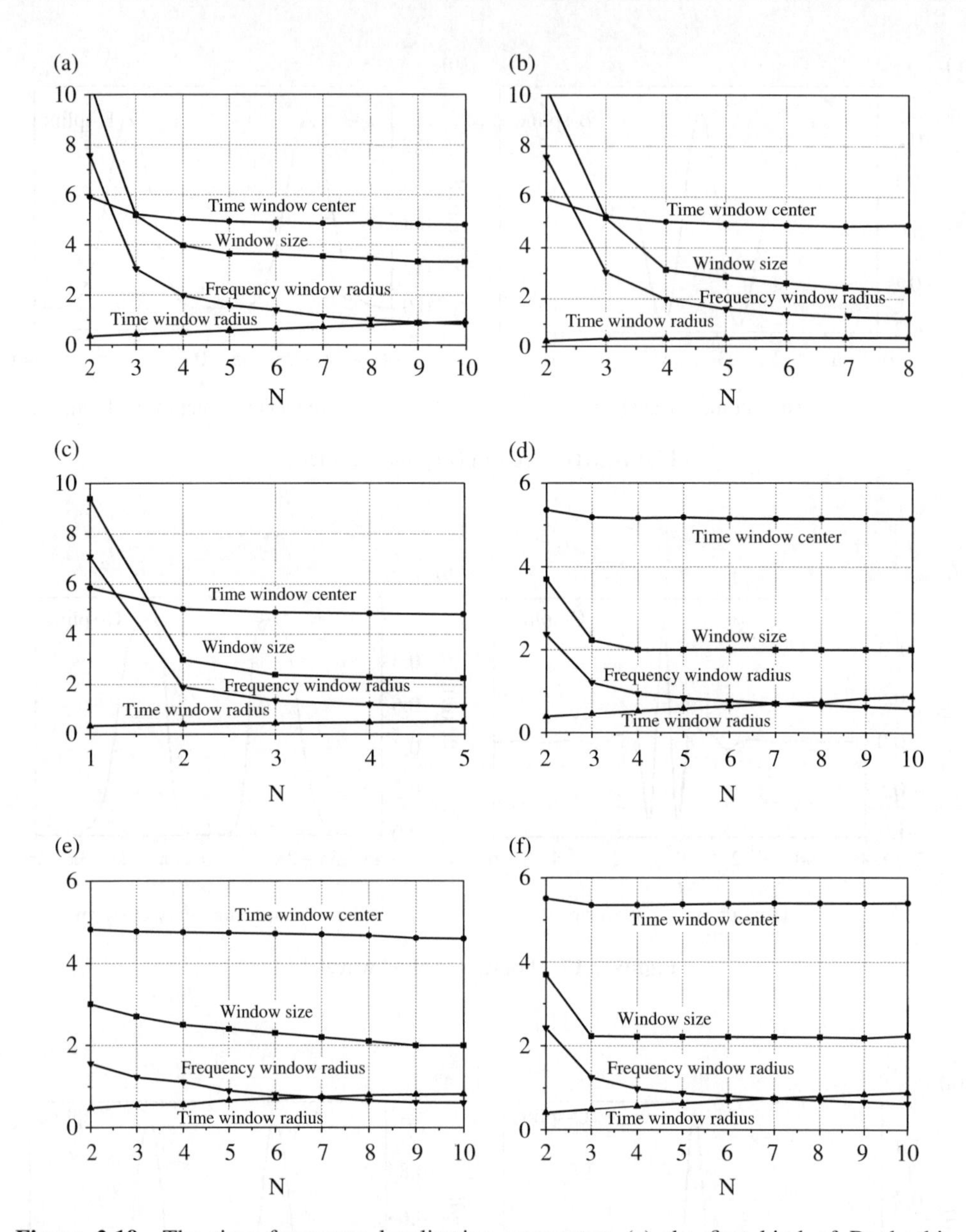

Figure 2.19 The time–frequency localization parameter: (a) the first kind of Daubechies wavelets, (b) the second kind of Daubechies wavelets, (c) the third kind of Daubechies wavelets), (d) B-spline wavelets, (e) O-spline wavelets, (f) C-spline wavelets

References

[1] Daubechies I., Ten lectures on wavelets. *Society for Industrial Mathematics*, 1992.
[2] Hu G. S., Modern signal processing guide. *Tsinghua University Press*, 2004.
[3] Pinsky M. A., Introduction to Fourier analysis and wavelets. *China Machine Press*, 2003.
[4] Mallat S., A wavelet tour of signal processing. *China Machine Press*, 2003.
[5] He Z. Y., The electric power system transient signal wavelet analysis method and its application on the non-unit transient protection principle. *Southwest Jiaotong University*, 2000.

3

Wavelet Analysis and Signal Singularity

Signals carry overwhelming amounts of data in which relevant information is often more difficult to find. Signal analysis and signal processing help people extract useful information from collected or recorded signals. It is evident that the information contained in signals is primarily extracted from the transient points or areas. For example, a time-varying DC signal does not provide much information other than the amplitude of the signal. In the same way, a slowly varying signal contains very little information. It is well acknowledged that white noise contains the richest information for that white noise is a random signal that two arbitrary points are irrelevant to each other.

3.1 Lipschitz exponents

Lipschitz exponents provide precise quantification of the local regularity of functions. Assume that function $x(t)$ is of the following property:

$$| x(t_0+h)-p_n(t_0+h) |=A| h |^{\alpha}, n<\alpha<n+1 \tag{3.1}$$

Then, α in Equation (3.1) is defined as the Lipschitz exponent of $x(t)$ at t_0, as shown in Figure 3.1.

In (3.1), h is a sufficiently small value, and $p_n(t)$ is a polynomial of degree n, $n \in \mathbb{Z}$, which are in fact the first n items in the Taylor series expansion of $x(t)$ at t_0.

$$T_n(t)=x(t_0)+a_1h+a_2h^2+.....+a_nh^n+o(h^{n+1})=p_n(t)+o(h^{n+1}) \tag{3.2}$$

Obviously, α is not necessarily equal to $n+1$, but it should be larger than n, and may be less than $n+1$. Assume

$$x(t)=x(t_0)+a_1h+a_2h^2+a_{2.5}h^{2.5}$$

Wavelet Analysis and Transient Signal Processing Applications for Power Systems, First Edition. Zhengyou He.

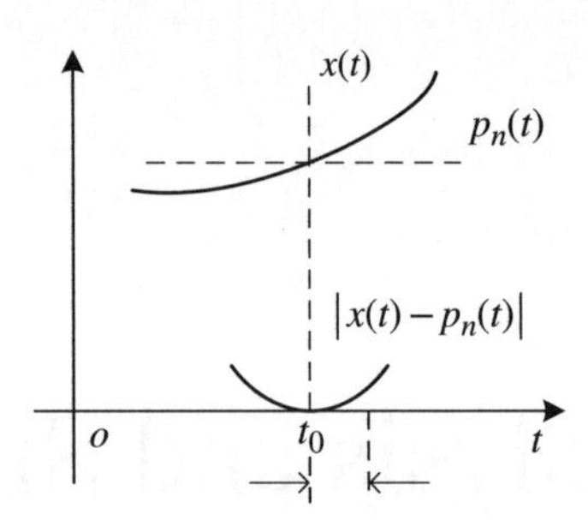

Figure 3.1 The graphic of Lipschitz exponent

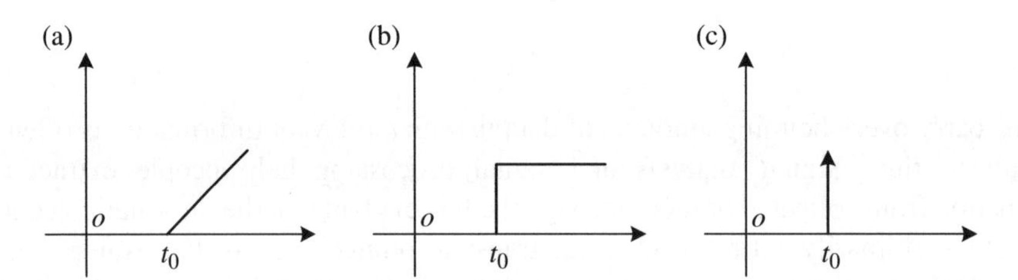

Figure 3.2 The Lipschitz exponents of common singularities: (a) ramp function, (b) step function, (c) impulse function

Then, $2<\alpha = 2.5<3$, thus α must be larger than n, but could be less than $n+1$. If $x(t)$ is n times differentiable, but fails to be $n+1$ times differentiable, then α is larger than n and smaller than $n+1$. If α is the Lipschitz exponents of $x(t)$, then $\alpha+1$ will be the Lipschitz exponents of $\int_0^t x(l)dl$. That is to say, the Lipschitz exponents of the function will increase by 1 if the function is integrated once; conversely, the Lipschitz exponents will decrease by 1 if the function is differentiated once.

There are three kinds of singularities from left to right in Figure 3.2.

The singularity in Figure 3.2a is indifferentiable, whose Lipschitz exponent is 1; the singularity in Figure 3.2b is an incontinuous step singularity, whose Lipschitz exponent is 0; the singularity in Figure 3.2c is a Dirac, whose Lipschitz exponents is –1; and the Lipschitz exponents of white noise singularity is $-1/2-\omega$, where $\omega > 0$. The Lipschitz exponent is defined at $x(t)$ of t_0. When expanding the definition to the whole segment of [a,b], two arbitrary points at t_0 and $t_0 + h$ need to satisfy (3.1). Hence, the segment is defined as regularly or uniformly Lipschitz (α). Accordingly, the following statements can also be given:

1. If and only if the integral of $x(t)$ is uniformly Lipschitz (α+1) in the segment of [a,b], then $x(t)$ is Lipschitz (α) in the segment of [a,b].
2. If and only if the derivative of $x(t)$ is uniformly Lipschitz (α–1) in the segment of [a,b], then $x(t)$ is Lipschitz (α) in the segment of [a,b].

Generally speaking, the Lipschitz exponents measure the singularity of the function at arbitrary points. That is to say, if Lipschitz exponents increase, the singularity of the function will be reduced, and consequently, the smoothness at this point will increase; conversely, if singularity increases, the smoothness of the function at the point will decrease. If $x(t)$ is differentiable at t_0, the Lipschitz exponents will be larger than 1; if $x(t)$ is indifferentiable at t_0, the Lipschitz exponents will be 1; and if $x(t)$ is incontinuous at t_0 and has limited value, the Lipschitz exponents will be greater than 0 and less than 1.

To continue the topic in detail, we will discuss the relationship between signal singularity and wavelet transformation in Section 3.2.

3.2 Characterization of signal singularity based on wavelet transform

According to the discussion in Section 3.1, Lipschitz exponents can describe the singularity of signals. We have already known that the smaller Lipschitz exponent leads to higher signal singularity, and the larger Lipschitz exponent leads to lower signal singularity.

Theorem 3.1: Let $x(t) \in L^2(\mathbb{R})$, and let [a,b] be a closed interval of $\mathbb{R}$, $0<\alpha<1$. For an arbitrary $\omega>0$, $x(t)$ is the function of uniformly Lipschitz (α) over $(a+\omega, b-\omega)$, if and only if there exist constant A_s and $t \in (a+\omega, b-\omega)$ satisfying the following statement for any arbitrary $s>0$:

$$\left|WT_x(t,s)\right| \le A_S s^{\alpha} \ or \log\left|WT_x(t,s)\right| \le \log A_s + \alpha \log s \tag{3.3}$$

The theorem shows the relationship between the decaying property of wavelet transformation coefficients and the singularity of local Lipschitz exponents.

In order to extend Theorem 3.1 to situations of Lipschitz exponents α greater than 1, we assume that the wavelet function $\psi(t)$ has great enough vanishing moments. A wavelet function $\psi(t)$ is said to have n vanishing moments, if and only if for all positive integer $k<n$, it satisfies

$$\int_{-\infty}^{+\infty} t^k \psi(t)\,dt = 0 \tag{3.4}$$

If the wavelet $\psi(t)$ has n vanishing moments, then Theorem 3.1 remains valid for any noninteger Lipschitz exponents $0<\alpha<n$; however, Theorem 3.1 is not necessarily valid for any integer Lipschitz exponents.

If $\alpha = 2^j$, then Equation (3.3) turns to be

$$\left|WT_{2^j}x(t)\right| \le A_S \left(2^j\right)^{\alpha} \ or \log_2\left|WT_{2^j}x(t)\right| \le \log_2 A_S + j\alpha \tag{3.5}$$

$j\alpha$ in Equation (3.5) relates the scale features j of wavelet transform with Lipschitz exponent α, and Equation (3.5) demonstrates the changing rules of wavelet transform values against j and α. As can be seen in Equation (3.5), when $\alpha>0$, the maximum of wavelet transform increases as the scale j increases; when $\alpha<0$, the maximum of wavelet transform increases as the scale j decreases. As for the step function ($\alpha = 0$), the maximum of the wavelet transform does not change as the scale changes. In Figure 3.3, the signals have different singularities: in the left neighborhood of $t = 0.16$, the local behavior of the signal will be $-t - 0.161$, and the Lipschitz exponent will be 1; when $t = 0.44$, the signal is a discrete Dirac (its Lipschitz exponent is -1); when $t = 0.5$ and $t = 0.7$, the signal is a step function (its Lipschitz exponent is 0); and, when $t = 0.8$, the signal is a Gaussian noise (its Lipschitz exponent is less than -0.5).

In fact, we can only process discrete signals with finite resolution, and the sampling rate shall be normalized to 1. Strictly speaking, it makes no sense to discuss the singularity, derivability, discontinuity, and Lipschitz exponents of discrete signals, because we are not able to calculate the asymptotic decay of wavelet transform at a scale smaller than 1. That is to say, we cannot prove Equation (3.5) at a scale smaller than 1. Although limited by the sampling resolution, we still want to use the mathematical tool – differential singularity – in real applications. Suppose that $x(t)$ could be approximated by a set of samples $(x_n)_{n \in z}$ at the resolution of 1, which also satisfy

$$\begin{cases} x_n = 0, n < n_0 \\ x_n = 1, n \geq n_0 \end{cases} \tag{3.6}$$

Obviously, $x(t)$ could be a continuous signal that has a sharp transition at n_0; however, since the resolution is too high, the continuity could not be observed. At this resolution, it is fair to say that $x(t)$ is singular at $t = n_0$; hence, the aforementioned singularity is closely related to scales. Therefore, the singularity obtained from the changing rules of scales by Equation (3.5) is the accurate measurement of discontinuity at the resolution of 1. By analyzing the changing rules of wavelet transform at a scale larger than 1 and finding the coefficient α_0, we can reach the optimal approximation of the relation between $\left|WT_{2^j} x(t)\right|$ (the wavelet transform of) and the scale by $A_s(2^j)^\alpha$ at the scale larger than 1.

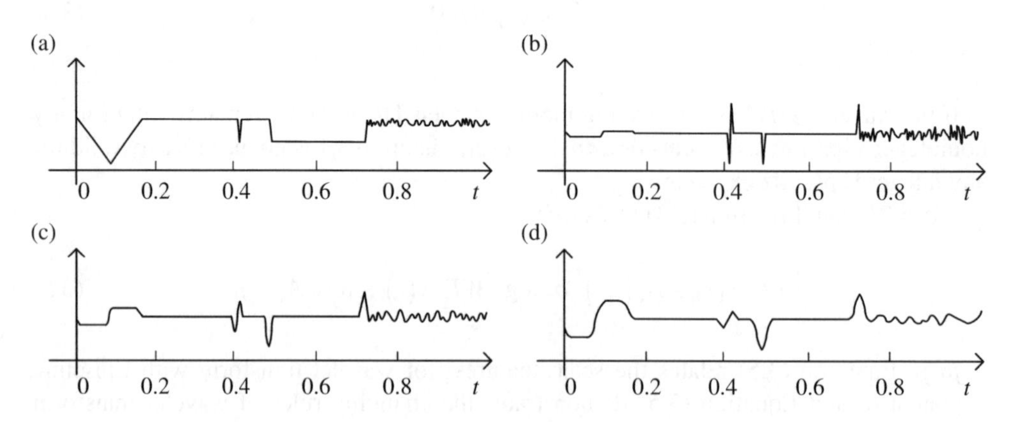

Figure 3.3 The singular signal and its triple decomposition: (a) Singular signal, (b) decomposition at scale 1, (c) decomposition at scale 2, (d) decomposition at scale 3

From above, if signal $x(t)$ has a sharp transition at t_0, then at each scale, there will be a local maximum of $|WT_{2^j}x(t)|$ in the neighborhood of t_0, which is changing with the scale and ultimately converges at t_0. Accordingly, we can make use of the wavelet modulus maximum feature to detect signal singularities and to calculate the Lipschitz exponents of singularities by the relation of modulus maxima against scales.

Usually, if signal $x(t)$ has a sharp transition at a point, it is not necessarily a singularity. Figure 3.3 shows an example of a smooth edge at $t = 0.44$; however, the edge could be just a noise, indicating that estimating the smoothness of a certain point is also a crucial step. We can assume that the observed signal is smoothed by a smoothing function, which is approximated by a Gaussian function with variance σ^2. Hence, σ becomes an important parameter that characterizes the edge, called the *smoothing factor*.

Assume $x(t)$ is the convolution of two functions: a function $h(t)$ with a singularity at t_0, and a Gaussian function with variance σ, which is given by

$$x(t) = h * g_\alpha(t) \tag{3.7}$$

where $g_\alpha(t) = \frac{1}{\sqrt{2\pi}\sigma}\exp\left(-\frac{t^2}{\sqrt{2\sigma^2}}\right)$. Assume the Lipschitz exponent of $h(t)$ at $t = t_0$ is $\alpha = \alpha_0$, and the wavelet function $\psi(t) = \frac{d\theta(t)}{dt}$, then we get

$$WT_{2^j}x(t) = 2^j\frac{d}{dt}\left(x * \theta_{2^j}\right)(t) = 2^j\frac{d\left(h * g_\sigma * \theta_{2^j}\right)(t)}{dt} \tag{3.8}$$

When $\theta(t)$ approximates the Gaussian function,

$$\theta_{2^j} * g_\sigma = \theta_{s_0}(t) \text{ and } s_0 = \sqrt{2^{2j} + \sigma^2} \tag{3.9}$$

Then Equation (3.8) can be written as

$$WT_{2^j}x(t) = 2^j\frac{d}{dt}\left(x * \theta_{2^j}\right)(t) = 2^j\frac{d}{dt}\left(h * \theta_{s_0}\right)(t) = \frac{2^j}{s_0}WT_{s_0}h(t) \tag{3.10}$$

Equation (3.10) indicates that the wavelet transform of the function $x(t)$, smoothed by a Gaussian function of variance σ^2, is equivalent to that of the nonsmoothed singular signal $h(t)$ at the scale of $s_0 = \sqrt{2^{2j} + \sigma^2}$. As a result, for an arbitrary scale of $s>0$, there exists a constant A_j that validates Equation (3.11) in the neighborhood of t_0:

$$\text{If } \left|WT_s h(t)\right| < A_s s^\alpha, \text{ then } \left|WT_{2^j}x(t)\right| < A_j 2^j S_0^{\alpha-1}, \text{ where } s_0 = \sqrt{2^{2j} + \alpha^2} \tag{3.11}$$

If $x(t)$ is multiplied by a constant C and gives $Cx(t)$, then σ and α will remain unchanged, and A_s will become $C\,A_s$; if the signal is smoothed again by another Gaussian function with variance σ_1, then σ^2 will become $\sigma^2 + \sigma_1^2$, while A_s and α remain unchanged. As seen from above, α and A_s represent different features of the singularity at t_0.

In the following part, we will discuss how to calculate Lipschitz exponents and smoothing factors with the behavior of wavelet maxima across scales. If we can detect modulus maxima at all scales rather than at only dyadic scales, we can plot the modulus maxima curve, also called "fingerprints" by Winkin because the maxima curves of some signals look like human fingerprints. As an example, the maxima curve of a noisy sinusoidal signal is depicted in Figure 3.4.

If the modulus maxima of two adjacent scales 2^j and 2^{j+1} are in the same maxima curve, we can consider that the maxima of scale 2^j propagate to the scale 2^{j+1}. As shown in Figure 3.3, the sharp transition point at $t = 0.5$ corresponds to modulus maxima in the neighborhood of $t = 0.5$. However, there also exists another sharp transition at $t = 0.44$, which corresponds to two modulus extrema (one is a maximum, and the other is a minimum). In general, we can estimate the positions and intensity of extrema of the next scale by observing that of the last scale through a simple algorithm. If there exists a relatively large maximum at scale 2^j, and its position is close to the modulus maximum of the same sign at scale 2^{j+1}, then we can conclude that the two maxima correspond to the same singularity. Although the algorithm is not precise enough, we do not have to perform wavelet transform at all scales but use the wavelet modulus maxima at dyadic scales to calculate the Lipschitz exponent and the smoothing factor; thus, the algorithm has some practical value.

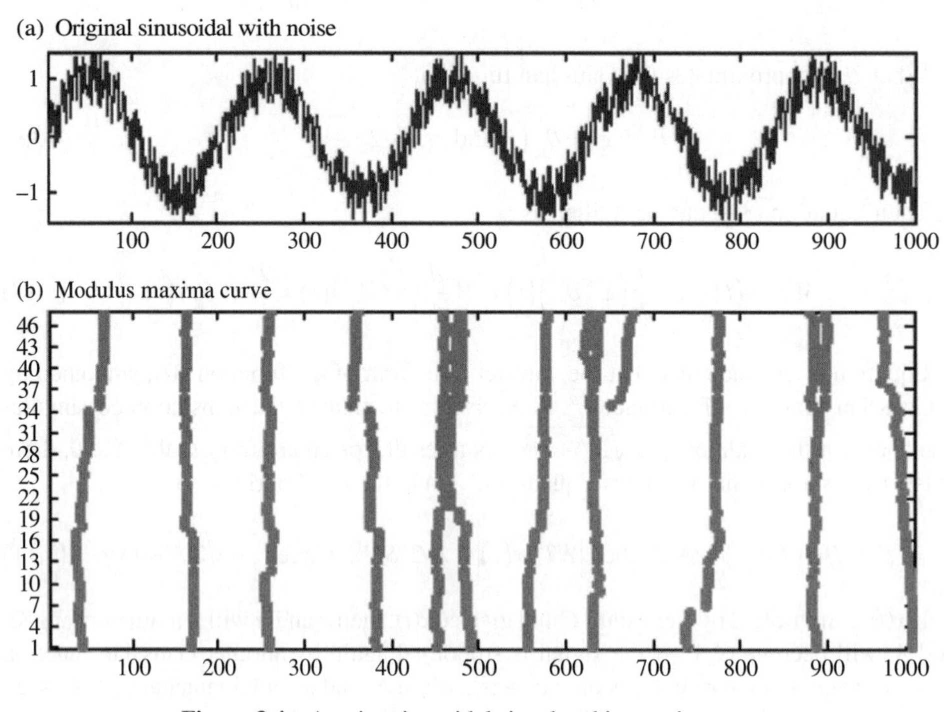

Figure 3.4 A noisy sinusoidal signal and its maxima curve

3.3 The wavelet transform property of random noise

From the discussion in Section 3.2, we know that the singularity of signals is closely related to their wavelet transform. In this section, we discuss the wavelet transform features of noise. According to different wavelet transform features of noise and signals, we can separate noise from signals and eliminate the influence of false edges in edge detection.

Assume $n(t)$ is a white noise of variance σ^2, and $WT_{2^j}n(t)$ is the wavelet transform of $n(t)$, then $WT_{2^j}n(t)$ will be a random process. If we assume that the wavelet function $\psi(t)$ is a real function, then the variance of $WT_{2^j}n(t)$ could be given by

$$\sigma_j = E\left[\left|WT_{2^j}n(t)\right|^2\right] = \int_{-\infty}^{+\infty}\int_{-\infty}^{+\infty} E\left[n(v)n(u)\right]\psi_{2^j}(t-v)\,dudv \tag{3.12}$$

Thus, $\sigma_j = E\left[\left|WT_{2^j}n(t)\right|^2\right] = \frac{\|\psi\|^2}{2^j}\sigma^2$. Because $\psi(t)$ is compactly supported, and $\|\psi\|$ and σ^2 are both constant, thereby $E\left[\left|WT_{2^j}n(t)\right|^2\right]$ will decrease as j increases.

For a given scale 2^{j_0}, the wavelet transform of $WT_{2^{j_0}}n(t)$ is a random process of variable t. Meanwhile, if $n(t)$ is a white noise, then $WT_{2^{j_0}}n(t)$ is also a Gaussian process. The average density of zero crossings of a derivable Gauss process is $\sqrt{-\frac{R^{[2]}(0)}{\pi^2 R(0)}}$, where $R^{(n)}(\tau)$ is the nth derivative of $R(\tau)$ and $R(\tau)$ is the autocorrelation function of a derivable Gauss process. If the process is twice differentiable, the density of its local extrema is equal to the density of its zero crossings, which is given by $\sqrt{-\frac{R^{[4]}(0)}{\pi^2 R^{[2]}(0)}}$.
The autocorrelation function of the Gaussian process $WT_{2^j}n(t)$ is defined by

$$R(\tau) = E\left[WT_{2^j}(t+\tau)WT_{2^j}(t)\right] = \int_{-\infty}^{+\infty}\int_{-\infty}^{+\infty} E\left[n(v)n(u)\right]\psi_{2^j}(t+\tau-u)\psi_{2^j}(t-v)\,dudv \tag{3.13}$$

From Equation (3.13), we can derive

$$R^{[4]}(0) = 2^{-5j}\sigma^2\left\|\psi^{(2)}\right\|^2, R^{[2]}(0) = 2^{-3j}\sigma^2\left\|\psi^{(2)}\right\|^2 \tag{3.14}$$

By substituting Equation (3.14) into the aforementioned equation, the extrema density of the process $WT_{2^j}n(t)$ can be obtained as $\frac{\left\|\psi^{(2)}\right\|}{2^j\pi\left\|\psi^{(1)}\right\|}$. Let m be the number of modulus maxima, e be the number of local extrema, and Z_0 be the number of zero crossings. It would be easy to verify the relation that follows:

$$m = \frac{e}{2} + Z_0 \tag{3.15}$$

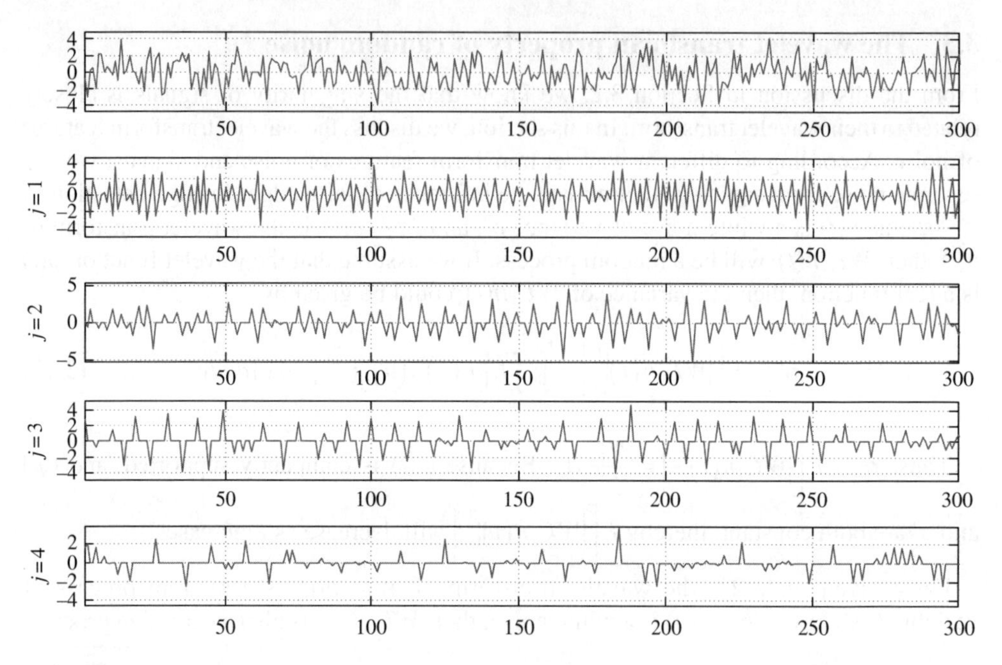

Figure 3.5 Gaussian white noise and its modulus maxima at different scales (1–4)

Given that the average density of zero crossings of $WT_{2^{j_0}}n(t)$ is $\frac{\|\psi\|^{(1)}}{2^j\pi\|\psi\|}$, the average density of modulus maxima is expressed by

$$ds = \frac{1}{2^j}\left(\frac{\|\psi^{(2)}\|}{\|\psi^{(1)}\|} + \frac{\|\psi^{(1)}\|}{\|\psi\|}\right) \tag{3.16}$$

where $\psi^{(1)}$ and $\psi^{(2)}$ are the first and second derivatives of ψ. As the equation shows, the average density of Gaussian noise maxima is reversely proportional to the scale, that is to say, when the scale increases, the density decreases. All in all, as the scale increases, $E\left[\left|WT_{2^j}n(t)\right|^2\right]$ and ds are both decreasing, hence $E\left[\left|WT_{2^j}n(t)\right|^2\right]$ and ds are important features that distinguish noises from signals. Experimental results can be seen in Figure 3.5.

Further reading

Daubechies I., Ten lectures on wavelets. *Society for Industrial Mathematics*, 1992.

He Z. Y., The electric power system transient signal wavelet analysis method and its application on the non-unit transient protection principle. *Southwest Jiaotong University*, 2000.

Hu G. S., Modern signal processing guide. *Tsinghua University Press*, 2004.

4

Sampling Techniques in Wavelet Analysis of Transient Signals

Before performing wavelet transform to continuous signals using computers, we need to first sample a continuous-time signal to obtain a series of discrete values and to sample the wavelet basis. The basic requirement for the sampling is to get a series of discrete values that contain all the information from the original signal and the wavelet basis. In fact, there are some differences between the sampling in wavelet transform and in Fourier transform. On one hand, the sampling intervals of Fourier transform in the time domain and space domain are both constants, whereas the sampling intervals of wavelet transform have a tendency to finer scales in both domains. On the other hand, after the first sampling of signal and wavelet bases, the secondary sampling of signal and wavelet bases is carried out. In addition, the sampling values in the sample space are obtained by calculation in the multiresolution analysis–based pyramidal decomposition, which is also known as the *discrete approximation* in the initial space of multiresolution analysis. In this chapter, we are going to introduce the sampling techniques in detail for wavelet analysis of transient signals [1].

4.1 Wavelet sampling in direct wavelet transform

Direct wavelet transform of signals refers to the projection of signal $x(t)$ on the wavelet basis, which is computed with the wavelet base function $\psi(t)$. Generally speaking, after the sampling of signals, the sampling interval of the signal is set to be the interval of the whole transform process; thus, the sampling values during the whole process are unchanging. Once the scale a changes, the sampling interval of the wavelet basis will be readjusted according to the scale a. There are various sampling methods for a and τ grids sampling. In engineering applications, scale a is usually set to be $a = 2^j$. Basic sampling methods include dyadic grids sampling and dyadic extraction sampling, as shown in Figure 4.1.

Wavelet Analysis and Transient Signal Processing Applications for Power Systems, First Edition. Zhengyou He.

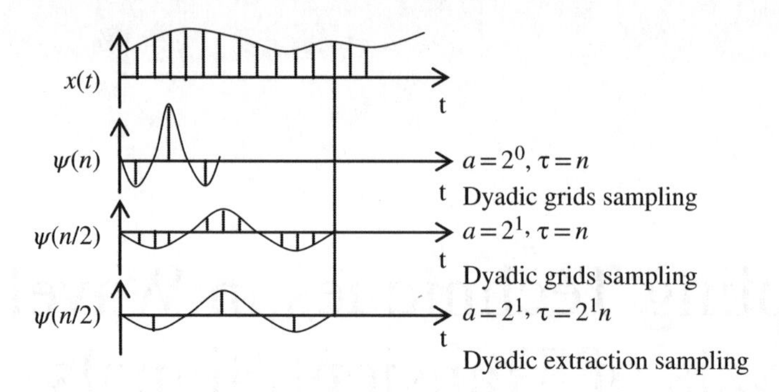

Figure 4.1 The sampling process using direct wavelet transform

1. Dyadic grids sampling is defined in Equation (4.1), where $a = 2^j$, $\tau = n$, $j = 0 \cdots J$, and $n \in \mathbb{Z}$.

$$DWT(n) = 2^{-j/2} \sum_k x(k) \psi\left[2^{-j}(k-n)\right] \tag{4.1}$$

2. Dyadic extraction sampling is defined in Equation (4.2), where $a = 2^j$, $\tau = n$, $j = 0 \cdots J$, and $n \in \mathbb{Z}$.

$$DWT(n) = 2^{-j/2} \sum_k x(k) \psi\left(2^{-j}k - n\right) \tag{4.2}$$

At scale $a = 2^j$, $\psi(2^{-j}t)$ is 2^j times as wide as $\psi(t)$, which means the frequency has decreased by 2^j times. Therefore, the sampling interval could be expanded by 2^j times, that is to say, the information contained will not be lost if the sampling interval along Axis τ is set to be 2^j uniformly. Thus, dyadic extraction sampling can satisfy the sampling requirements.

4.2 Pre-sampling in indirect wavelet transform

Indirect wavelet transform is a concept comparable to direct wavelet transform. Given that the computing of wavelet functions of $L^2(\mathbb{R})$ is much more complicated than computing the basic scaled function, indirect wavelet transform is performed from multiresolution analysis space to implement pyramidal decomposition through filter banks. In this way, one obtains the discrete wavelet transform of the digital signal. However, one performs the pyramidal decomposition method by first projecting the sampling signal onto the sample space of multiresolution analysis, and then employing the decomposition algorithm at every resolution to obtain the wavelet components. The traditional Mallat algorithm directly decomposes the sample signals, which lacks theoretical foundation. Thus, we need to research the projection of signals onto the initial multiresolution analysis space, in other words, the pre-sampling of signals on V_N.

Assume the projection of $x(t)$ on V_N is $x_N(t)$, then, in terms of the multiresolution analysis condition, we get

$$\lim_{N\to\infty}\left\|x_N\right\| = x \lim_{N\to-\infty}\left\|x - x_N\right\| = 0 \tag{4.3}$$

Obviously, $x_N(t)$ is the approximation of $x(t)$. Let $x_N(t) = \sum_{k=-\infty}^{\infty} c_{N,k}\phi(2^N t - k)$, then $c_{N,k}$ is called the sampling value on V_N. The computing of sampling values is also the computing of $c_{N,k}$ when $\| x - x_N \|$ reaches the minimum. For real signal $x(t)$, assume $C(\omega) = \sum_{k=-\infty}^{\infty} c_{N,k} e^{-k\omega i}$, then we get

$$\begin{aligned}
\left\|x - x_N\right\| &= \frac{1}{2\pi}\int_{-\infty}^{\infty}\left|X(\omega) - 2^{-N}X\left(2^{-N}\omega\right)\Phi(2^{-N}\omega)\right|^2 d\omega \\
&= \frac{2^N}{2\pi}\int_{-\infty}^{\infty}\left|X\left(2^N\omega\right) - 2^{-N}C(\omega)\Phi(\omega)\right|^2 d\omega \\
&= \frac{2^N}{2\pi}\sum_{k=-\infty}^{\infty}\int_{-\pi+2k\pi}^{\pi+2k\pi}\left|X\left(2^N\omega\right) - 2^{-N}C(\omega)\Phi(\omega)\right|^2 d\omega \\
&= \frac{2^N}{2\pi}\int_{-\pi}^{\pi}\left(\sum_{k=-\infty}^{\infty}\left|X\left[2^N(\omega + 2k\pi)\right]\right|^2 + 2^{-2N}\sum_{k=-\infty}^{\infty}\left|C(\omega + 2k\pi)\Phi(\omega + 2k\pi)\right|^2\right. \\
&\quad - 2^{-N}\sum_{k=-\infty}^{\infty}X\left[2^N(\omega + 2k\pi)\right]\bar{C}(\omega + 2k\pi)\bar{\Phi}(\omega + 2k\pi) \\
&\quad \left. - 2^{-N}\sum_{k=-\infty}^{\infty}\bar{X}\left[2^N(\omega + 2k\pi)\right]C(\omega + 2k\pi)\Phi(\omega + 2k\pi)\right)d\omega
\end{aligned} \tag{4.4}$$

Let $[X,G](\omega) = \sum_{k=-\infty}^{\infty} X(\omega + 2k\pi)\bar{G}(\omega + 2k\pi)$ and $X_{2^N}(\omega) = X(2^N\omega)$. To minimize the above functional, let

$$C(\omega) = \frac{2^N\left[X_{2^N},\Phi\right](\omega)}{\left[\Phi,\Phi\right](\omega)} \tag{4.5}$$

Assuming $x_0(n)$ is the discrete sampling signal of $x(t)$, we can consider $x_0(n)$ is the sampling signal of $x(t)$ with a sampling period of 1 if the timeline unit is properly selected. According to Shannon's theorem, we get

$$x(t) = \sum_{k=-\infty}^{\infty} x_0(n)\frac{\sin\pi(t - n)}{\pi(t - n)} \tag{4.6}$$

Let $u_{[a,b]}(\omega)=\begin{cases}1, a\le\omega\le b\\0\end{cases}$, then $X(\omega)=X_0(\omega)u_{[-\pi,\pi]}(\omega)$. So, from Equation (4.5), we can derive

$$C(\omega)=2^N X_0\left(2^N\omega\right)u_{\left[-2^{-N}\pi,2^{-N}\pi\right]}(\omega)\frac{\bar{\Phi}(\omega)}{\left[\Phi,\Phi\right](\omega)} \tag{4.7}$$

In the expression of $C(\omega)$, $2^N X_0(2^N\omega)u_{\left[-2^{-N}\pi,2^{-N}\pi\right]}(\omega)$ is equivalent to the sampling rate conversion of $x_0(n)$, and $x_0(k/2^N)$ is obtained by means of reducing its frequency band. The term $u_{[-\pi,\pi]}(\omega)\frac{\bar{\Phi}(\omega)}{\left[\Phi,\Phi\right](\omega)}$ is used to implement low-pass filtering. Let

$$q(n)=\int_{-\pi}^{\pi}u_{[-\pi,\pi]}(\omega)\frac{\bar{\Phi}(\omega)}{\left[\Phi,\Phi\right](\omega)}e^{k\omega i}d\omega=\int_{-\infty}^{\infty}\sin c\pi t\tilde{\phi}(t-n)dt \tag{4.8}$$

$$c_{N,k}=\sum_{k=-\infty}^{\infty}x_0\left(k/2^N\right)q(k-n) \tag{4.9}$$

where $\tilde{\Phi}(\omega)=\frac{\bar{\Phi}(\omega)}{\left[\Phi,\Phi\right](\omega)}$ is defined as the sampling function of V_N. Generally, we can first perform the sampling rate conversion of $x_0(n)$, then obtain the digital sampling of $c_{N,k}$ on V_N by pre-filtering on the basis of $q(n)$. When sampling band-limited analog signals, the sampling rate is usually set to be more than two times as high as the sampling rate limit. Therefore, $X(\omega)$ is naturally constrained within the range of $[-\pi/2,\pi/2]$, and N is usually set to be 0. At this point, there is no need to perform any sampling rate conversion. In fact, $\tilde{\phi}(t)$ is a dual scaling function of $\phi(t)$, which means when $[\Phi,\Phi](\omega)=1$, then $\tilde{\phi}(t)=\bar{\phi}(t)$. For field applications, it is excessively complicated to choose the proper $q(n)$ through Equation (4.8). Other alternatives can be used as practical selection algorithms for pre-sampling filters in different applications, as introduced in Section 4.3.

4.3 Selection algorithms for pre-sampling filters

4.3.1 Direct method

In practical applications, when we neither plot the time–frequency analysis nor extract the fractal exponent, but only decompose signals of different frequencies, we usually let the pre-filter $q(n)=\delta(n)$, and regard the sampling signal $x_0(n)$ as the measurement on the sampling space V_N. This is the commonly used Mallat direct decomposition based on multiresolution. In general situations, $x_0(n)$ does not denote the coefficients of the scale function $x(t)$; thus, it does not clarify any issues if we use the components $\{c_{j,k}\}, \{d_{j,k}\}$ obtained by direct decomposition to perform the time–frequency analysis. To be aware of the components in a chosen frequency band, we can perform spectrum analysis using fast Fourier transform (FFT). However, it is not enough to use Mallat pyramidal decompositions in the spectrum analysis. The graphical display algorithm is also needed to avoid too few data points.

4.3.2 *Wavelet transform–based method*

Replacing the Shannon sampling function $\sin c(\pi t)$ with $\delta(t)$, we can assume $\operatorname{Sup}\tilde{\phi} = [0, L]$; therefore, we get

$$q(n) = \tilde{\phi}(n), n = 1, L \tag{4.10}$$

Herein,

$$c_{N,k} = \int_{-\infty}^{\infty} x_0(t)\tilde{\phi}(t-n)dt = \sum_j x_0(k)\tilde{\phi}(k-n) \tag{4.11}$$

In fact, $c_{N,k}$ is the expansion coefficients of $x_0(n)$ on the scale function V_0. The algorithm is called the wavelet transform–based method for this reason. The calculation of pre-filter $q(n)$ in Equation (4.9) requires the integer values of the dual of the scale function $\tilde{\phi}$, which could be obtained by solving the system of linear equations established by the two-scale relation of the multiresolution analysis.

4.3.3 *Sampling function–based method*

The sampling function method is defined as the derivation of $q(n)$ in Equation (4.10), where $q(n)$ could be previously obtained from $\operatorname{sinc}\pi t$ and $\tilde{\phi}(t-n)$ with the Romberg method. The pre-filters used in the sampling function–based method and the wavelet transform–based method are of infinite length, which should be truncated in field applications. The truncation will unavoidably generate errors, which makes the measurements over the sampling space different from the optimal approximation of the signal on the sampling space.

4.3.4 *The optimal method based on minimum error of wavelet coefficients*

According to multiresolution analysis theory, if we assume that the sequence $c_{N,n}$ is the expansion coefficients of $x(t)$ on the scale function, then the discrete wavelet coefficients will be the exact decomposition of the signal. Now, let us consider the error between $c_{N,n}$ and the discrete wavelet transform (DWT) coefficients of $c'_{N,n}$ filtered by Equation (4.7).

$$d_{j,k} - d_{j,k}^{(s)} = DWT\left\{c_{N,n} - c'_{N,n}; j, k\right\} \tag{4.12}$$

The DWT coefficients error could be obtained by two steps:

1. First, we can obtain this equation (the process is omitted):

$$c_{N,n} = \int_{-\infty}^{\infty} x\left(t/2^N\right)\tilde{\phi}\left(2^N t - n\right)dt = \frac{2^N}{2\pi}\int_{-\infty}^{\infty} X\left(-2^N\omega\right)\widetilde{\Phi}\left(\omega\right)e^{-i\omega n}d\omega \tag{4.13}$$

$$c'_{N,n} = \sum_m x\left(m/2^N\right)q\left(m-n\right) = \frac{2^N}{2\pi}\int_{-\infty}^{\infty} X\left(-2^N\omega\right)Q\left(\omega\right)e^{-i\omega n}d\omega \tag{4.14}$$

Therefore,

$$c_{N,n} - c'_{N,n} = \frac{2^N}{2\pi}\int_{-\infty}^{\infty} X\left(-2^N\omega\right)\left[\widetilde{\Phi}\left(\omega\right) - Q\left(\omega\right)\right]e^{-i\omega n}d\omega \tag{4.15}$$

2. Second, corresponding to discrete orthogonal wavelet decomposition, the norm error of DWT is described according to the energy conservation principle by

$$\sum_{j\le N-1}\sum_k \left|d_{j,k} - d_{j,k}^{(s)}\right|^2 = \sum_n \left|c_{N,n} - c'_{N,n}\right|^2 \tag{4.16}$$

Particularly, if $x(t)$ is the $2^N\pi$ band-limited signal, Equation (4.16) yields

$$\sum_{j\le N-1}\sum_k \left|d_{j,k} - d_{j,k}^{(s)}\right|^2 = \frac{2^{N-1}}{\pi}\int_{-\pi}^{\pi}\left|X\left(-2^N\omega\right)\right|^2\left|Q\left(\omega\right) - \widetilde{\Phi}\left(\omega\right)\right|^2 d\omega \tag{4.17}$$

Letting $C_{x,\varphi}(q) = \int_{-\pi}^{\pi}\left|X(-2^N\omega)\right|^2\left|Q(\omega) - \widetilde{\Phi}(\omega)\right|^2 d\omega$, and assuming that the norm error of DWT is set to be minimal (i.e., let $C_{x,\varphi}(q)$ equal its minimal value), we can get the optimized pre-filter $Q(\omega)$. For biorthogonal wavelet transform, we can obtain the same $C_{x,\varphi}(q)$. Expanding $C_{x,\varphi}(q)$ to $q(n)$, we can get

$$\begin{aligned} C_{x,\varphi}\left(q\right) &= \int_{-\pi}^{\pi}\left|X\left(-2^N\omega\right)\right|^2\left|\sum_n q\left(n\right)e^{-i\omega n} - \widetilde{\Phi}\left(\omega\right)\right|^2 d\omega \\ &= A\sum_n q^2\left(n\right) - \sum_{n<m} B_{n,m}q\left(n\right)q\left(m\right) + \sum_n C_n q\left(n\right) + D \end{aligned} \tag{4.18}$$

$$\begin{aligned} A &= \int_{-\pi}^{\pi}\left|X\left(-2^N\omega\right)\right|^2 d\omega, \\ B_{n,m} &= 2\int_{-\pi}^{\pi}\left|X\left(-2^N\omega\right)\right|^2\cos\left(\left(n-m\right)\omega\right)d\omega, \\ C_n &= \int_{-\pi}^{\pi}\left|X\left(-2^N\omega\right)\right|^2\operatorname{Re}\left(\widetilde{\Phi}\left(\omega\right)e^{-i\omega n}\right)d\omega, \\ D &= \int_{-\pi}^{\pi}\left|X\left(-2^N\omega\right)\right|^2\left|\widetilde{\Phi}\left(\omega\right)\right|^2 d\omega \end{aligned} \tag{4.19}$$

Hence, the solution of $q(n)$ will be a process of solving the equation $\partial C_{x,\varphi}(q)/\partial q(n) = 0$. Assuming the pre-filter is a finite impulse response (FIR) filter with a length of L, the optimized filter $q_0(n)$ $(L_1 \le n \le L_2 - 1)$ could be the solution to a system of linear equations. From the derivation above, we observe that when $Q(\omega) = \widetilde{\Phi}(\omega)$, $C_{x,\varphi}(q) = 0$. The optimized pre-filter is equal to the pre-filter obtained by the wavelet-based method. In addition, the computation of the optimized pre-filter depends on the Fourier spectrum of the signal; thus, the filter well matches the frequency characteristics of the signal. The length of the pre-filter is limited. The error between the DWT decomposition results and the theoretical decomposition results is partly reduced as well.

4.3.5 Simulation example

In order to analyze the errors of the DWT coefficients after the processing of pre-filters, which are calculated by the computational methods discussed above, we assume the signal $x(t)$ is a $2^6\pi$ band-limited signal. The Fourier transform of the signal is denoted by

$$X(\omega) = \begin{cases} e^{-(\omega/100)^2}, |\omega| < 2^6\pi \\ 0, \textit{others} \end{cases} \tag{4.20}$$

Daubechies orthogonal wavelet basis db2 is adopted to compute the pre-filter coefficients according to the aforementioned four methods and to perform the discrete wavelet transform after the pre-filtering process. The filter length is $L = 8$. The norm errors between the theoretical wavelet decomposition coefficients and the practical wavelet decomposition coefficients can be calculated by Equation (4.18). The results are shown in Figure 4.2, Figure 4.3, Table 4.1, and Table 4.2. The DWT coefficients error is an absolute error in Figure 4.2. For the convenience of plotting, the error of DWT is set to be dB = 10log (DWT errors) in Figure 4.3. Figure 4.2 shows that among the four pre-filter calculation methods, even though

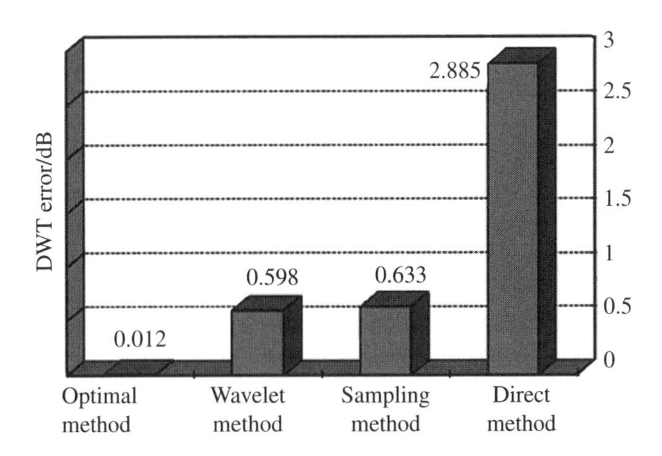

Figure 4.2 Comparison of the DWT errors using four different methods

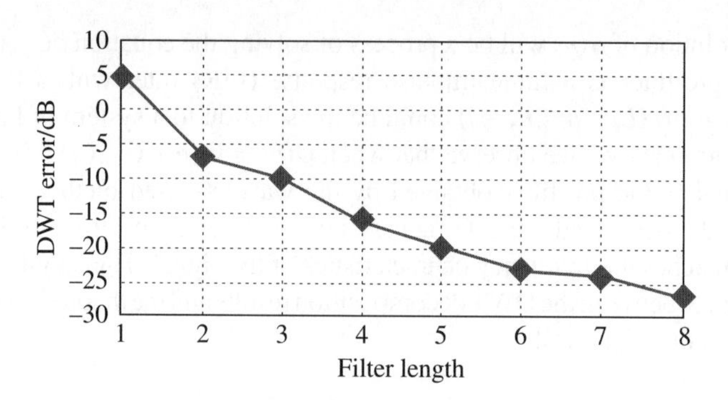

Figure 4.3 Variation of DWT errors (dB) under different filter lengths

Table 4.1 The filter using the orthogonal wavelet method

Length L	$q(n), -L/2 \le n \le L/2-1$							
8	−0.0435	0.0618	−0.1575	0.8097	0.3779	−0.1036	0.0613	−0.0437

Table 4.2 The filter using the optimized orthogonal wavelet method

Length L	$q_0(n), 0 \le n \le L-1$
1	1.0000
2	0.6291 0.5524
3	0.6178 0.5884 −0.1873
4	0.6154 0.5899 −0.1974 0.0980
5	0.6162 0.5918 −0.2005 0.0992 −0.0589
6	0.6131 0.5945 −0.2029 0.1056 −0.0654 0.0418
7	0.6123 0.5957 −0.2116 0.1077 −0.0683 0.0508 −0.0320
8	0.6122 0.5958 −0.2116 0.1093 −0.0688 0.0523 −0.0409 0.0253

the wavelet-based method is the precise decomposition of discrete wavelet transform, the error is great owing to the truncation of the pre-filter. For the Mallat algorithm, the error between discrete wavelet decomposition and theoretical decomposition is maximum because the Mallat algorithm adopts direct sampling for the decomposition process. The optimized method produces the lowest DWT errors after the optimized filtering process because it matches the frequency characteristics of the signal. Moreover, the DWT error is reversely proportional to the length of the filter (when L=1, it is equivalent to the Mallat algorithm).

4.4 Problems in the implementation of wavelet transform

In the fast algorithm of discrete wavelet transform, a different selection algorithm of the pre-filter refers to different discrete wavelet transform methods and application fields. The direct method cannot be used directly in the illustration of time–frequency curves and the extraction of fractal exponents because it results in the highest norm errors of the discrete wavelet coefficients and the theoretical wavelet coefficients. The decompositions of the wavelet-based selection method correspond to components of different frequencies and reflect the discrete wavelet coefficients precisely. The sampling function method obtains the measurement values by adding interpolations to the sampling signal according to the sampling theorem, which better reveals the features of the analog signal. However, when the pre-filter is truncated, higher errors will be produced. The optimized pre-filter algorithm has features such as signal frequency tracking and low error, and it is partly flexible because the length of the filter could be selected in accordance with the accuracy and decomposition rate in different applications. Consequently, we can choose different pre-filters according to different applications and requirements. It is worth noting that the influence of different pre-filters on the phase of signals needs further investigation.

Reference

[1] He Z. Y., The electric power system transient signal wavelet analysis method and its application on the non-unit transient protection principle. *Southwest Jiaotong University*, 2000.

Further reading

Chen Z. X., The analysis algorithm and application of wavelet. *Xi'an Jiaotong University Press*,1998.

Chen Z. X., Li S. G., Numerical approximations and numerical solutions of differential equations. *Xi'an Jiaotong University Press*,2000.

He Z. Y., Dai X. W., Qian Q. Q., Prefilter selection methods in fast algorithm of discrete wavelet transform. *Journal of Southwest Jiaotong University*, vol. 35, no. 2, pp. 183–187, 2000.

Tewfik A. H., Sinha D., Jorgensen P., On the optimal choice of a wavelet for signal representation. *IEEE Transactions on Information Theory*, vol. 38, no. 2, pp. 747–765, 1992.

5

Wavelet Basis Selection for Transient Signal Analysis of Power Systems

Wavelet transform is different from Fourier transform, which has only one function or transformation kernel. Theoretically, there are infinite wavelet bases or transformation kernels to meet the requirements of various problems. However, one should select the proper wavelet base for a specific problem, or the wavelet base will not perform satisfactorily. For example, some wavelet bases are good for sharp variation signals but bad for nonsharp variation signals. We can see that many characteristics of wavelet bases are correlative and restrict each other from the analysis in Chapter 2. Therefore, the selection of wavelet bases cannot isolate from the application object and the key points to the problem.

Up to now, there is no unified model to make the wavelet coefficients on the selected wavelet base have the most concentrated energy and good detection and location performance. Generally speaking, in the wavelet transform of power transient signals, the selection of wavelet bases should consider the following factors in terms of the transient sources and analysis purposes. These factors include computation speed, reconstruction precision, wavelet base attenuation, phase characteristics, time–frequency resolution, de-noising ability, orthogonality, vanishing moment, and so on.

5.1 The sources and features of power system transient signals

The power system transients can be categorized into two groups. The first group are the disturbances that do not affect the normal operation of power systems, lines, or devices, such as oscillation transients, impulsive transients that influence the power quality, and the noises generated by the power electronic switch and the nonlinear devices. The second group are the fault transients induced by abnormal operation, such as fault-generated

Wavelet Analysis and Transient Signal Processing Applications for Power Systems, First Edition. Zhengyou He.

Table 5.1 The common disturbance types

Types	Causes	Duration time
Oscillation transients	Load switching, capacitor switching Transformer excitation	μs–ms
Impulsive transients	Inductive circuit switching Lightning strike	μs–s
Voltage sag/swell	Remote motor start and stop	μs–ms
Noises	Abnormal ground connection Solid-state switching transients	–

traveling waves and the fault arc. These disturbances have different duration times according to the causes of disturbances, as shown in Table 5.1.

The fault transients include transmission line transients and electrical equipment transients. The transmission line fault-generated high-frequency components contain abundant information, which is from DC components to the hundreds or thousands Hz. There are mainly the three following reasons for high-frequency fault transients' occurrence. First, the occurrence of a fault is equivalent in magnitude and opposite in sign to the pre-fault voltage at the fault point. Thus, the voltage at the fault point suddenly reduces to a low value. This sudden change produces a high-frequency electromagnetic impulse called a traveling wave. Second, the reflection and refraction of traveling waves result in the distortion of transient voltage and current. The distortion corresponds to the broadband high-frequency components in the frequency domain. Third, the nonlinear characteristic of a fault arc produces high-frequency components with a wide frequency band.

According to the measured record and analysis of some power transient signals, the power system transients can be expressed mathematically by damped transients shown in Equation (5.1) and by oscillatory damped transients shown in Equation (5.2). The noise is considered as Gaussian white noise.

$$I_1 = I_m e^{-t/\tau} \text{ or } I_1 = I_{m1} e^{-t/\tau_1} - I_{m2} e^{-t/\tau_2} \tag{5.1}$$

$$I_2 = I_m \sin(2\pi f t) e^{-t/\tau} \text{ or } I_2 = I_m \sin(2\pi f t)\left(e^{-t/\tau_1} - e^{-t/\tau_2}\right) \tag{5.2}$$

In an extra-high voltage transmission line, the rise time of traveling waves is very short, whereas the fall time is three times longer than the rise time. The rise time is about 0.7–8.3 μs. Therefore, in the following wavelet analysis study, τ_1 is 2 μs and τ_2 is 6 μs.

5.2 Detection of a high-order singular signal

A function $x(t) \in L^2(\mathbb{R})$ is said to be of high-order singularity at t_0, if $x^{(k)}(t_0 - 0) \neq x^{(k)}(t_0 + 0)$. $x^{(k)}(t_0 - 0)$ and $x^{(k)}(t_0 + 0)$ are, respectively, the left kth derivative and right kth derivative. Since the power system transients are continuous signals and have high-order singularity, wavelet transform is able to detect the sharp variation points of signals.

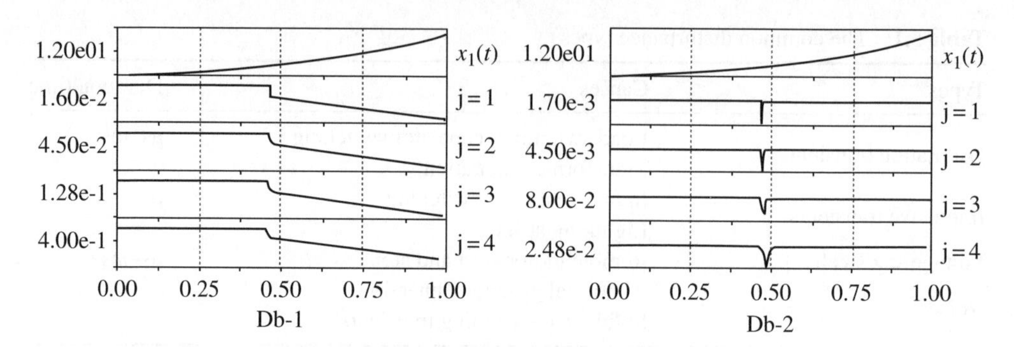

Figure 5.1 Wavelet decomposition of signal $x_1(t)$

Moreover, wavelet transform is capable of detecting the high-order singular points. Consider the following three signals.

$$x_1(t)=\begin{cases}5t, t\in[0,0.46875)\\ t^2/0.09375, t\in[0.46875,1]\end{cases}\tag{5.3}$$

$$x_2(t)=\begin{cases}2.5t^2, t\in[0,0.46875)\\ t^3/0.28125+2.5\times0.46875^2-0.46875^3/0.28125, t\in[0.46875,1]\end{cases}\tag{5.4}$$

$$x_3(t)=\begin{cases}5t^3/6, t\in[0,0.46875)\\ t^4/1.125+\left(2.5\times0.46875^2-0.46875/0.2815\right)t\\ +5\times0.46875^3/6, t\in[0.46875,1]\end{cases}\tag{5.5}$$

where $x_1(t)$, $x_2(t)$, and $x_3(t)$ are continuous signals. $x_3^{(3)}(t_0+0)=x_2^{(2)}(t_0+0)=x_1^{(1)}(t_0+0)=10$, and $x_3^{(3)}(t_0-0)=x_2^{(2)}(t_0-0)=x_1^{(1)}(t_0-0)=5$. The Haar wavelet (db1 wavelet), db2 wavelet, and db3 wavelet are, respectively, used as the wavelet basis functions to decompose $x_1(t)$, $x_2(t)$, and $x_3(t)$. The decomposition results are shown in Figures 5.1, 5.2, and 5.3. It can be seen that the Haar wavelet cannot detect the high-order singular points $x_2(t_0)$ and $x_3(t_0)$ because it has only one vanishing moment. The wavelet basis db2 has a low-order vanishing moment, so it is unable to detect the high-order singular point $x_3(t_0)$. The db3 wavelet is able easily to detect the high-order singular points since it has three vanishing moments. Of course, the db3 wavelet is also capable of detecting the low-order singular points, which coincides with the wavelet theory of singular signals. Besides, a large number of simulation experiments find that the wavelet with low-order vanishing moments will be able to detect the higher order singular points if the wavelet transform scales increase. However, the location ability weakens. Therefore, we should select the wavelet basis with some vanishing moments for power system transient detection and feature extraction.

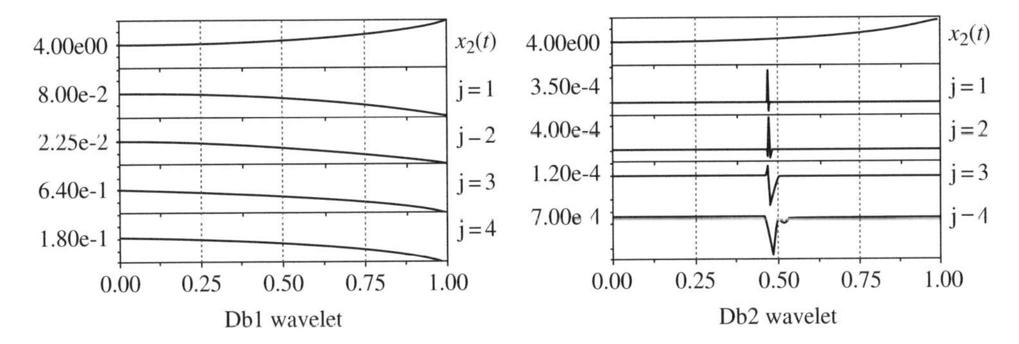

Figure 5.2 Wavelet decomposition of signal $x_2(t)$

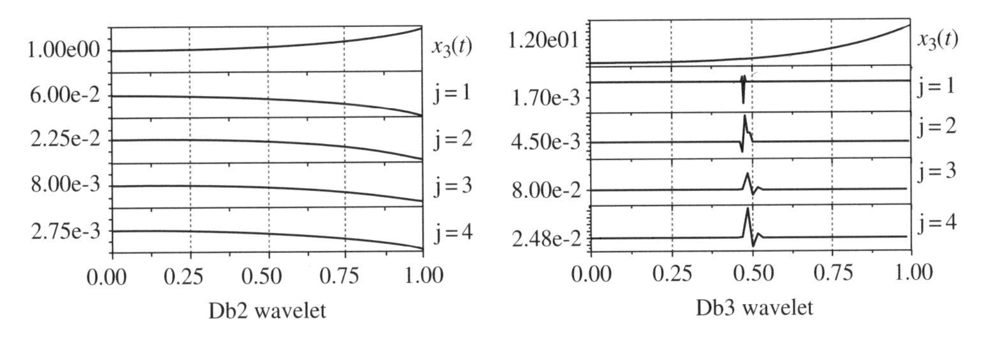

Figure 5.3 Wavelet decomposition of signal $x_3(t)$

5.3 Detection of weak transients of low-frequency carriers

The four simulated transient signals shown in Equations (5.1) and (5.2) are superimposed on the mixed carrier, which includes multiple frequencies. The frequencies of the mixed carrier are lower than the transient signals. The maximum amplitude ratio between the transient signals and carrier is 1:10. The signal $x(t)$ synthesized with transient signals and the mixed carrier is shown in Figure 5.4. The unilateral damped transient and bilateral damped transient are, respectively, superimposed at sampling points 50 and 150. The unilateral oscillatory damped transient and bilateral damped transient are, respectively, superimposed at sampling points 150 and 250. Figure 5.4a–d shows the wavelet decomposition results of $x(t)$ with a db4 wavelet, Mexican hat wavelet, Morlet wavelet, and Meyer wavelet. Some conclusions drawn from the simulation results are as follows:

1. At the same scale, the Mexican hat wavelet has the worst detection ability. The db4 wavelet takes second place. The Morlet wavelet and Meyer wavelet are good at detecting weak transients. From the analysis in Chapter 2, we know that the relation among center frequencies of four wavelets is $\omega^*_{db4} > \omega^*_{morlet} > \omega^*_{Meyer} > \omega^*_{Mex}$. That is to say, at the same scale the center frequency of a Mexican wavelet is small and will be mixed with the low-frequency carrier components. The center frequencies of a db4 wavelet and

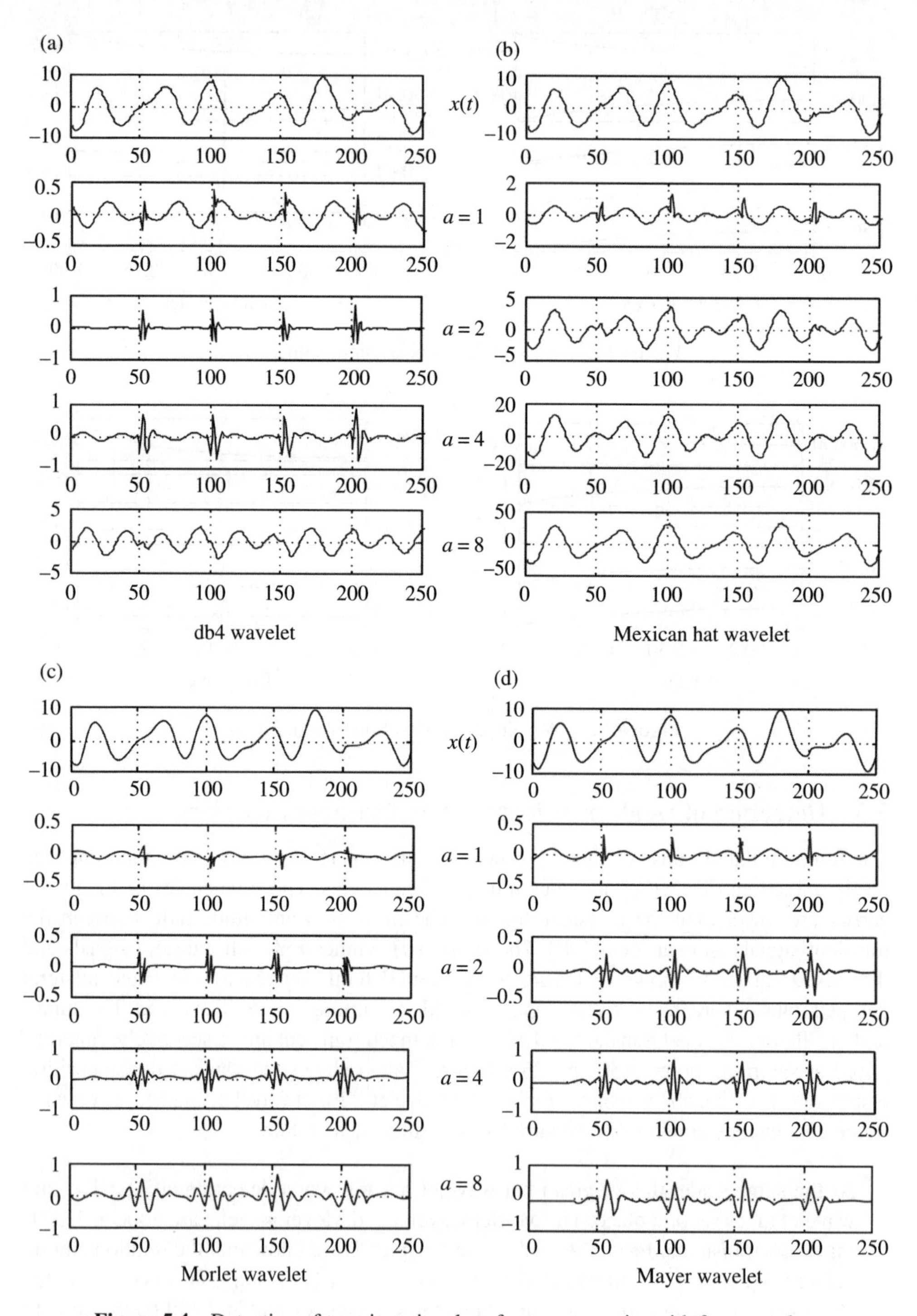

Figure 5.4 Detection of transients in a low-frequency carrier with four wavelets

Morlet wavelet are high, which makes it easy to extract the transient signals and restrain the low-frequency carrier with these two wavelets. Hence, we should select a wavelet basis with a high center frequency to detect the transients in the carrier.

2. The relation among the frequency window radii of four wavelets is $\Delta\omega_{db4} > \Delta\omega_{Meyer} > \Delta\omega_{morlet} > \Delta\omega_{Mex}$. Figure 5.4b shows that the frequency window radius of a Mexican hat wavelet is too small to match the wide-band transients. Thus, the Mexican hat wavelet goes against the extraction of transient signals at arbitrary scales. On the other hand, the db4 wavelet has a wide frequency window to mix abundant carrier components at scale $a = 2^4$. The Morlet wavelet and Meyer wavelet can detect and extract the transient features remarkably at multiple scales since they have adequate frequency window width to match the wide-band transient signals. The detection of weak transients in a strong carrier is related to the width of the frequency window. Therefore, the selection of wavelet bases and scales should consider the effect of low-frequency components.
3. The simulation results above indicate that the detection ability has a bearing on the amplitude-frequency characteristics of wavelets, but basically has nothing to do with wavelets' symmetry, orthogonality, compact support, and so on.
4. The damped transients and oscillatory damped transients perform different symmetries at different scales, which is helpful for classification of transient signals. The continuous wavelet transform could observe and analyze the transform results at any scale. However, the detection abilities are different at different scales. Hence, the proper wavelet base and detection scale should be selected to depress the low-frequency carrier.

5.4 Detection of transients in narrow-band interference

On-line partial discharge (PD) detection is a method to detect the insulating property of power equipment. One of the difficult problems is how to eliminate various interferences, especially narrow-band interference. When applying continuous wavelet transform to detect PD signals from narrow-band interference, the selection of wavelet bases is the same as in Section 5.3. That is to say, the wavelet base with a proper time–frequency center, frequency window width, and scale should be selected to depress the narrow-band interference. Additionally, this section will illustrate the effects of scales and wavelet filter properties on transient detection from the discrete wavelet transform standpoint.

PD signals also can be simulated by damped signals, and oscillatory damped signals as described in Equations (5.1) and (5.2). The narrow-band interference is synthesized with multiple sinusoidal signals that have different frequencies. This book uses 10 sinusoidal signals whose frequencies are from 50 kHz to 500 kHz to express the narrow-band interference. The interference frequency is usually 30 kHz to 300 kHz. Four impulsive transients are superimposed at sampling point 150 to express PD signals. The time interval is $20\,\mu s$. $f_c = 300\,kHz$, $\tau = 5\mu s$. The maximum amplitude ratio between transient signals and interference is 1:10. The sampling frequency is 1 MHz. Figures 5.5 and 5.6 show the discrete wavelet transform results of signal $x(t)$. In Figure 5.5, $x(t)$ is synthesized with damped transient and narrow-band interference. In Figure 5.6, $x(t)$ is synthesized with oscillatory damped transient and narrow-band interference.

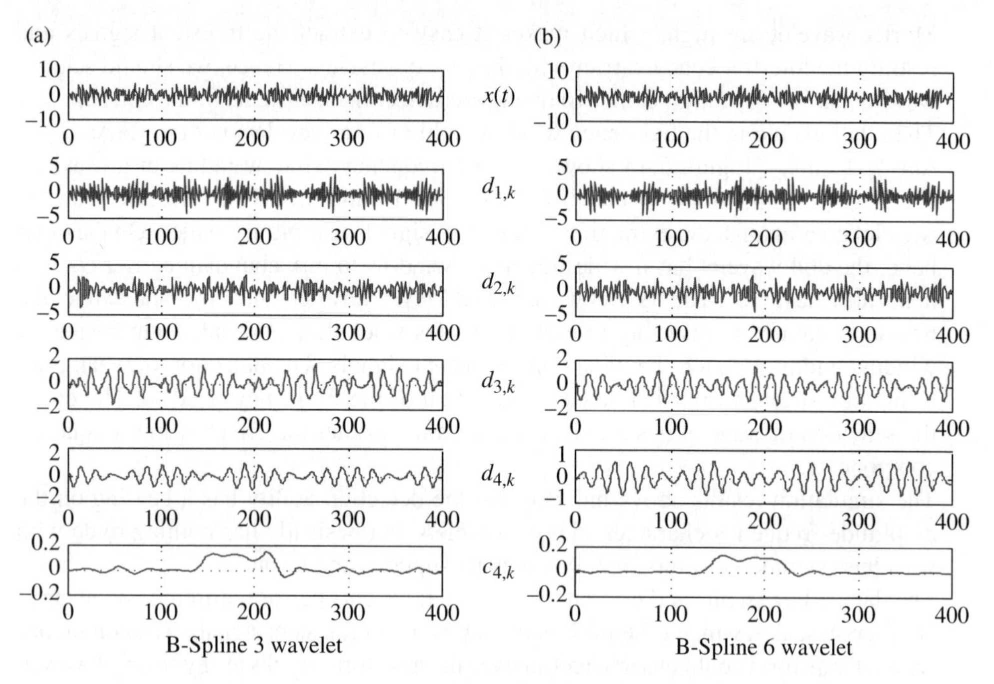

Figure 5.5 Detection of transients in narrow-band interference

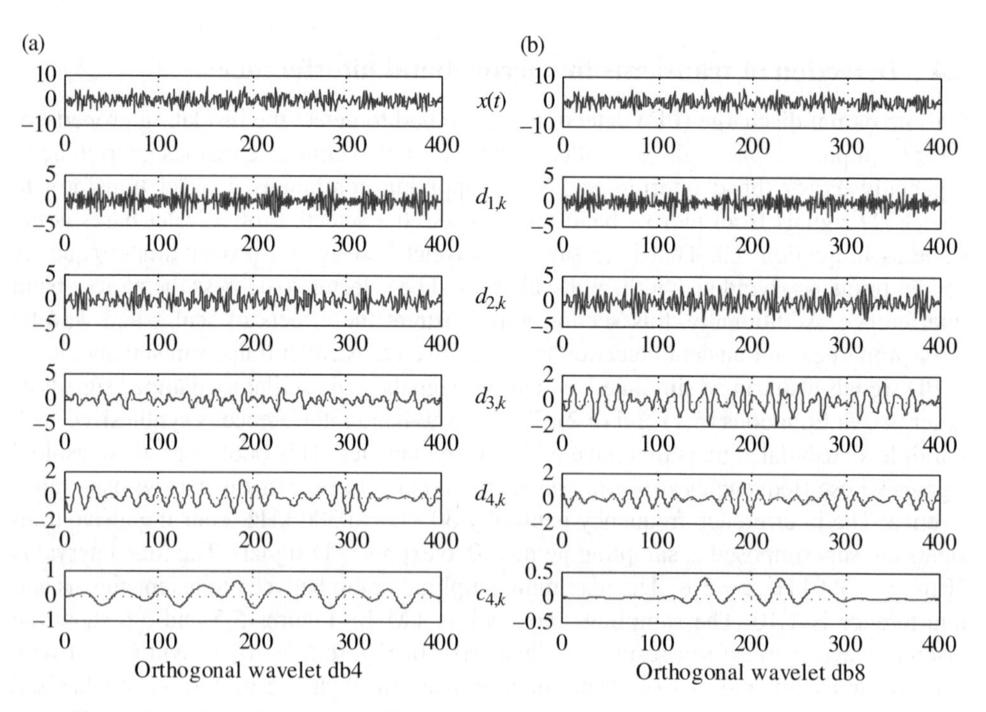

Figure 5.6 Detection of an oscillatory damped transient in narrow-band interference

1. The transient signal can be detected successfully by approximation at scale 4. The detection performance of a high-order B-spline wavelet and high-order db wavelet is better than that of a low-order B-spline wavelet and low-order db wavelet. The amplitude-frequency characteristics of the B-spline wavelet and db wavelet demonstrate that the low-pass filter $H(\omega)$ corresponding to the scaling function and high-pass filter $G(\omega)$ corresponding to the wavelet function of a high-order wavelet tend to 0 faster and have a narrower transition band than those of a low-order wavelet. The high-order wavelet therefore has stronger detection ability due to its frequency division ability and high-frequency elimination. However, a high-order wavelet corresponds to longer filter length. In other words, the scaling function and wavelet function have long support and weak space localization properties. We should consider both time resolution and frequency resolution to obtain better feature extraction and detection performance.
2. Comparing Figure 5.5 with Figure 5.6, we know that when the high-order wavelet is selected, the energy concentrated in the low-frequency band of a damped transient is greater than that of an oscillatory damped transient. The feature of a damped transient can be extracted well, and the four damped transients can be separated. When increasing the sampling frequency, the four damped transients also can be separated. Besides, the damped coefficient will influence the feature extraction. The smaller the damped coefficient is, the weaker the extraction performance is.

5.5 Data compression and de-noising of transients

De-noising, filtering, and data compression all require signal reconstruction. In order to ensure the signal reconstruction is stable and undistorted, it is necessary to select the wavelet that has a linear phase. This section emphatically studies the de-noising ability of wavelets generated by Gauss functions and its derivatives and B-spline functions.

As known in Chapter 3, the zero-crossing density of a differentiable Gauss process is $\sqrt{\left|R^{(2)}(0)/\pi^2 R(0)\right|}$, where $R(\tau)$ is the autocorrelation function of a derivable Gauss process and $R^{(n)}(\tau)$ is the nth derivative of $R(\tau)$. If the process is twice differentiable, the density of its local extrema is equal to the density of its zero crossings, which is given by

$$d = \sqrt{\left|R^{(4)}(0)/\pi^2 R(0)\right|} \tag{5.6}$$

We can prove that $R^{(4)}(0) = n_0^2 \left\|x^{(2)}\right\|^2$, $R^{(2)}(0) = n_0^2 \left\|x^{(1)}\right\|^2$. Therefore,

$$d = \frac{\left\|x^{(2)}\right\|}{\pi \left\|x^{(1)}\right\|} \tag{5.7}$$

For first-order Gauss-H wavelet and quadratic B-spline wavelet, the calculated average densities are $d_G \approx 1.58/(\pi\sigma)$ and $d_B \approx 2.74/(\pi s)$. The window of a first-order Gauss-H wavelet is $[-3\sqrt{2}\sigma, 3\sqrt{2}\sigma]$. The window of a quadratic B-spline wavelet at scale s is $[-2s+1, 2s+1]$. If the two windows are the same, we have $s = 2.12\sigma$. Thus, $d_B = 1.29/(\pi\sigma)$. From the analysis above, we know $d_G > d_B$. The modulus maximum density of noise obtained by a first-order Gauss-H wavelet is greater than that of noise obtained by a B-spline wavelet. That is to say, the B-spline wavelet has better noise suppression capability.

5.6 Location of transients

In fault location based on high-frequency transients, the location accuracy is crucial and affects the location error. It is therefore of great significance to study the effect of a wavelet base on location accuracy. The location property is inversely proportional to the time width of the wavelet at the same scale. A Haar wavelet has two filtering coefficients and has the best time domain localization capability. However, the frequency domain division capability of a Haar wavelet is worst. The accurate location of sharp variation points requires that the wavelet basis have a linear phase. Here, we compare the location accuracy of the commonly used Gaussian wavelet and B-spline wavelet.

Let the detected and located signal be $x(t) = u(t+d/2) - u(t-d/2)$, where d is the bilateral sharp variation distance. The derivative of $X(t)$, which is the response of $x(t)$, is described as

$$X'(t) = \left(\phi(t) * x(t)\right)' = \phi'(t+d/2) - \phi'(t-d/2) \tag{5.8}$$

At $t = -d/2$, $X'(t)$ can be accurately approximated in the form of a Taylor series. After omitting the high-order infinitesimal, we obtain

$$X'(t) = \phi'(0) + \phi''(0)(t+d/2) - \phi'(-d) - \phi''(-d)(t+d/2) \tag{5.9}$$

From the symmetry of $\phi(t)$, we know that $\phi'(0) = 0$. Letting t_0 be the zero-crossing point of $X'(t)$, we have

$$t_0 = \frac{\phi'(-d)}{\phi''(0) - \phi''(-d)} - \frac{d}{2} \tag{5.10}$$

where t_0 corresponds to the maximum. After wavelet transform, t_0 can be used to locate the signal. When $\phi'(-d) \neq 0$, $t_0 \neq -d/2$. At this time, the sharp variation point shifts. Assume the support of a filtering function is $[-w/2, w/2]$. When $d > w/2$, $\phi'(-d) = 0$, the sharp variation point can be located accurately. When the scaling function $\phi(t)$ of a Gauss-H1 wavelet is a Gaussian function, $w = 6\sqrt{2}\sigma$, where σ is scale. When the scaling function $\phi(t)$ of a B-spline 2 wavelet is a quadratic B-spline function, $w = 3s$, where s is scale. Correspondingly, the scale for the accurate location of a sharp variation point is $\sigma < d/3\sqrt{2}$, $s < d/3$. As shown in Figure 5.7, when the filtering window widths are the same, a first-order Gauss-H wavelet has better location properties than a quadratic B-spline wavelet.

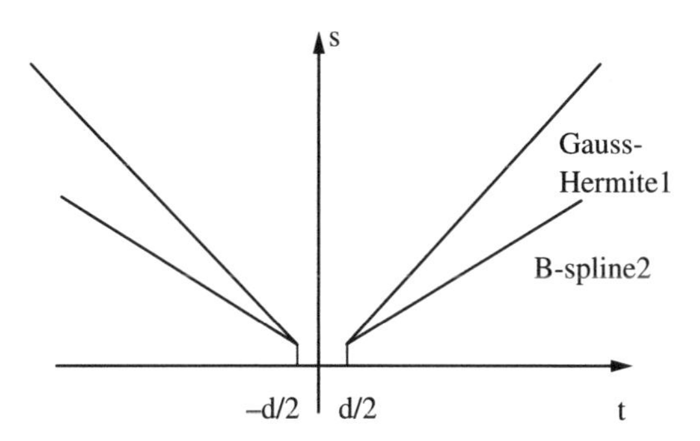

Figure 5.7 The property comparison of a Gauss-H wavelet and B-spline wavelet

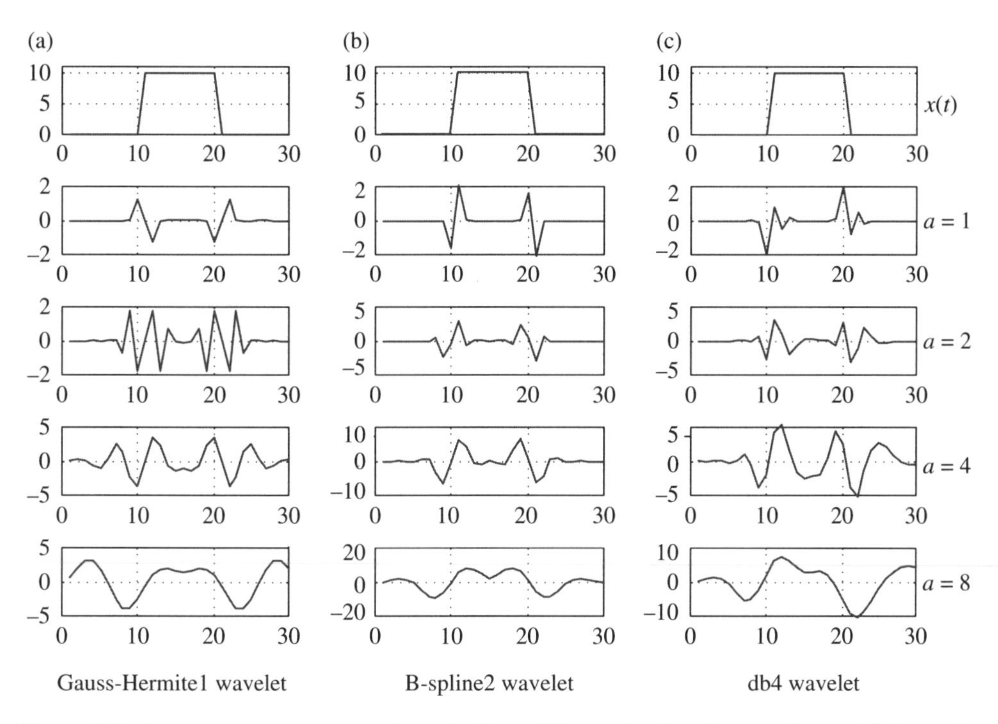

Figure 5.8 Location property comparison of a Gauss-H1 wavelet, B-spline wavelet, and db4 wavelet

Figure 5.8 shows the wavelet decomposition results of a rectangular signal using a Gauss-H1 wavelet, B-spline 2 wavelet, and db4 wavelet. In transient signal locations, one finds the following conclusions:

1. The first-order Gauss-H wavelet and quadratic B-spline wavelet have better location properties for a sharp variation point than a db4 wavelet. This is because a db4 wavelet does not have a linear phase. Hence, the symmetry wavelets or antisymmetry wavelets should be selected for sharp variation point location since they have a linear phase to ensure the location accuracy.

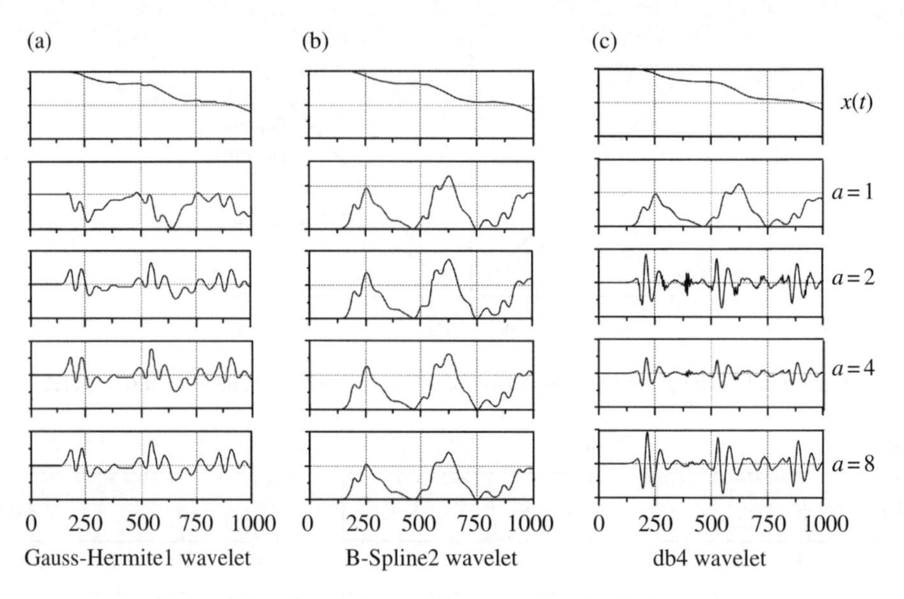

Figure 5.9 Comparison of three wavelets for fault location

2. Gauss-H series wavelets have better location properties than B-spline series wavelets, which is in accordance with theoretical analysis. Besides, the location of sharp variation points sometimes corresponds to the zero-crossing points of the wavelet transform, and sometimes they correspond to the modulus maxima. Figure 5.8 shows that the sharp variation points correspond to zero-crossing points for symmetric wavelets but to extrema points for asymmetric wavelets. This is contrary to the sharp variation at peak point. Due to the fact that the zero-crossing points are subjected to noise, the edge location should select the asymmetric wavelets while the peak location should select the symmetry wavelets.
3. The power system transients are close to sharp variation at peak points. We therefore usually select symmetric wavelets. Figure 5.9 shows the fault location results for the transmission line in Appendix A.2. The fault occurred at the maximum of the phase-A voltage (t = 15 ms). As can be seen in Figure 5.9, a quadratic B-spline wavelet is capable of locating the wave fronts of an initial traveling wave and the reflected traveling wave from fault points. The fault location can be achieved using the time difference between the first-arrival traveling wave and the second-arrival traveling wave.
4. If other conditions are all satisfied, we should select the wavelet basis with a small time radius and few filtering coefficients as much as possible.

5.7 Selection of wavelet decomposition scales

The selection of decomposition scales is very important for power system transient analysis. The decomposition scales are selected according to the frequency division characteristics, as shown in Figure 5.10.

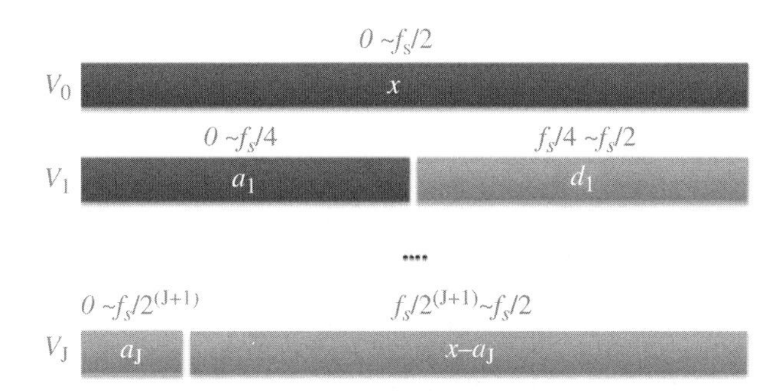

Figure 5.10 Frequency band diagram of discrete wavelet decomposition (f_s is the sampling frequency)

Different research objects need different decomposition scales. Assume that the studied signal is a low-frequency oscillation signal. The energy of a low-frequency oscillation signal is mostly concentrated over a frequency band that is lower than 50 Hz. Consequently, we hope that the approximation coefficients carry the information over a frequency band that is lower than 50 Hz. The best decomposition scale can be obtained by converting the sampling frequency to the main frequency of the studied object. In most cases, the frequency band carrying the most energy of the signal is unknown. At this time, other methods should be used to determine the best decomposition scale.

The decomposition scale must satisfy some basic conditions. Assume that ω_c is the sampling frequency of the wavelet basis, ω_0 is the center frequency of wavelet ψ, and $\Delta\omega$ is the frequency width of the wavelet basis. The highest frequency of a wavelet basis should be less than half of the sampling frequency. The scale a should satisfy

$$2(\omega_0 + \Delta\omega)/a \le \omega_c, \text{ i.e. } a \ge 2(\omega_0 + \Delta\omega)/\omega_c \tag{5.11}$$

On the other hand, the sampling number of wavelets should be less than the sampling number of signals. Assuming that the sampling number of the wavelet basis is N_ψ and the sampling number of the signal is N_x, we have

$$aN_\psi < N_x, \text{ that is, } a < N_x / N_\psi \tag{5.12}$$

If the dominant frequency or frequency band of a transient signal is prior known, the scale can be determined by the time–frequency window characteristics of continuous wavelet transform or the frequency division characteristics of discrete wavelet transform. However, most of the transients are uncertain. It is difficult to select a proper scale factor a to make the modulus maximum greater. The research shows that, under the condition of no prior knowledge or lack of knowledge, the scale spectrum of a wavelet transform can solve this problem. The scale spectrum is defined as the distribution of the square of the wavelet transform modulus with time and scales. Figure 5.11 shows the scale spectra of damped transients and oscillatory damped transients, which indicate the evolution of the wavelet transform modulus across scales. The best detection scale can be obtained by the peaks of these moduli. For example, the Morlet wavelet is utilized to decompose

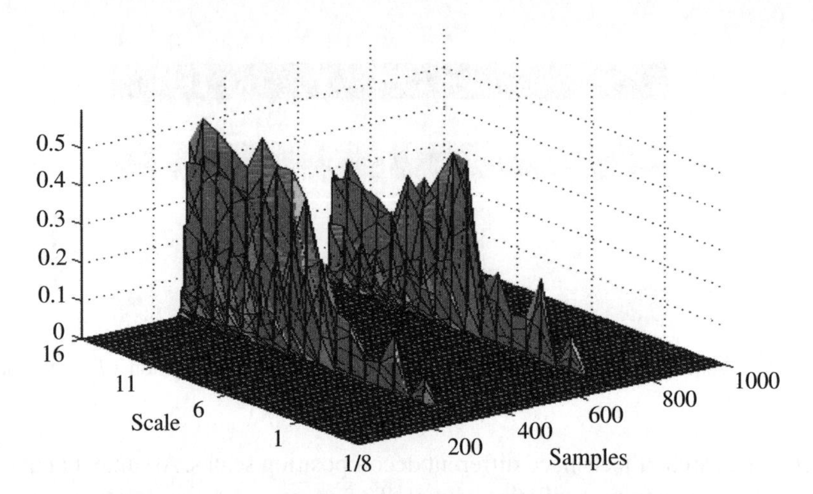

Figure 5.11 Timescale spectra of two transients

transients in Figure 5.11. The best detection scale is $a = 6$ for damped transients and $a = 8$ for oscillatory damped transients. The noise or interference will certainly be considered in the practical application.

One alternative method is to calculate the correlation coefficients of the reconstructed signal and the original signal. Then, the best decomposition scale can be achieved by the setting threshold. Let x be the original signal and y be the reconstructed signal by approximation coefficients. The correlation coefficients between x and y are calculated as

$$\rho(x,y) = \frac{\Sigma x_i y_i - \left(\frac{1}{N}\right)\Sigma x_i \,\Sigma y_i}{\left[\Sigma x_i^2 - \left(\left(\frac{1}{N}\right)\Sigma x_i\right)^2\right] - \left[\Sigma y_i^2 - \left(\left(\frac{1}{N}\right)\Sigma y_i\right)^2\right]} \tag{5.13}$$

where N is the signal length, $i = 1,2,\ldots,N$.

The correlation coefficients are calculated for each scale. The best decomposition scale is obtained by comparing the calculated coefficients with the setting threshold.

The correlation coefficient is only an alternative evaluation criterion. Some scholars have proposed other evaluation criteria to determine the best decomposition scale, such as root-mean-square error, average gray value, entropy, and spatial frequency. The basic principles of these evaluation criteria are coincident. Different evaluation criteria are compared with the threshold to determine the best decomposition scale.

To sum up, one has to select the appropriate wavelet basis for the application of power system transient signals to achieve the satisfied decomposition and feature extraction. The high-order singular transient signals should select the wavelet basis that has a number of vanishing moments. In detection of transients in low-frequency carriers, the wavelet basis with a higher center frequency should be selected to prevent low-frequency components infiltration. When detecting transients in narrow-band interference, the high-order wavelet basis with a narrow transition band and good frequency division capability is needed to eliminate the high-frequency interference and extract the wide-band transient. In the application of de-noising, filtering, and data compression, a B-spline wavelet basis has better noise suppression capabilities than a Gauss-H wavelet because the density of the modulus maxima of a B-spline wavelet is smaller than that of a Gauss-H wavelet. In location of transient signals, a symmetric wavelet basis is recommended. Apart from the selection of a wavelet basis, the appropriate decomposition scale should be considered in terms of the analyzed signal as well.

Further reading

Dugan R.C., Mcgranaghan M.F., Beaty H.W., Electrical power systems quality. *New York: McGraw-Hill*, 1996.

He Z.Y., Qian Q.Q., Mother wavelet option method in the transient signal analysis of electric power systems. *Automation of Electric Power Systems*, vol. 27, no. 10, pp. 45–48, 76, 2003.

Kopf U., Feser K., Rejection of narrow-band noise and repetitive pulses in on-site PD measurements. *IEEE Transactions on Dielectrics and Electrical Insulation*, vol. 2, no, 3, pp. 433–466, 1995.

Kaewpijit S., Le Moigne J., El-Ghazawi T., Automatic reduction of hyperspectral imagery using wavelet spectral analysis. *IEEE Transactions on Geoscience and Remote Sensing*, vol. 41, no. 4, pp. 863–871, 2003.

6

Construction Method of Practical Wavelets in Power System Transient Signal Analysis

For a specific signal, we need to find an appropriate wavelet basis to decompose a signal in the process of its wavelet analysis; sometimes, we also need to construct special wavelet bases for different application purposes. Just as it is important in wavelet theory, the construction of an effective wavelet basis is the foundation of the decomposition of power system transient signals, in the sense that a good wavelet basis helps to converge the energy of each decomposed transient signal on its wavelet basis.

In essence, Mallat multiresolution analysis is a kind of wavelet transform on a "dyadic band" foundation. Mallat multiresolution analysis possesses a relatively wide bandwidth for high-frequency channels and a relatively narrow bandwidth for low-frequency channels. This feature makes it suitable for the analysis of low-frequency signals mixed with pulses or short-time transient signals. However, if the target signal is a high-frequency signal mixed with relatively narrow pulses or transients, dyadic wavelets would be inappropriate for the analysis, because transient signals cannot be well extracted from high-frequency power system carrier signals by dyadic wavelets (e.g., the extraction of partial discharge signals).

To meet the needs of real-time processing of power system signals, various fast transform algorithms are developed. As one of these algorithms, the recursive algorithm possesses unique advantages for its high real-time performance. Besides, because the traveling wave and the high-frequency transient arc signals caused by an EHV (extra-high-voltage) transmission line fault that take up relatively wide bandwidth, the extraction of transient components from a narrow-frequency band seems less effective. Therefore, it is necessary to develop a general construction method of recursive wavelets and an effective algorithm for recursive wavelet transform to extract signal characteristics from specified frequency bands.

Wavelet Analysis and Transient Signal Processing Applications for Power Systems, First Edition. Zhengyou He.

Transient fault signals in power systems are nonstationary signals primarily composed of the fundamental component and lower harmonics. They are also accompanied by instantaneous wide-band high-frequency signals. Thus, it requires effort to develop a method for constructing an optimal wavelet basis for the requirement of power system transient data compression, and for the requirement of fault recording and data transmission in integrated automatic power systems.

6.1 The construction and application of a class of M-band wavelets [1–2]

6.1.1 Construction conditions of M-band wavelets

Assume $\phi(t)$ is a scaling function so that $\phi_{jk} = \left\{M^{-j/2}\phi(M^{-j}t-k)\right\}_{k\in\mathbb{Z}}$ could be employed as the orthonormal basis of the multiresolution space V_j. The corresponding wavelets denoted by $\psi^{(i)}(t)$ have a total number of $M-1$, where $i = 1,2,\cdots M-1$. The scaling function and the wavelet function comply with the double-scaling equation:

$$\begin{cases} \phi(t) = \sqrt{M}\sum_k h_0(k)\phi(Mt-k) \\ \psi^{(i)}(t) = \sqrt{M}\sum_k h_i(k)\phi(Mt-k) \end{cases} \tag{6.1}$$

where $h_i(k)$ is the impulse response sequence of the filter corresponding to the scaling function or the wavelet function. $\left\{\psi_{jk}^{(i)}(t) = M^{-j/2}\psi^{(i)}(M^{-j}t-k)\right\}_{k\in\mathbb{Z}}$ is employed as the orthonormal basis of the ith orthogonal complement $W_j^{(i)}$ of V_j in V_{j-1}.

$$V_j = V_{j+1} \oplus \left(\bigoplus_{i=1}^{M-1} W_{j+1}^{(i)} \right) \tag{6.2}$$

where $\oplus$ signifies the direct sum operation, hence $\left\{M^{-j/2}\psi^{(i)}(M^{-j}t-k)\right\}_{j,k\in\mathbb{Z}}$ forms the orthonormal basis of $L^2(\mathbb{R})$ [3]. The construction of an M-band orthogonal wavelet is equivalent to the design of a perfect reconstruction quadrature mirror filter (PR-QMF) [4], which is expressed by

$$\sum_k h_0(k) = \sqrt{M} \tag{6.3}$$

$$\sum_k h_i(k)h_j(Ml+k) = \delta(l)\delta(i-j)\; i,j = 0,1,2,\cdots M-1 \tag{6.4}$$

$$\sum_l (-1)^l l^m h_0(l) = 0\; m = 1,2,\cdots K \tag{6.5}$$

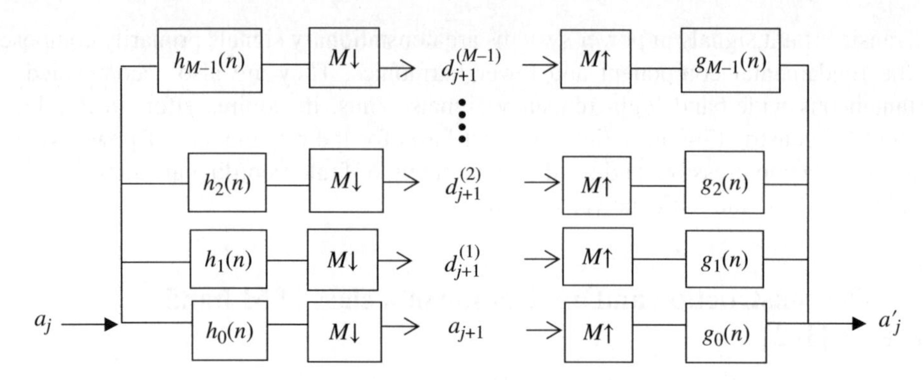

Figure 6.1 The decomposition and reconstruction of an M-band wavelet

Suppose a_j is the jth component of wavelet approximation, then the M-band wavelet decomposition is expressed as

$$\begin{cases} a_{j+1}(n) = \sum_k \bar{h}_0(Mn-k)a_j(k) \\ d_{j+1}^{(i)}(n) = \sum_k \bar{h}_i(Mn-k)a_j(k) \end{cases} \tag{6.6}$$

where $\bar{h}_i(n) = h_i(-n)$. This equation indicates that the approximation component and the detail component of lower resolution could be obtained by filtering the higher resolution approximation component and then sampling the filtered signal M times. Hence, the approximation component at the jth resolution level could be decomposed to an approximation component at the $(j+1)$th resolution level and $M-1$ detail components $d_{j+1}^{(i)}$. Just like the dyadic wavelet transform, in the M-band wavelet transform, we can reconstruct the original signal with the approximation components and detail components. The processes of decomposition and reconstruction are shown in Figure 6.1.

6.1.2 The construction of an M-band wavelet basis based on cosine modulation [5–8]

The reconstruction structure of an M-band wavelet is given in Figure 6.1, where h_i is the decomposition filter and g_i is the synthesis filter. If the input signal and output signal are identical, it will become a perfect reconstruction (PR) filter bank. With the satisfaction of $g_i(n) = h_i(-n)$, the perfect reconstruction filter bank will become a unitary filter bank that also satisfies the quadrature-mirror filter condition (QMF condition).

$$\sum_i \sum_k h_i(Mk+n_1)h_i(Mk+n_2) = \delta(n_1 - n_2) \tag{6.7}$$

$$\sum_k h_i(k)h_j(Ml+k) = \delta(l)\delta(i-j) \tag{6.8}$$

Therefore, the low-pass filter in the unitary filter group satisfies Equations (6.3) and (6.4); if it also satisfies Equation (6.5), the orthogonal wavelet basis could be derived. Here we only discuss the situation when the regularity $K = 1$.

According to Reference [8], the basic idea behind the cosine-modulated PR-QMF filter bank is to employ the prototype low-pass linear filters $g(n)$ to construct, after cosine modulation, the decomposition filter $h_i(n)$ that satisfies

$$h_i(n) = C_{in} g(n) \tag{6.9}$$

where $g(n)$ is a linear low-pass filter with even symmetry and of length $N = 2Mm$ (where m is a nonnegative integer). The cosine modulation factor C_{in} is defined by

$$C_{in} = 2\cos\left[\frac{\pi}{2M}(2i+1)\left(n - \frac{N-1}{2}\right) + (-1)^i \frac{\pi}{4}\right] \tag{6.10}$$

which also satisfies $C_{i(n+2Ml)} = (-1)^l C_{in}, l \in \mathbb{Z}$. The polyphase components [7] of the prototype filter are defined as

$$G(z) = \sum_{n=0}^{2M-1}\left[\sum_{l-0}^{m-1} g(2Ml+n) z^{-2Ml}\right] z^{-n} = \sum_{n=0}^{2M-1} z^{-n} G_n\left(z^{2M}\right) \tag{6.11}$$

where $G_n(z)$ is a finite impulse response filter of finite length m. Because of the symmetry of $G(z)$, it can be proved that

$$G_n(z) = z^{-(m-1)} G_{2M-1-n}\left(z^{-1}\right) \tag{6.12}$$

Reference [6] has proved that if the cosine-modulated filter bank is a unitary filter group, it must satisfy

$$G_n\left(z^{-1}\right) G(z) + G_{M+n}\left(z^{-1}\right) G_{M+n}(z) = \frac{1}{2M} \quad n = 0, 1, \cdots M-1 \tag{6.13}$$

If we signify the integer arithmetic operation of $M/2$ as $J = [M/2]$, then, according to the parity of M, we can rewrite Equation (6.13) as follows.

1. If M is an even number, the number of the equations could be reduced to J. And Equation (6.13) could be rewritten as

$$G_n\left(z^{-1}\right) G_n(z) + G_{M+n}\left(z^{-1}\right) G_{M+n}(z) = \frac{1}{2M} \quad n = 0, 1, \cdots J-1 \tag{6.14}$$

2. If M is an odd number, Equation (6.13) could be rewritten as

$$G_n\left(z^{-1}\right) G_n(z) + G_{M+n}\left(z^{-1}\right) G_{M+n}(z) = \frac{1}{2M} \quad n = 0, 1, \cdots J-1 \tag{6.15}$$

$$G_J(z) G_J\left(z^{-1}\right) = \frac{1}{4M} \tag{6.16}$$

From the above equation, the finite impulse response filter $G_J(z)$ could be presumably defined as a delay filter $G_J(z) = \pm\frac{1}{\sqrt{4M}}z^{-k}, k \in \mathbb{Z}$. Now we parameterize $G_n(z)$ and $G_{M+n}(z)$ with cosine modulation (the same approach as stated in Reference [6]) and denote them with m angular parameters θ_{nl} $(l = 0,1,2,\cdots m-1)$. Assume

$$\begin{bmatrix} G_n(z) \\ G_{M+n}(z) \end{bmatrix} = \lambda \prod_{l=m-1}^{1} \left\{ \begin{bmatrix} \cos\theta_{nl} & \sin\theta_{nl} \\ \sin\theta_{nl} & -\cos\theta_{nl} \end{bmatrix} \begin{bmatrix} 1 & 0 \\ 0 & z^{-1} \end{bmatrix} \right\} \begin{bmatrix} \cos\theta_{n0} \\ \sin\theta_{n0} \end{bmatrix} \tag{6.17}$$

where λ is an undetermined parameter.

On one hand, the construction of an M-band orthogonal wavelet basis requires the condition $H_0(1) = 1$. On the other hand, the Z transform of the decomposition filter is expressed by

$$H_i(z) = \sum_{n=0}^{2M-1} C_{in} z^{-n} G_n\left(-z^{2M}\right) \tag{6.18}$$

If the cosine modulation used in PR-QMF satisfies the wavelet basis construction condition $H_0(1) = 1$, Equation (6.18) becomes

$$\sum_{n=0}^{2M-1} C_{0n} G(-1) = \sqrt{M} \tag{6.19}$$

According to Equation (6.17) and the property of trigonometric functions, we can prove

$$\begin{bmatrix} G_n(-1) \\ G_{M+n}(-1) \end{bmatrix} = \lambda \begin{bmatrix} \cos\theta_n \\ \sin\theta_n \end{bmatrix} \tag{6.20}$$

where $\theta_n = \sum_{l=1}^{m-1} \theta_{nl}$. From the filter construction condition expressed in Equation (6.13), λ is determined by $\lambda = 1/\sqrt{2M}$. With Equation (6.12), we can also prove that

$$\begin{bmatrix} G_{2M-1-n}(-1) \\ G_{M-1-n}(-1) \end{bmatrix} = \begin{bmatrix} (-1)^{m-1} G_n(-1) \\ (-1)^{m-1} G_{M+n}(-1) \end{bmatrix} \frac{1}{\sqrt{2M}} \begin{bmatrix} (-1)^{m-1}\cos\theta_n \\ (-1)^{m-1}\sin\theta_n \end{bmatrix} \tag{6.21}$$

If M is an odd number, considering the even symmetry of $g(n)$, we get

$$\begin{bmatrix} G_J(-1) \\ G_{M+J}(-1) \end{bmatrix} = \pm\sqrt{\frac{1}{4M}} \begin{bmatrix} (-1)^k \\ (-1)^{k+1-m} \end{bmatrix} \tag{6.22}$$

Supposing $\beta_n = \frac{\pi}{2M}(n - Mm + 1/2) + \frac{\pi}{4}$, it can be proved by the periodicity of C_{in} that $C_{0n} = 2\cos\beta_n$, $C_{0(n+M)} = -2\sin\beta_n$, $C_{0(M-1-n)} = 2(-1)^{m-1}\cos\beta_n$, and $C_{0(2M-1-n)} = -2(-1)^{m-1}\sin\beta_n$.

If M is an odd number, we obtain

$$\begin{bmatrix} C_{0J} \\ C_{0(J+M)} \end{bmatrix} = 2\begin{bmatrix} \cos\left[\frac{\pi}{2}(1-m)\right] \\ -\sin\left[\frac{\pi}{2}(1-m)\right] \end{bmatrix} \tag{6.23}$$

Combining Equations (6.21), (6.22), and (6.23), the solution of Equation (6.19) could be obtained. According to the M-band wavelet construction condition, whether Equation (6.19) has a solution relates to the construction of the M-band wavelet basis. Two different conditions are discussed here.

1. If M is an even number, we can prove that

$$\sum_{j=0}^{2M-1} C_{0n} G_j(-1) = \frac{2}{\sqrt{M}} \sum_{j=0}^{J} \sin\left(\beta_n + \theta_n + \frac{\pi}{4}\right) \tag{6.24}$$

Hence $\theta_n = \frac{\pi m}{2} - \frac{\pi}{4M}(2n+1)$.

If M is an odd number, we can also prove that

$$\sqrt{\frac{4}{M}} \sum_{n=0}^{J-1} \sin\left(\beta_n + \theta_n + \frac{\pi}{4}\right) + C_{0J} G_J(-1) + C_{0(M+J)} G_{M+J}(-1) = \sqrt{M} \tag{6.25}$$

As is known from Equation (6.10), if m is a fixed number, then either C_{0J} or $C_{0(M+J)}$ is equal to 0. Consequently, we can rewrite Equation (6.25) as

$$\sum_{n=0}^{J-1} \sin\left(\beta_n + \theta_n + \frac{\pi}{4}\right) = \frac{M \pm 1}{2} \tag{6.26}$$

Apparently, if the right side of Equation (6.26) is equal to $M + 1$, there would be no solution to the formula, that is to say, the filter could not be constructed effectively.

Note that $G_J(z)$ is a delay filter. The selection of $G_J(z)$ is somehow a random process. Therefore, to facilitate the construction of the filter, $G_J(z)$ could be selected in terms of the existence of the solution to Equation (6.26). We assume the filter satisfies

$G_J(z)=\sqrt{1/4M}z^{-k}$ (where k is an even number); thus, we can transform Equation (6.26) to the form of Equation (6.24):

$$\sum_{n=0}^{J-1}\sin\left(\beta_n+\theta_n+\frac{\pi}{4}\right)=J \tag{6.27}$$

Hence, if M is an odd number, the formula will still have a solution, so that the orthogonal wavelet bases could be constructed.

6.1.3 The application of an M-band wavelet in the analysis of power system transient signals

As was discussed in this chapter, the acquisition of an M-band orthogonal wavelet basis could be implemented by (1) fulfilling the construction condition of M-band orthogonal wavelet; (2) solving the prototype filter $g(n)$; and (3) solving the corresponding M-band wavelet and the scaling function. Now we take $M=6$ as an example to demonstrate the process above: if $M=6$, then $N=2M=12$, that is to say, the filter is of length 12, $m=1$, $J=[M/2]=3$. From Equation (6.24), we get $\theta_0=(11/24)\pi$, $\theta_1=(9/24)\pi$, and $\theta_2=(7/24)\pi$; then, the prototype low-pass filter could be given by

$$\begin{aligned}g(n)=\sqrt{1/12}\{&\cos\theta_0,\cos\theta_1,\cos\theta_2,\sin\theta_2,\sin\theta_1,\sin\theta_0,\\&\sin\theta_0,\sin\theta_1,\sin\theta_2,\cos\theta_2,\cos\theta_1,\cos\theta_0\}\end{aligned} \tag{6.28}$$

Hence, we can achieve the cosine modulation factor and the filter bank by combining Equations (6.9) and (6.10). The amplitude-frequency characteristic of the filter bank is depicted in Figure 6.2 when $M=6$. The time domain and frequency domain characteristics are demonstrated in Figures 6.3 and 6.4. From the simulation result, we can say that the cosine modulation method could effectively construct an M-band wavelet basis. It has provided a useful and concise tool for the analysis of power system transient signals.

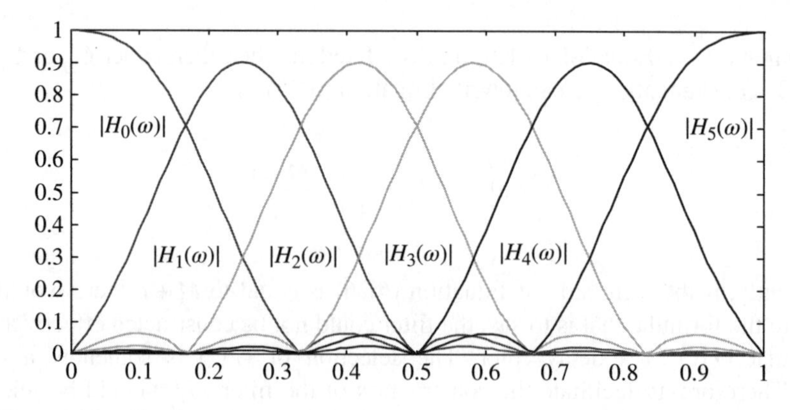

Figure 6.2 The frequency domain characteristic of a six-band wavelet filter bank

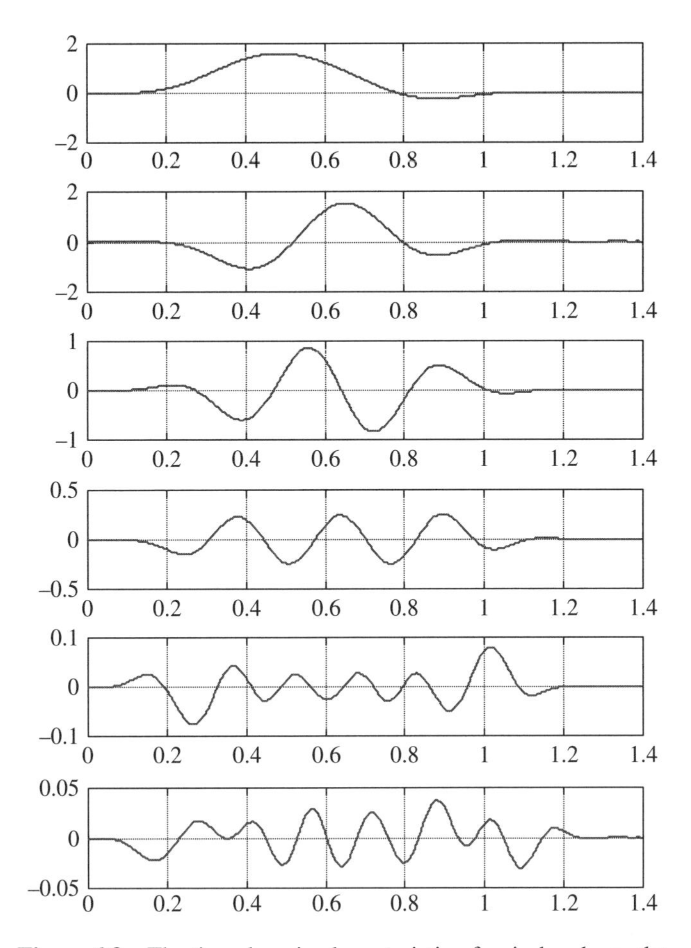

Figure 6.3 The time domain characteristic of a six-band wavelet

In Figures 6.5 and 6.6, a comparison is conducted between M-band wavelet transform and dyadic wavelet transform. The curves show the results of the two transforms dealing with the same signal mixed with a short-time transient signal in the high-frequency band. The high-frequency signal is simulated with multiple sinusoidal signals in the range of 10~90 kHz. The short-time impulse transient signals are simulated with oscillation damping, and mixed with high-frequency signals from the 200th to the 600th point. The sampling frequency is set to be $f_s = 200$ kHz. (The simulation parameters are the same as those of Chapter 5. The maximum amplitude is 1:10.)

The M-band wavelet analysis of the single-phase grounding fault in an EHV transmission line is shown in Figure 6.7, and it is discussed here:

1. After the M value and the regularity level of the M-band wavelet basis have been determined, the prototype filter and the modulation factor could be easily obtained. Then the low-pass and high-pass filter sequence corresponding to the wavelet scaling

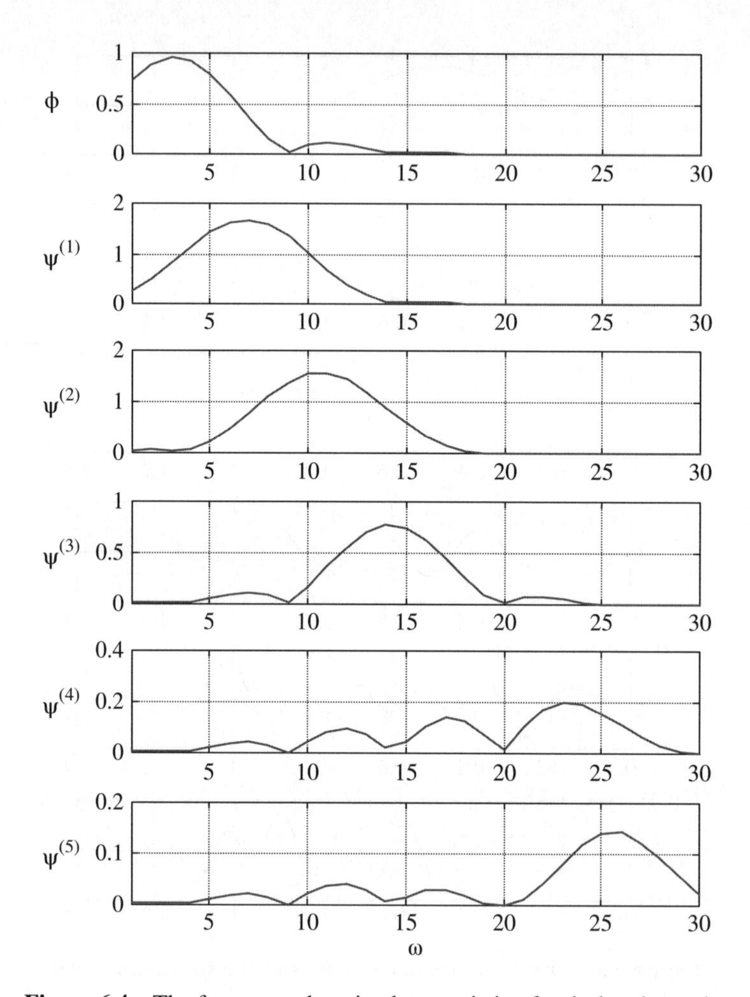

Figure 6.4 The frequency domain characteristic of a six-band wavelet

function and the wavelet function could be obtained. Thus, it is easy and feasible for the construction of M-band wavelets based on cosine modulation.

2. Compared with the dyadic wavelet analysis, the decomposition velocity of M-band wavelet analysis is faster while the decomposition subband is kept identical. This is because the target signal is decomposed in multiple channels in M-band wavelet transform, which features a parallel computational structure. Thus, it is highly practical to employ M-band wavelet analysis for retrieving and processing transient signal characteristics.
3. As we can see from Figure 6.6, the frequency bands are comparatively coarsely divided in dyadic wavelet analysis, and the transient signals are not significantly retrieved at all scales. In M-band wavelet transform, however, the frequency domain is equally divided, and the high-frequency part is divided with finer resolution. As seen in Figure 6.5, after M-band wavelet transform, two short-time transient signals could be effectively retrieved or detected in the high-frequency part (Figure 6.6e,f).

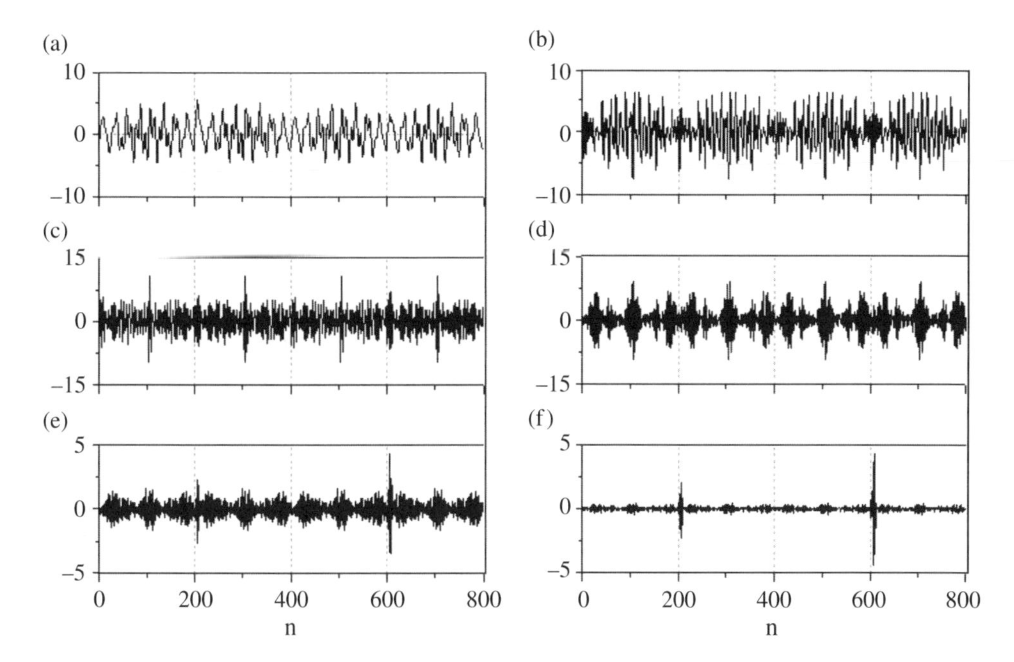

Figure 6.5 The six-band wavelet decomposition of a high-frequency signal mixed with short-time transient signals: (a) $c_1(n)$, (b) $d^{(1)}{}_1(n)$, (c) $d^{(2)}{}_1(n)$, (d) $d^{(3)}{}_1(n)$, (e) $d^{(4)}{}_1(n)$, (f) $d^{(5)}{}_1(n)$

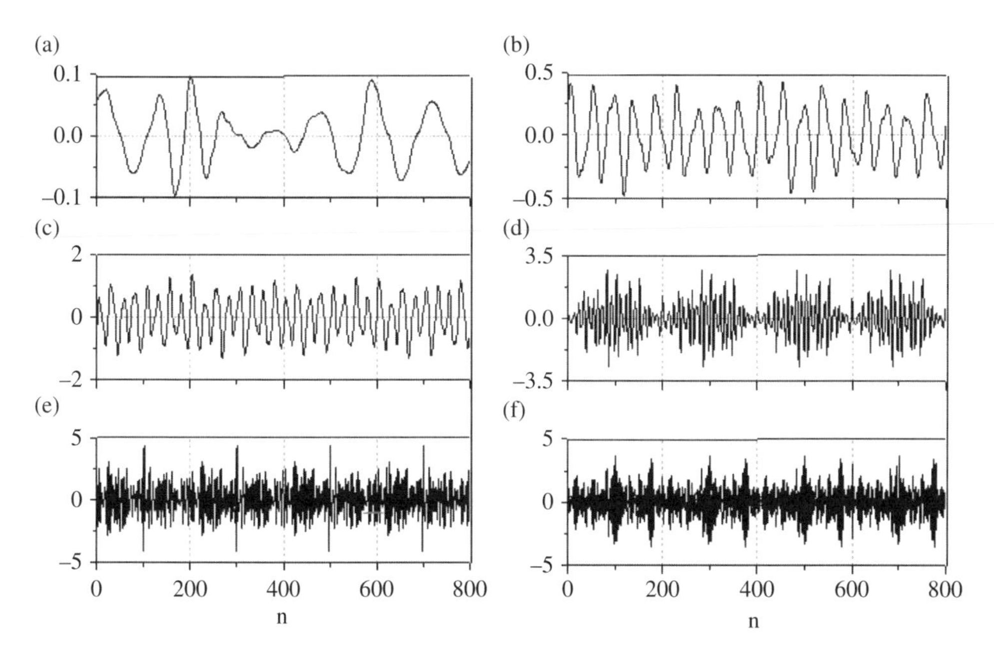

Figure 6.6 The dyadic wavelet transform of a high-frequency signal mixed with short-time transient signals: (a) $c_1(n)$, (b) $d^{(1)}{}_1(n)$, (c) $d^{(2)}{}_1(n)$, (d) $d^{(3)}{}_1(n)$, (e) $d^{(4)}{}_1(n)$, (f) $d^{(5)}{}_1(n)$

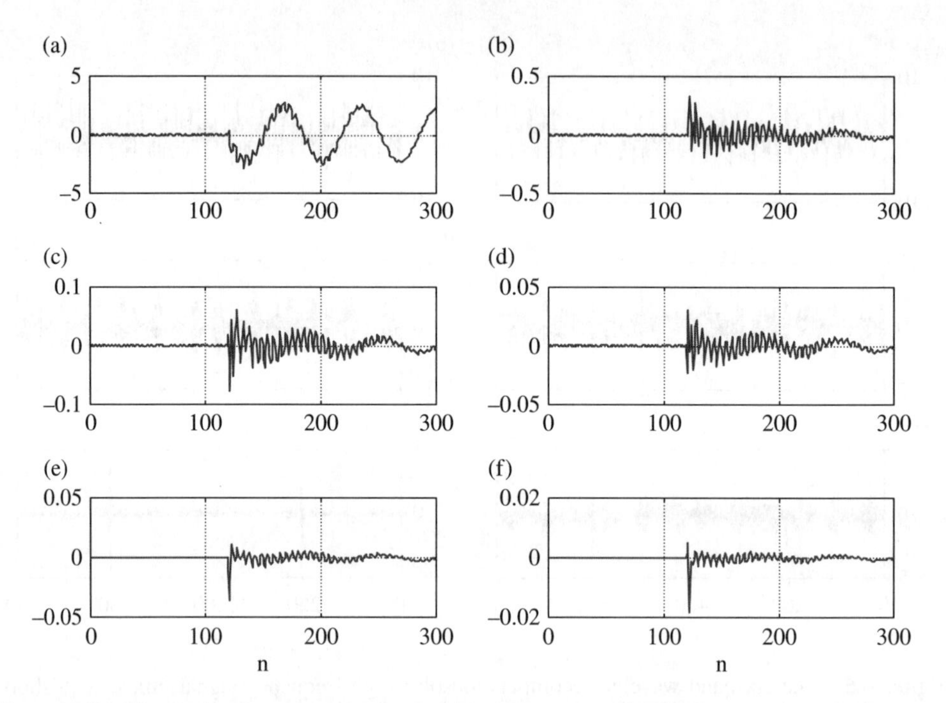

Figure 6.7 The six-band wavelet decomposition of a zero-mode fault current in an EHV transmission line: (a) $c_1(n)$, (b) $d^{(1)}{}_1(n)$, (c) $d^{(2)}{}_1(n)$, (d) $d^{(3)}{}_1(n)$, (e) $d^{(4)}{}_1(n)$, (f) $d^{(5)}{}_1(n)$

Hence, the M-band wavelet transform is particularly suitable for the analysis of high-frequency signals mixed with comparatively narrow impulse signals. As is depicted in Figure 6.7, if the M value is properly chosen, the M-band wavelet transform is very effective for retrieving traveling wave and electric arc fault transient characteristics.

4. M-band wavelet analysis and M-band wavelet packet analysis have good application prospects for retrieving the fundamental and harmonic components of a signal, and especially for retrieving and estimating higher harmonics.

6.2 The construction and application of a class of recursive wavelets

6.2.1 *The Construction condition of recursive wavelets*

A class of super-Gaussian function is defined in Equation (6.29):

$$g_\sigma^m(t) = \exp\left(-|\sigma t|^m\right), \sigma > 0, m \geq 1 \tag{6.29}$$

where σ determines the shape of the function, and m is the order of the super-Gaussian function. The Fourier transform of the super-Gaussian function could be given by

$$G_\sigma^m(\omega) = \frac{1}{\sigma}\sum \frac{(-1)^k}{(2k+1)!} 2^{\frac{2k+1}{m}} \Gamma\left(1 + \frac{2k+1}{m}\right)\left(\frac{\omega}{2\sigma}\right)^{2k} \tag{6.30}$$

where $\Gamma(m)$ is a chi-square function. Obviously, if $m = 2$, $\Gamma(m)$ will be a Gaussian function, we can construct a series of Gaussian wavelets, such as Mexican-hat wavelets, Morlet wavelets, and so on. Inspired by this, this book presents the construction method of a class of wavelets by multiplying polynomials with the super-Gaussian function (or the exponential function) where $m = 1$ and is supplemented with a frequency-shifting operator. As the derivation in Reference [9] shows, if the wavelet function can be written in the form of $\psi(t) = A(t)\exp(bt)$ (where $A(t)$ is a polynomial about t, and b is a constant value), the Z transform of the wavelet could be explicitly expressed as a rational function about z^{-1}. The recursive algorithm could be implemented by wavelet transform; thus, we give the name *recursive wavelet* (RW) to the class of wavelet that has a recursive algorithm.

6.2.2 The construction method of recursive wavelets

In Equation (6.30), when $m = 1$, there is

$$G_{\sigma}^{1}(\omega) = \frac{\sigma}{\sigma^2 + (\omega)^2} \tag{6.31}$$

Obviously, the super-Gaussian function $g_{\sigma}^{1}(t)$ does not satisfy the mother wavelet condition. Therefore, we must reconstruct $g_{\sigma}^{1}(t)$. Considering the relation between the Mexican-hat wavelet and the Gaussian wavelet, we calculate the differential of $G_{\sigma}^{1}(\omega)$ and introduce the frequency-shifting factor $\exp(i\omega_0 t)$. Letting $g_{\sigma}(t) = e^{-\sigma t}e^{i\omega_0 t}u(t)$, then the Fourier transform of $g_{\sigma}^{1}(t)$ is expressed by

$$F(s)\Big|_{s=i(\omega-\omega_0)} = G_{\sigma}(\omega) = \frac{1}{s+\sigma}\Big|_{s=i(\omega-\omega_0)} \tag{6.32}$$

The linear combination of the differential of each order could be seen as the Fourier transform of $\theta(t)$, which is expressed by

$$\vartheta(\omega) = \frac{\alpha_1}{s+\sigma} + \frac{\alpha_2\sigma}{(s+\sigma)^2} + \cdots + \frac{\alpha_N\sigma^{N-1}}{(s+\sigma)^N},\ s = i(\omega - \omega_0) \tag{6.33}$$

The corresponding primitive function is given by

$$\theta(t) = \left[\alpha_1 e^{-\sigma t} + \alpha_2\sigma t e^{-\sigma t} + \cdots + \frac{\alpha_N}{(N-1)!}(\sigma t)^{N-1} e^{-\sigma t}\right] e^{\omega_0 t} u(t) \tag{6.34}$$

Letting $\psi(t) = \bar{\theta}(-t)$, by adjusting the relation of ω_0, σ, $\psi(t)$ could satisfy the mother wavelet condition; thus, it can be used as the mother wavelet of forward recursive wavelet transform. When applying different α_i and N, we can achieve a series of recursive

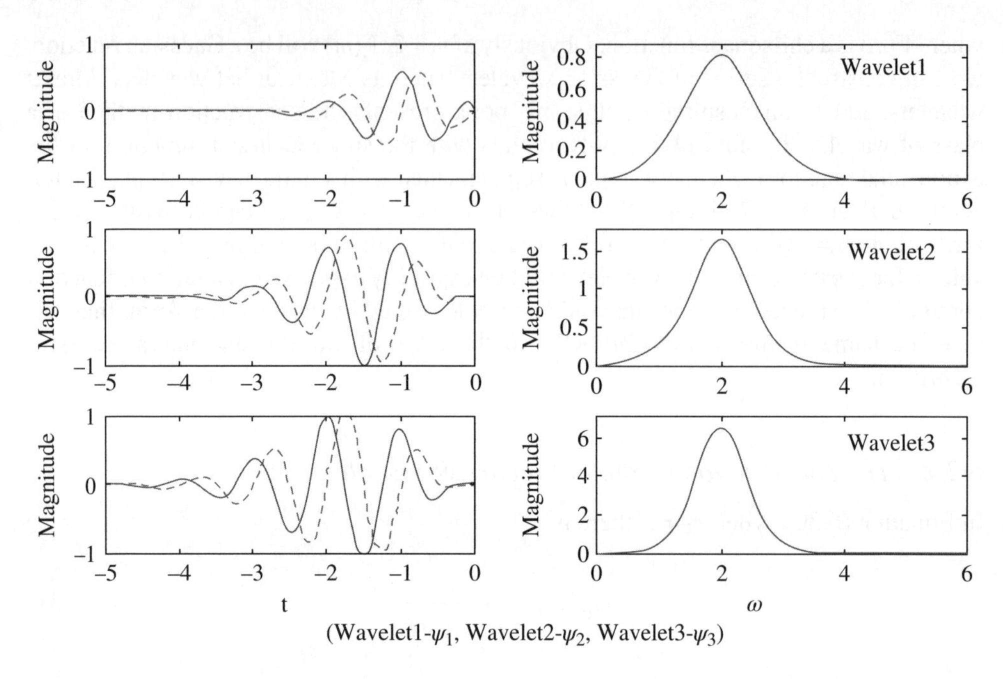

Figure 6.8 The time domain and frequency domain waveforms of three kinds of recursive wavelets

Table 6.1 Wavelet localization parameters (wavelet1 is ψ_1, wavelet2 is ψ_2, wavelet3 is ψ_3)

Wavelet	Time domain center t^*	Time domain width $\Delta\psi$	Frequency domain center ω^*	Frequency domain width $\Delta\hat{\psi}$	The window area $4\,\Delta\psi\,\Delta\hat{\psi}$
Wavelet1	−0.9924	0.3224	6.2832	2.1623	2.8128
Wavelet2	−1.5349	0.4702	6.2832	1.1948	2.2472
Wavelet3	−1.7632	0.6670	6.2832	0.8887	2.3712

wavelets, such as three classes of wavelet ψ_1, ψ_2, ψ_3; the time–frequency characteristics of the three wavelets are shown in Figure 6.8 and Table 6.1, in which ψ_2 is applied in Reference [10], where $3\sigma^2 = \omega_0^2$. As N increases, the frequency localization ability strengthens and the window area decreases as it approximates the optimal window (the window area is 2). Thus, we can construct recursive wavelets based on different time–frequency localization characteristics according to different requirements. The recursive wavelets constructed with this method are nonorthogonal wavelets that do not possess compactly supported sets. The same method could be applied to construct a series of bidirectional recursive wavelets based on bidirectional exponential function according to Reference [9].

$$\psi_1(t)=\left(-\sigma t-\sigma^2t^2-\frac{2}{3}\sigma^3t^3\right)e^{(\sigma+i\omega_0)t}u(-t)$$
$$\psi_2(t)=\left(-\frac{1}{3}\sigma^3t^3-\frac{1}{6}\sigma^4t^4-\frac{1}{15}\sigma^5t^5\right)e^{(\sigma+i\omega_0)t}u(-t) \tag{6.35}$$
$$\psi_3(t)=\left(-\frac{1}{15}\sigma^5t^5-\frac{1}{90}\sigma^6t^6-\frac{1}{630}\sigma^7t^7\right)e^{(\sigma+i\omega_0)t}u(-t)$$

Let T be the sampling period, $f=1/a$ be the wavelet transform scale, and k be the translation of wavelet functions in the time domain. According to the definition of wavelet transform, the discrete wavelet transform of signal $x(t)$ is given as

$$WT_\psi(k)=\sqrt{f}T\sum_{n=1}^{\infty}x(n)\bar{\psi}(n-k)=\sqrt{f}T\left[x(n)*\theta(n)\right] \tag{6.36}$$

Perform a Z transform on both sides of the wavelet function:

$$WT_\psi(z)=\sqrt{f}T\left[x(z)\cdot\theta(z)\right] \tag{6.37}$$

where $WT_\psi(z), x(z), \theta(z)$ are the Z transform of $WT_\psi(k), x(k), \theta(k)$, respectively. Letting $B=f\sigma T$, $A=(if\omega_0T-f\sigma T)$, if $\theta(t)$ could be written in the form of $\theta(t)=A(t)\exp(bt)$, we can get

$$\begin{aligned}\theta(z)&=\sum_{0}^{\infty}\left[\alpha_1+\alpha_2Bk+\cdots+\frac{\alpha_N}{(N-1)!}(Bk)^{N-1}\right]e^{Ak}\cdot z^{-k}\\&=\frac{\delta_1z^{-1}+\delta_2z^{-2}+\cdots+\delta_kz^{-k}\cdots+\delta_{N-1}z^{-N+1}}{1+\lambda_1z^{-1}+\lambda_2z^{-2}+\cdots+\lambda_kz^{-k}\cdots+\lambda_Nz^{-N}}\end{aligned} \tag{6.38}$$

where δ_k $(k=1,\cdots N-1)$ and λ_k $(k=1,\cdots N)$ are constant values that could be rearranged from Equation (6.38) to the form of

$$WT_\psi(z)\left(1+\lambda_1z^{-1}+\lambda_2z^{-2}+\cdots+\lambda_Nz^{-N}\right)=\sqrt{f}Tx(z)\left(\delta_1z^{-1}+\delta_2z^{-2}+\cdots+\delta_{N-1}z^{-N+1}\right) \tag{6.39}$$

According to the property of the Z transform, the forward recursive algorithm could be obtained from Equation (6.39):

$$\begin{aligned}WT_\psi(k)=&\sqrt{f}T\left[\delta_1x(k-1)+\delta_2x(k-2)+\delta_3x(k-3)+\cdots+\delta_{N-1}x(k-N+1)\right]\\&-\lambda_1WT_\psi(k-1)-\lambda_2WT_\psi(k-2)-\cdots-\lambda_NWT_\psi(k-N)\end{aligned} \tag{6.40}$$

6.2.3 The application of recursive wavelets in power transient signal analysis

Real-time phasor measurement of each node current and voltage is the foundation of safe monitoring. Therefore, phasor estimation and frequency estimation algorithms with finer resolutions are significant for judging grid status and predicting the development of power systems and fast and proper protection measurement. As for the transient signals in power systems, because the Fourier transform lacks time localization ability, it could not effectively measure phasors in a precise and real-time manner. This defect could be compensated with wavelet transform as analyzed in Chapter 2. Thus, in this chapter, a kind of recursive wavelet where $N=7$ is taken as an example for the application of transient phasor measurement.

The chosen recursive wavelet is expressed by

$$\psi(t)=\left(\frac{\sigma^4 t^4}{12}-\frac{\sigma^5 t^5}{30}+\frac{\sigma^6 t^6}{90}\right)\cdot\exp(\sigma t+j\omega_0 t)\cdot u(-t) \tag{6.41}$$

To satisfy the permission condition of mother wavelets, we can assign $\sigma=2\pi/\sqrt{3}$, $\omega_0=2\pi$. According to Equation (6.40), the forward recursive algorithm chosen in this chapter is

$$WT_\psi(a,k)=\frac{1}{\sqrt{a}}T\left(\sum_{i=1}^{6}\delta_i x(k-i)\right)-\left(\sum_{j=1}^{7}\lambda_j W(k-j)\right) \tag{6.42}$$

where δ_i and λ_j are the parameters of the recursive wavelet in Z transform. The parameters satisfy

1. $[\delta_1\ \delta_2\ \delta_3\ \delta_4\ \delta_5\ \delta_6]^T=K_1\times[B_1\ B_2\ B_3]^T\times[C^1\ C^2\ C^3\ C^4\ C^5\ C^6]$;
2. $[\lambda_1\ \lambda_2\ \lambda_3\ \lambda_4\ \lambda_5\ \lambda_6\ \lambda_7]^T=K_2\times[C^1\ C^2\ C^3\ C^4\ C^5\ C^6\ C^7]$

where

$$C=e^{-fT(\sigma-j\omega_0)},\ K_1=\begin{bmatrix}1 & -1 & 1\\ 9 & -25 & 57\\ -10 & 40 & 302\\ -10 & 40 & 302\\ 9 & 25 & 57\\ 1 & -1 & 1\end{bmatrix},\ K_2=\begin{bmatrix}-7\\ 21\\ -35\\ 35\\ -21\\ 7\\ -1\end{bmatrix},$$

$$[B_1\quad B_2\quad B_3]=\begin{bmatrix}\frac{(\sigma fT)^4}{12} & \frac{(\sigma fT)^5}{30} & \frac{(\sigma fT)^6}{90}\end{bmatrix}$$

Suppose the real component of the wavelet transform $WT_\psi(a,k)$ is $R(k)$, the imaginary component is $I(k)$, the phasor estimate of the instantaneous phase angle is $\theta(k)$, the instantaneous frequency is $f(k)$, the precise estimate of the phasor amplitude is $A_x(k)$, and the amplitude gain coefficient is e_a. Thus, the relation of these parameters could be expressed below:

$$\theta(k)=\tan^{-1}\left(I(k)/R(k)\right),\quad f(k)=\frac{1}{2\pi}\frac{\theta(k)-\theta(k-1)}{T},$$
$$A_x(k)=e_a\sqrt{\left(R^2(k)+I^2(k)\right)}$$

In Chapter 7, we are going to present the experiment, which measures two transient signals with a frequency-hopping or decaying direct current (DC) component, and then compare the results with discrete Fourier transform (DFT) to validate the advantages of recursive wavelet transform in the analysis of transient signals.

6.2.3.1 The simulation of phasor measurement of signals with frequency hopping

Equation (6.43) shows a Gaussian white-noise signal with frequency hopping. Before the moment of 1 s, the signal keeps at the frequency of 50 Hz. And when the time is no less than 1 s, the signal is 49 Hz. Figures 6.9, 6.10, 6.11, and 6.12 show the frequency measurement curve, amplitude error curve, phase angle error curve, and total measurement error curve, respectively.

$$\begin{cases} x(t)=20e^{j*2\pi ft}+3\cos(6\pi ft)+\cos(10\pi ft)+\cos(2\times 5.51ft)+\varepsilon & (0<t<1s) \\ x(t)=20e^{j*2\pi ft}+3\cos(6\pi ft)+\cos(10\pi ft)+\cos(2\times 5.51ft)+\varepsilon & (t\geq 1s) \end{cases} \tag{6.43}$$

where $\begin{cases} f=50\,Hz & 0<t<1s \\ f=49\,Hz & t\geq 1s \end{cases}$, ε is a Gaussian white-noise signal; its SNR = 50 dB, and its signal amplitude is 20 kV.

6.2.3.2 The simulation of phasor measurement of signals with decaying DC components

Figure 6.13 shows the fault phase current when a single-phase grounding fault happens. As is seen in the figure, the current is synthesized with a DC component since the moment of 1 s; in other words, a ramp is introduced in the current curve. The amplitude curve measured with DFT and recursive wavelet transform is demonstrated in Figure 6.14.

From the phasor measurement curve of the two signals, a recursive wavelet brings fast stabilization of the measurement curve. At the same time, in the enlarged view of Figures 6.10 and 6.11, we can see that the recursive wavelet has better performance in antiharmonic and antinoise measures compared with the Fourier transform method.

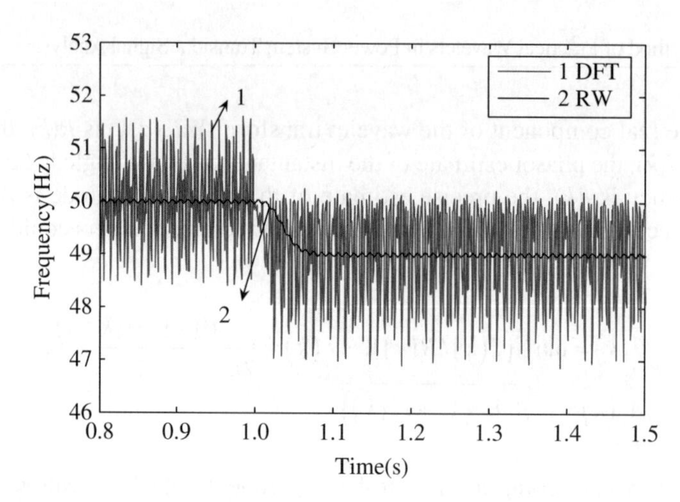

Figure 6.9 The frequency measurement curve performed by DFT and the recursive wavelet method when the frequency hops

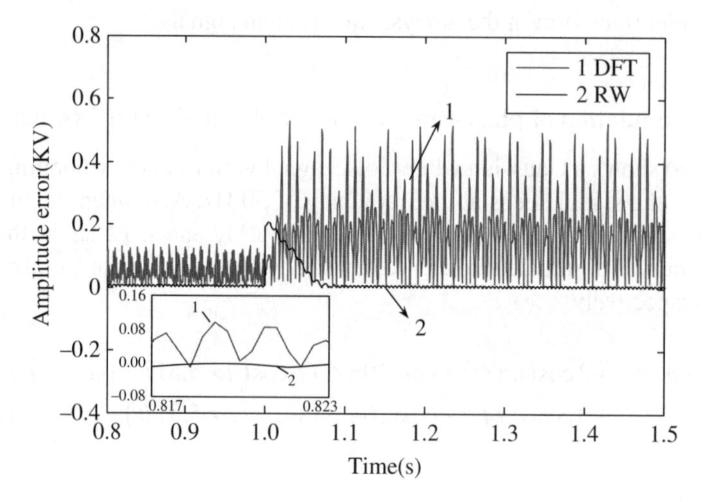

Figure 6.10 The phasor measurement curve of amplitude error performed by DFT and the recursive wavelet method when the frequency hops

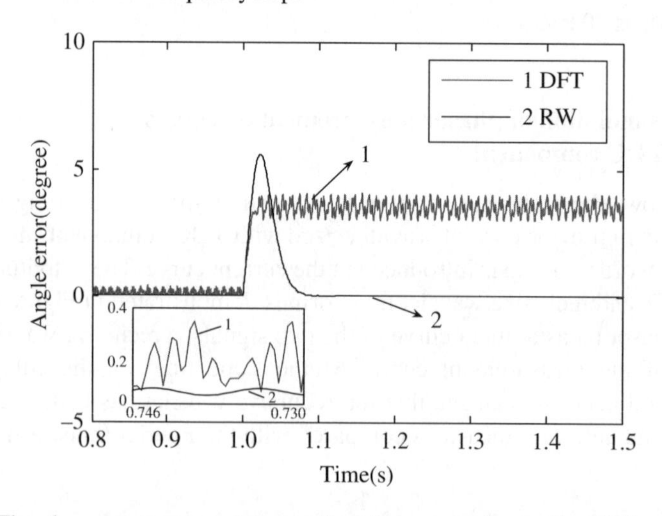

Figure 6.11 The phasor measurement curve of angle error performed by DFT and the recursive wavelet method when the frequency hops

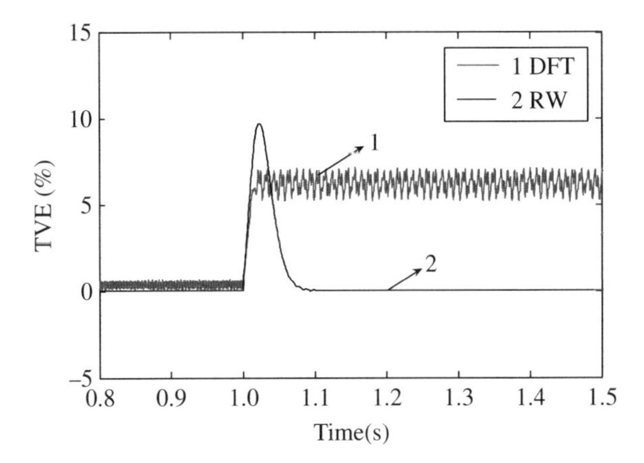

Figure 6.12 The phasor measurement curve of the total error performed by DFT and the recursive wavelet method when the frequency hops

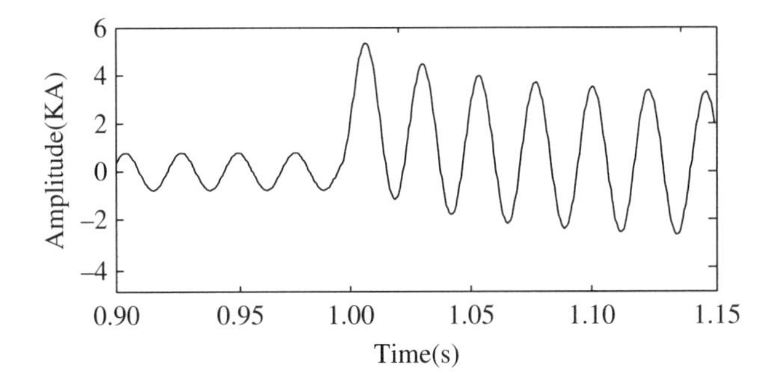

Figure 6.13 The fault current measurement of a signal with decaying DC component

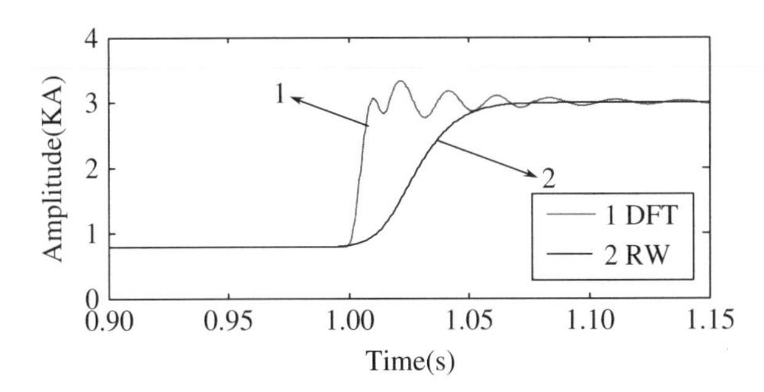

Figure 6.14 The amplitude measurement curve of a signal with declining DC component using the wavelet transform method and recursive wavelet method

In power systems, although the influence factors such as fault and disturbance may happen in various transient situations, the recursive wavelet method still has favorable precision in phasor measurement. Wide development is expected in the transient phasor measurement area.

6.3 The application and structure of an optimal wavelet basis

6.3.1 The construction method of an optimal wavelet basis

In applications such as data compression, the optimization of a wavelet basis is a necessary process. This is because $x(t)$ is replaced with $x^J(t)$ expressed at a finite resolution, which is equivalent to cutting off certain frequencies of the signal, causing decomposition and reconstruction error to some extent. Naturally, the optimization of wavelet basis $\psi(t)$ and scale function $\phi(t)$ minimizes the frequency domain norm of $x^J(t)$ and $x(t)$; because of this feature, it is wise to optimize the wavelet basis and the discrete wavelet transform pre-filter [11, 12]. As is discussed in Chapter 2, $\psi(t)$ and $\phi(t)$ could be exclusively determined by the dyadic scale sequence $h(k)$; thus, the optimization of the wavelet basis becomes the optimization of $h(k)$ $(0 \leq k < K)$ for the purpose of minimizing the cost function.

$$L = \left\| X(\omega) - 2^{-J} A(\omega/2^J) \Phi(\omega/2^J) \right\|_2 = \left\| 2^J X(2^J \omega) - A(\omega)\Phi(\omega) \right\|_2 \tag{6.44}$$

where $X(\omega)$, $\Phi(\omega)$ are the Fourier transform of $x(t)$, $\phi(t)$, $A(\omega) = 2^{J/2} \sum_{m=-\infty}^{\infty} a(J,m) e^{-im\omega}$, and $a(j,m)$ is the discrete wavelet approximation at scale j. Because the orthogonal scale function corresponding to $h(k)$ must satisfy Equations (2.39) and (2.44), it is also called a constrained optimization task. In addition, because the implicit function with an objective function $h(k)$ is not a strict concave-convex function, the optimization of Equation (6.44) is a very difficult task [12].

For the real-time requirement of data compression of power system fault information, we turn to consider the determination of the filter coefficient $h(k)$, to achieve the best performance of discrete wavelet approximation at resolutions from scale J to scale $J+1$. That is to say, the norm of wavelet approximation approaches the highest value in the time domain. It could also be seen as the optimization of the cost function in Equation (6.45) with the condition of Equation (2.39) being satisfied. (For computational simplicity, the norm is considered to be the Euclidean norm of $L^2(\mathbb{R})$.)

$$\max \sum_m \left| a(J,m) \right|^2 = \max \sum_m \left| \sum_k h(k) a(J+1, k+2m) \right|^2 \tag{6.45}$$

It is proved in Reference [12] that the sequence $h(k)$ $(0 \leq k < K)$ satisfying Equation (2.39) could be obtained by mapping the sequence $\theta_k \in [0, 2\pi], 0 \leq k < K/2-1$. For example, $h(k)$ $(0 \leq k \leq 3)$ and $h(k)$ $(0 \leq k \leq 5)$ could be expressed as

$$\begin{aligned} h(0) &= (1-\cos\theta_1 + \sin\theta_1)/\sqrt{8} \quad h(1) = (1+\cos\theta_1 + \sin\theta_1)/\sqrt{8} \\ h(2) &= (1+\cos\theta_1 - \sin\theta_1)/\sqrt{8} \quad h(3) = (1-\cos\theta_1 - \sin\theta_1)/\sqrt{8} \end{aligned} \tag{6.46}$$

where $\theta_1 \in [0, 2\pi]$; and as

$$\begin{aligned}
h(0) &= \left[1+\cos\theta_1 - \cos\theta_2 + \sin\theta_1 - \sin\theta_2 - \cos(\theta_1-\theta_2) - \sin(\theta_1-\theta_2)\right]/\sqrt{32} \\
h(1) &= \left[1-\cos\theta_1 + \cos\theta_2 + \sin\theta_1 - \sin\theta_2 - \cos(\theta_1-\theta_2) + \sin(\theta_1-\theta_2)\right]/\sqrt{32} \\
h(2) &= \left[1+\cos(\theta_1-\theta_2) + \sin(\theta_1-\theta_2)\right]/\sqrt{8} \\
h(3) &= \left[1+\cos(\theta_1-\theta_2) - \sin(\theta_1-\theta_2)\right]/\sqrt{8} \\
h(4) &= \left[1-\cos\theta_1 + \cos\theta_2 - \sin\theta_1 + \sin\theta_2 - \cos(\theta_1-\theta_2) - \sin(\theta_1-\theta_2)\right]/\sqrt{32} \\
h(5) &= \left[1+\cos\theta_1 - \cos\theta_2 - \sin\theta_1 + \sin\theta_2 - \cos(\theta_1-\theta_2) + \sin(\theta_1-\theta_2)\right]/\sqrt{32}
\end{aligned} \tag{6.47}$$

where $\theta_1, \theta_2 \in [0, 2\pi]$.

Therefore, the optimization problem becomes the optimization of θ_k, which is a relatively simple computational method.

6.3.2 *An instance of wavelet basis optimization*

The waveforms in Figures 6.15a and 6.16a show the sampled fault phase voltage and zero-mode current signal when a single-phase grounding fault happens in the simulation system as described in the Appendix. Figure 6.15b,c and Fig. 6.16b,c show the curves of corresponding cost functions described by $\sum_m |\sum_k h(k)a(J+1, k+2m)|^2$, where in (b) of each figure, *h*(*k*) is a four-coefficient function, with its first parameter being θ_1. In (c) of each figure, *h*(*k*) is a six-coefficient function, with its first and second parameters being θ_1, θ_2.

Table 6.2 shows the optimized dyadic sequence *h*(*k*), from which we can calculate the wavelet sequence $g(k) = (-1)^k h(1-k)$. When constructing the wavelet, *h*(*k*) complies with the orthogonal-scale function requirements, and *g*(*k*) and *h*(*k*) satisfy the quadrature mirror condition; thus, the optimized wavelet would be an orthogonal wavelet that is a compactly supported asymmetrical wavelet. The optimal wavelets that correspond to the fault voltage and current are depicted in Figures 6.17 and 6.18. The amplitude-frequency characteristic of a six-coefficient *h*(*k*) is depicted in Figure 6.19.

6.3.3 *The application of an optimized wavelet basis in the data compression of power system transient information*

An effective method for compressing power system transient data is by reserving the details of wavelet approximation related to the transient signal, setting the wavelet coefficient of noise and redundant information to zero, and then reconstructing the signal. One way to identify useful and redundant components and noise is by configuring a threshold value η_s as the criterion. The configuration of the threshold value could be implemented in multiple ways; here, we adopt two adaptive configuration methods by setting η_s:

$$\eta_s = \left\{ |d(j,m_s)| \left| \begin{matrix} num\{|d(j,m)| \geq |d(j,m_s)|\} = u \times num\{d(j,m)\} \\ num\{|d(j,m)| < |d(j,m_s)|\} = (1-u) \times num\{d(j,m)\} \end{matrix} \right. \right\} \tag{6.48}$$

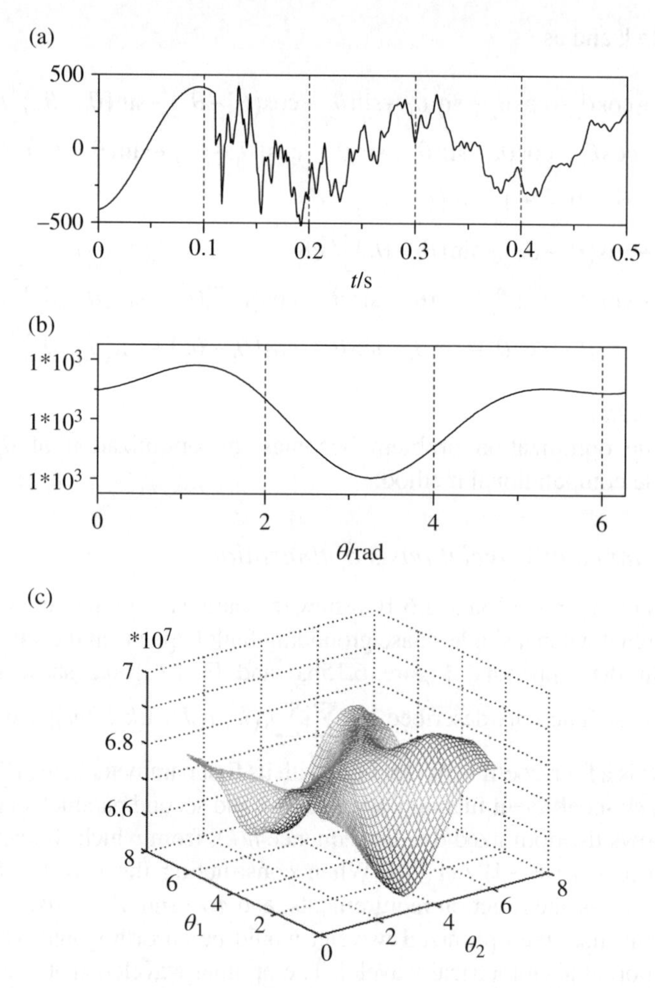

Figure 6.15 The optimization of the wavelet basis in the analysis of a transient fault phase voltage signal: (a) voltage, (b) four-coefficient, (c) six-coefficient

$$\eta_s = \lambda \times \max(|\, d(j,m\,|) \tag{6.49}$$

where $d(j, m)$ is the discrete wavelet detail at scale j; *num* represents the number of the sequence; and $0 \le u \le 1$ and $0 \le \lambda \le 1$ are the proportional coefficients. For example, $u = 0.1$ means that 10% of the coefficients are reserved in the wavelet detail, and $\lambda = 0.1$ means that the threshold value is set to be 0.1 times the maximum modulus of the wavelet detail. As a result, the wavelet detail of the reconstructed signal could be expressed by

$$\bar{d}(j,m) = \begin{cases} d(j,m), |d(j,m)| \ge \eta_s \\ 0, |d(j,m)| < \eta_s \end{cases} \tag{6.50}$$

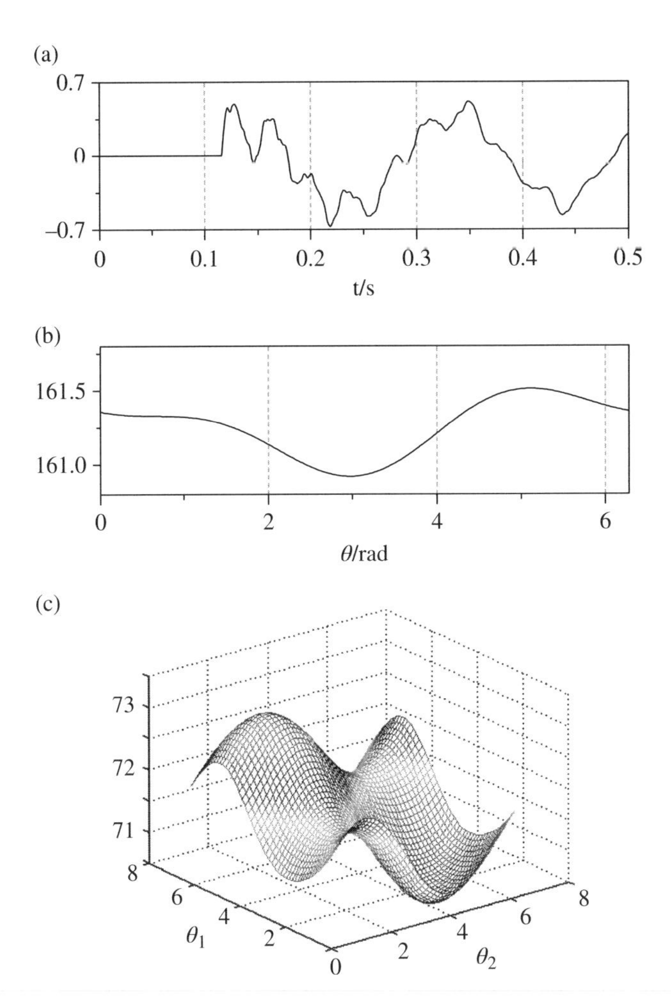

Figure 6.16 The optimization of a wavelet basis in the analysis of a transient fault zero-mode current signal: (a) voltage, (b) four-coefficient, (c) six-coefficient

Table 6.2 The optimized four-coefficient and six-coefficient dyadic sequence $h(k)$

Sequence	$h(k), 0 \le k \le 3$	$h(k), 0 \le k \le 5$
Voltage signal	0.5460 0.8150 0.1611 −0.1079	0.5433 0.7995 0.2140 −0.1266 −0.0503 0.0342
Current signal	−0.1098 0.1720 0.8169 0.5414	−0.0208 0.0235 −0.0215 0.0229 0.7494 0.6606

Therefore, the system only needs to store and transfer a small amount of data. The original signal could be recovered by reconstructing the wavelet approximation $a(J_1, m)$ and the wavelet detail $\bar{d}(j,m), j = J, J+1, \cdots J_1$.

Employing the aforementioned data compression method, we perform wavelet transform and data compression to the fault phase voltage and zero-mode signal in Figures 6.15a

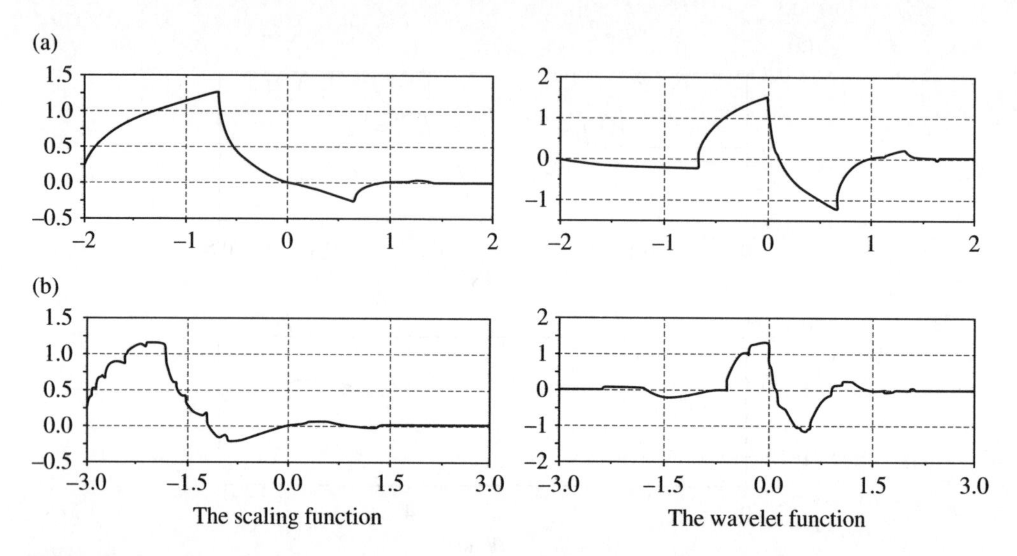

Figure 6.17 The optimal wavelets corresponding to the fault voltage signals: (a) four-coefficient, (b) six-coefficient

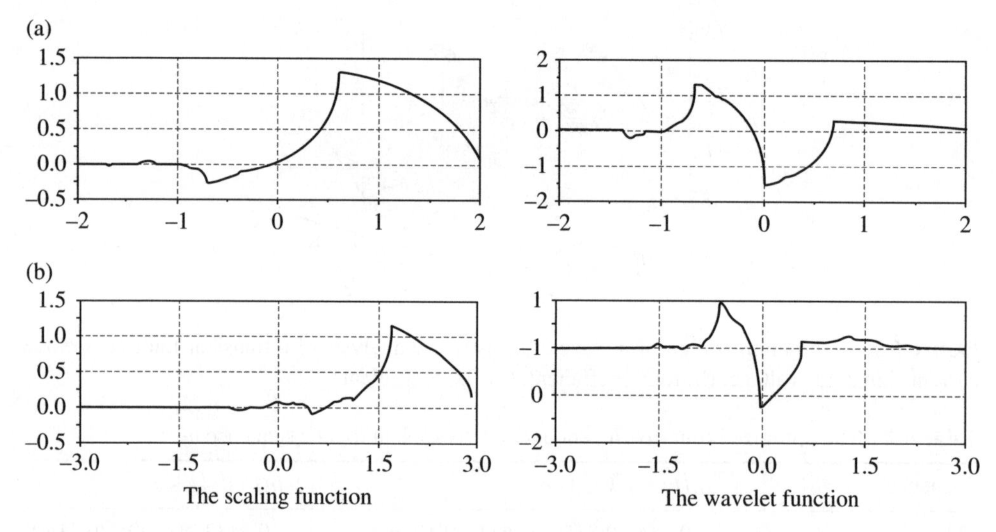

Figure 6.18 The optimal wavelets corresponding to the fault current signals: (a) four-coefficient, (b) six-coefficient

and 6.16b (by setting the sampling point as $N = 1024$, and the sampling frequency to 20 kHz). The wavelet basis is selected as the optimal filter sequence obtained by db3 and Table 6.2. The threshold value is set according to Equation (6.49). The results are shown in Figures 6.20, 6.21, 6.22 and 6.23. It is found that when the decomposition scale and threshold value are set to be fixed values, less data in the wavelet detail will be reserved.

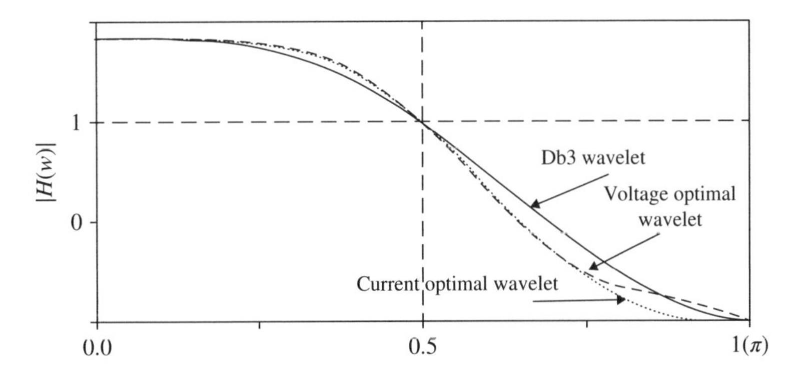

Figure 6.19 Comparison of three low-pass filters employing different wavelets

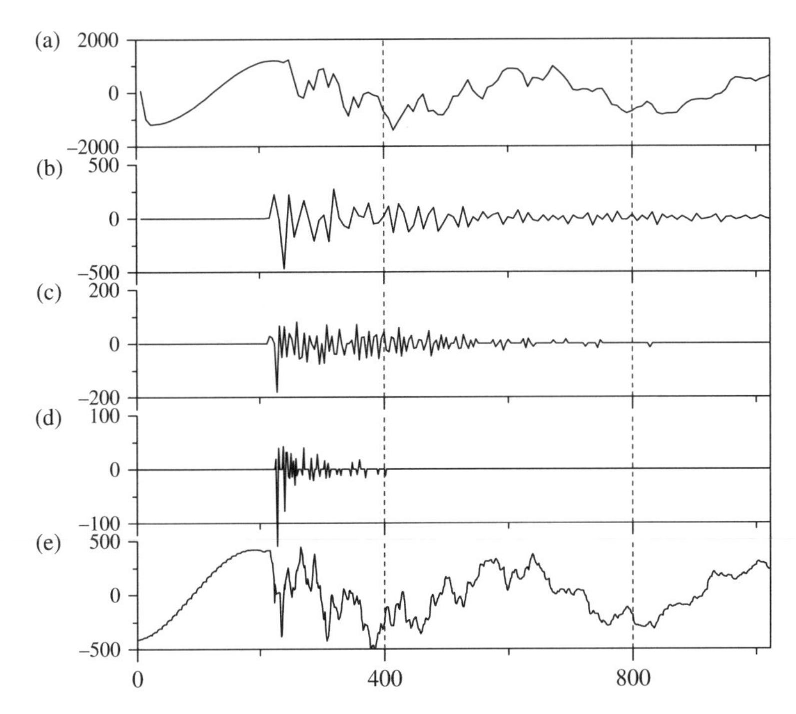

Figure 6.20 Voltage transient compression based on optimal wavelets: (a) discrete wavelet approximation $a(J-3,m)$, (b) wavelet detail $\bar{d}(J-3,m)$, (c) wavelet detail, (d) wavelet detail $\bar{d}(J-1,m)$, (e) reconstruction signal, as $\eta_s = 0.05 * \max(d(j,m))$

It is also evident that the reconstructed waveform better approximates the original voltage and current waveform, which reflects the effectiveness of data compression of the transient signal by optimizing the wavelet basis.

The compression rate r ($J_1 - J$ is the decomposition layer) and the reconstruction root mean square d_{mse} are defined in Equations (6.51) and (6.52), where $x_r(n)$ is a reconstruction signal and N is the number of discrete signal points.

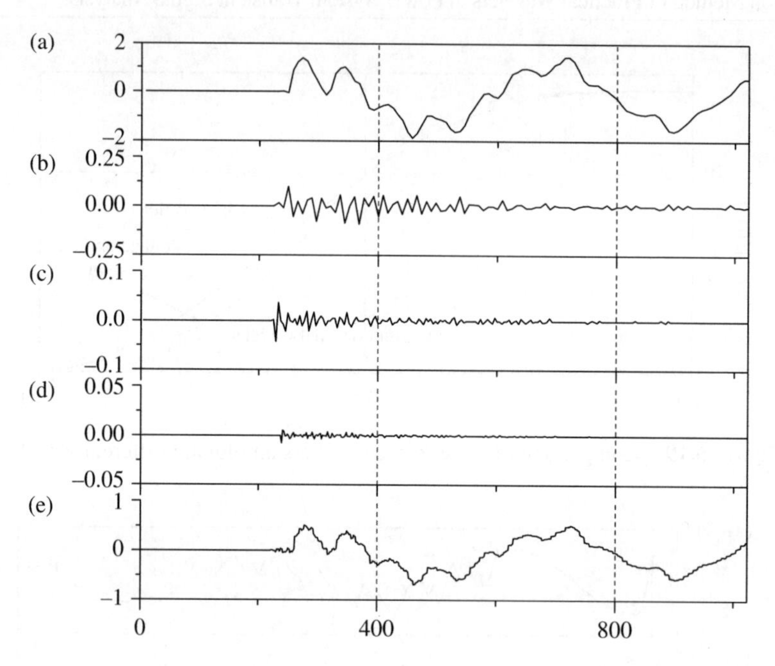

Figure 6.21 Current transient compression based on optimal wavelets: (a) discrete wavelet approximation $a(J-3,m)$, (b) wavelet detail $\overline{d}(J-3,m)$, (c) wavelet detail, (d) wavelet detail $\overline{d}(J-1,m)$, (e) reconstruction signal, as $\eta_s = 0.05 * \max(d(j,m)$

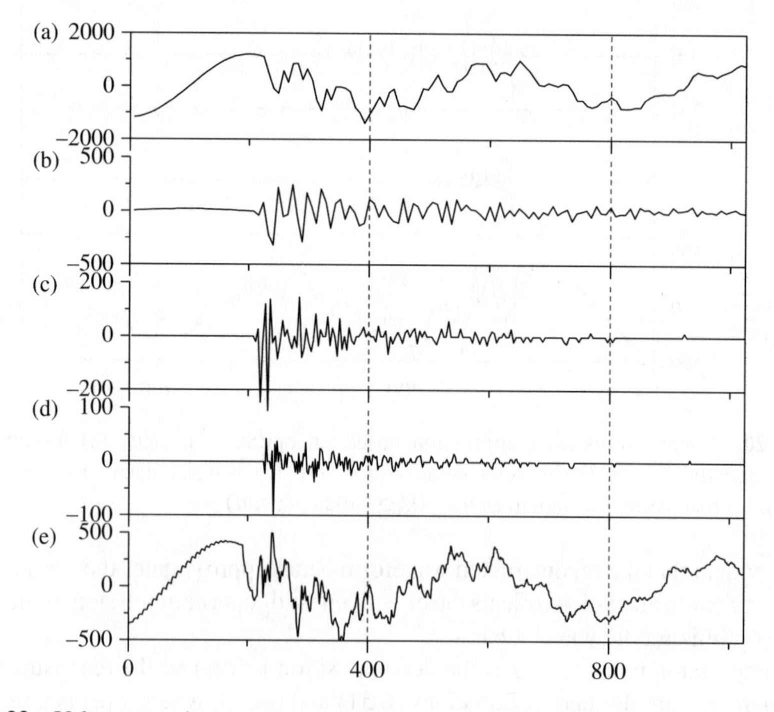

Figure 6.22 Voltage transient compression based on db3: (a) discrete wavelet approximation $a(J-3,m)$, (b) wavelet detail $\overline{d}(J-3,m)$, (c) wavelet detail, (d) wavelet detail $\overline{d}(J-1,m)$, (e) reconstruction signal, as $\eta_s = 0.05 * \max(d(j,m)$

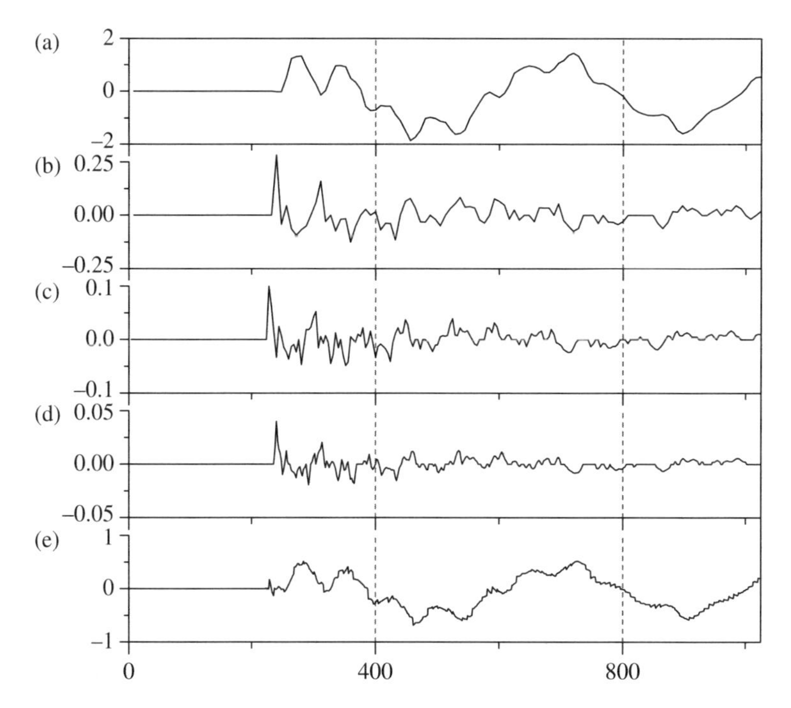

Figure 6.23 Current transient compression based on db3: (a) discrete wavelet approximation $a(J-3,m)$, (b) wavelet detail $\bar{d}(J-3,m)$, (c) wavelet detail, (d) wavelet detail $\bar{d}(J-1,m)$, (e) reconstruction signal, as $\eta_s = 0.05 * \max(d(j,m))$

$$r = N/\left(\frac{N}{2^{J_1-J}} + \frac{uN}{2^{J_1-J}} + \cdots + \frac{uN}{4} + \frac{uN}{2}\right) \tag{6.51}$$

$$d_{mse} = 10\log_{10}\frac{\sum_{n=1}^{N}\left|x(n) - x_r(n)\right|^2}{\sum_{n=1}^{N}\left|x(n)\right|^2} \tag{6.52}$$

To facilitate the calculation of the compression rate, the threshold configuration method is adopted as in Equation (6.48). The relation of the compression rate and reconstruction error between the two wavelet bases could be seen in Figure 6.24. We see that the reconstruction error increases when adopting db3 instead of the optimal wavelet if we want to guarantee the compression rates remaining the same. It is found by the amplitude-frequency characteristic of the filter in Figure 6.19 that the Daubechies wavelet inhibits useful information in low-frequency signals and allows more high-frequency noise to mix in. On the contrary, the employment of the optimal wavelet produces a smaller reconstruction error. Especially when $r = 1–4$ (in other words, more data are contained in the

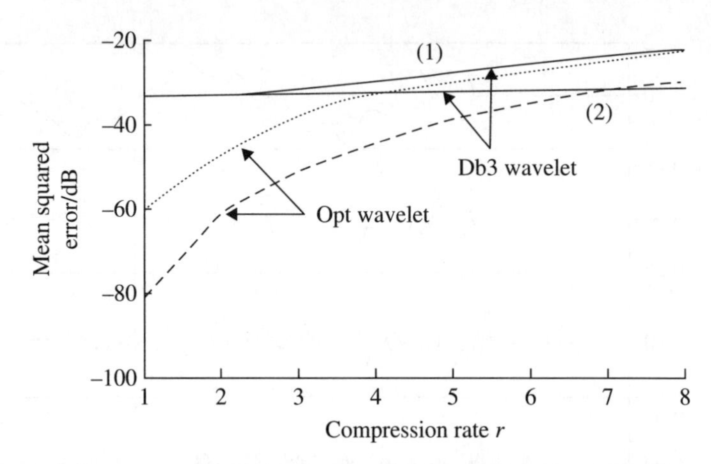

Figure 6.24 The relation of compression rate and reconstruction error between an optimal wavelet (Opt) and db3 wavelet (Db3): (1) fault phase voltage, (2) zero-mode current

wavelet detail), the reconstruction error is significantly reduced. This result coincides with the wavelet optimization mechanism that maximizes the principle component of the signal (the norm of wavelet approximation). The addition of decomposition layers results in a higher compression rate to meet the requirement of actual needs.

The wavelet transform based on the aforementioned optimal wavelet basis better suits the analysis of electric power transient signals, especially signals with low-frequency components as their principle component (fundamental component and lower components). Wideband transients are rich in these signals. Hence, the application of discrete wavelet transform based on optimal wavelets will lead to higher compression rates and smaller reconstruction errors in the compression of fault transient data.

References

[1] He Z. Y., The electric power system transient signal wavelet analysis method and its applicaton on the non-unit transient protection principle. *Southwest Jiaotong University*, 2000.

[2] Liu Z. G., Multiwavelet theory and its application study in power system fault signals processing. *Southwest Jiaotong University*, 2003.

[3] Zou H. H., Ahmed H. Tewfik., Discrete orthogonal M-band wavelet decompositions. *Acoustics, Speech, and Signal Processing, 1992*, ICASSP-92, 1992 IEEE International Conference on. Vol. 4. IEEE, 1992.

[4] Steffen P., Heller P. N., Gopinath R. A., et al., Theory of regular M-band wavelet bases. *IEEE Transactions on Signal Processing*, vol. 41, no. 12, pp. 3497–3511, 1993.

[5] Vetterli M., Herley C., Wavelets and filter banks: theory and design. *IEEE Transactions on Signal Processing*, vol. 40, no. 9, pp. 2207–2232, 1992.

[6] Koilpillai R. D., Vaidyanathan P. P., Cosine-modulated FIR filter banks satisfying perfect reconstruction. *IEEE Transactions on Signal Processing*, vol. 40, no. 4, pp. 770–783, 1992.

[7] Vaidyanathan P. P., Multirate digital filters, filter banks, polyphase networks, and applications: a tutorial. *Proceedings of the IEEE*, vol. 78, no. 1, pp. 56–93, 1990.

[8] Strichartz R. S., How to make wavelets. *The American Mathematical Monthly*, vol. 100, no. 6, pp. 539–556, 1993.

[9] Chaari O., Meunier M., Brouaye F., Wavelets: a new tool for the resonant grounded power distribution systems relaying. *IEEE Transactions on Power Delivery*, vol. 11, no. 3, pp. 1301–1308, 1996.
[10] Zhang C. L., Huang Y. Z., Ma X. X., et al., Study of relaying protection for transformer applying IRWT. *Automation of Electric Power Systems*, vol. 23, no. 17, pp. 20–22+53, 1999.
[11] He Z. Y., Dai X.W., Qian Q. Q., Prefilter selection methods in fast algorithm of discrete wavelet transform. *Journal of Southwest Jiaotong University*, vol. 35, no. 2, pp. 183–187, 2000.
[12] Tewfik A. H., Sinha D., Jorgensen P., On the optimal choice of a wavelet for signal representation. *IEEE Transactions on Information Theory*, vol. 38, no. 2, pp. 747–765, 1992.

7

Wavelet Postanalysis Methods for Transient Signals in Power Systems

The wavelet transform results contain abundant wavelet decomposition information. Processing of wavelet decomposition information should not be limited to showing several beautiful pictures. At present, in part with detection methods, the feature extraction needs manual intervention and assumption of specific conditions. In classification methods, the wavelet transform results are usually interpreted as the inputs of neural networks or fuzzy classification systems. The neural networks and fuzzy classification systems are usually huge due to the large amount of wavelet decomposition information. It is necessary to process the wavelet decomposition results further. This chapter aims to extract features from fault transients quantitatively. Several wavelet postprocessing methods for power system transients are introduced systematically, such as modulus maxima and the singularity analysis method, the energy analysis method, wavelet neural networks, wavelet coefficients statistics, and the clustering analysis method.

7.1 Modulus maxima and the singularity exponents method

7.1.1 Wavelet transform property of singular signals

Theorem 7.1 (Mallat) Let n be a strictly positive integer. Let $\psi(x)$ be a wavelet with n vanishing moments, n times continuously, and compactly supported. Let $x(t) \in L^1[c,d]$ ([c, d] be a real number zone. If there exists a scale $a_0 > 0$ such that for all scales $a < a_0$ and $t \in [c,d]$, $|WT_x(a,t)|$ has no local maxima, then for any $\varepsilon > 0$ and $\alpha < n$, $x(t)$ is uniformly Lipschitz α in $[c+\varepsilon, d-\varepsilon]$ (ε is an arbitrary small positive number).

This theorem proves that a function is not singular in any neighborhood where its wavelet transform has no modulus maxima. The singularity is related to modulus maxima at each scale. The modulus maxima converge to the singularity across scales. Therefore,

Wavelet Analysis and Transient Signal Processing Applications for Power Systems, First Edition. Zhengyou He.

we can decompose the signal at multiple scales using multiresolution analysis. The decomposed wavelet coefficients has modulus maxima at sharp variations.

Mallat proved that if the positions and values of the modulus maxima are at all scales, the original signal can be reconstructed sufficiently with alternative projection, which can restore the value of wavelet transform at each scale. That is to say, the positions and values of the modulus maxima at all scales carry enough information to reconstruct the corresponding function. As a result, extracting the modulus maxima of power system transient signals is one of the transient signals' wavelet postanalysis methods to obtain features. For example, the modulus maxima are used to detect the fault inception time and fault duration time. When applied in traveling-wave-based fault location, the fault distance is determined depending on the polarities and positions of modulus maxima. The modulus maxima perform well in a fault location application. However, there are multiple modulus maxima to be recorded at each scale. The selection of modulus maxima always needs to be completed gradually from rough selection to fine selection. Moreover, the features contain time shift parameters and amplitude parameters, which make the algorithm complex and require large amounts of computation and of storage volume. Therefore, the singularity in signals is usually used as another feature in practical applications. The singularity of a signal is often measured with Lipschitz exponents. In order to estimate Lipschitz exponents, it is necessary to find modulus maxima lines [2].

Nevertheless, it is difficult and complicated to search for strictly maxima lines of a certain singularity. In addition, using wavelets with more vanishing moments increases the number of maxima lines. The workload is heavy. The traditional Lipschitz exponents utilize a simple ad hoc algorithm to find the maxima lines. We suppose that a modulus maximum propagates from a scale 2^j to a coarser scale 2^{j+1}, if it has a large amplitude and its position is close to a maximum at the scale 2^{j+1} that has the same sign. Such an ad hoc algorithm is not exact but saves computations since we do not need to compute the wavelet transform at any other scale. The application results show that the singularities in actual signals are not isolated singularities. It is difficult, along with the noise interference, to apply this algorithm because it requires some prior knowledge of different signals. Besides, it takes quite a long time to find the maxima lines. For that reason, this book puts forward a window Lipschitz to characterize approximately the singularity according to Theorem 7.1.

Definition 7.1 If the signal $x(t) \in L^2(\mathbb{R})$ is uniformly Lipschitz α on the real axis, then α is the global Lipschitz exponent. β is defined as the window Lipschitz exponent at t_0, if $X(t) = x(t)$ is uniformly Lipschitz α over the intervals $[t_0 - \Delta t, t_0 + \Delta t]$.

According to Definition 7.1 the window Lipschitz exponent characterizes approximately the singularity at t_0. Let d_k^j be the discrete wavelet transform of a window signal at x_0 sampled from a discrete transient signal $x(n)$. The estimation of Lipschitz exponents thus consists of estimating the optimal c and α, which satisfy

$$\left|d_k^j\right| \leq c2^{-j(1/2+\alpha)}, j = 1,2,\cdots J \tag{7.1}$$

α can be solved in the following three steps:

1. Compute discrete wavelet decomposition to obtain the wavelet coefficients d_k^j at decomposition level j and instant k, as well as $d_j^* = \max\left|d_k^j\right| > 0$. Then, the problem consists of finding the optimal c and α that satisfy

$$d_j^* \le c2^{-j(1/2+\alpha)}, j = 1,2,\cdots J \tag{7.2}$$

2. Let $b_j^* = \log_2 d_j^*$, and $b = \log_2 c$; then, inequalities (Equation (7.2)) yield that

$$j\left(1/2+\alpha\right)+b_j^* \le b, j = 1,2,\cdots J \tag{7.3}$$

3. In order to solve inequalities (Equation (7.3)), let $j(1/2+\alpha)+b_j^*+\beta_j = b$. Then $\beta_j = b - j(1/2+\alpha) - b_j^*$. Thus, the problem consists of finding α and b that satisfy

$$\min\sum_j \beta_j^2 = \sum_j \left[b - j\left(1/2+\alpha\right) - b_j^*\right]^2, j = 1,2,\cdots J \tag{7.4}$$

Using the least-squares method, we obtain

$$\alpha = \frac{\Sigma j \Sigma b_j^* - M \Sigma j b_j^*}{J \Sigma j^2 - (\Sigma j)^2} - \frac{1}{2}, j = 1,2,\cdots J \tag{7.5}$$

This is the window Lipschitz exponent at t_0.

7.1.2 Analysis of the theoretical signals

Generally, in real power systems, the supplied current or voltage signal contains a fundamental component, harmonics, a sharp variation component, and some noise. In order to investigate the application of singularity exponents in power equipment fault detection, we first analyze the singularity exponents based on wavelet transform of theoretical signals that contained a power fault signal. We take the singular signals such as the step signal $u(t)$, impulse signal $\delta(t)$, white noise $n(t)$, fundamental component, integer harmonics, and their mixed signals as the objects of the research. The wavelet bases, such as the Haar wavelet, Daubechies-4 (db4) wavelet, and cubic B-spline wavelet, are chosen as the mother wavelets for discrete wavelet transform computed on three scales. The sampling frequency is 10 kHz. The analysis results appear in Table 7.1 and Figures 7.1, 7.2, and 7.3.

Some conclusions from Table 7.1 and Figures 7.1, 7.2, and 7.3:

1. The Lipschitz exponents of current signals decrease when the harmonic order increases. This indicates that the Lipschitz exponents are inversely proportional to the degree of the signal sharp variation. However, the Lipschitz exponents are positive.

Table 7.1 Lipschitz exponents of typical signals

	Haar wavelet	db4 wavelet	B-spline wavelet
Step signal	0.0000	0.0145	0.0182
Impulse signal	−1.0000	−0.8932	−0.8840
White noise	−0.5-ε (ε >0)	−0.5-ε (ε >0)	−0.5-ε (ε >0)
Sinusoidal signal	0.5000	0.4999	0.4998

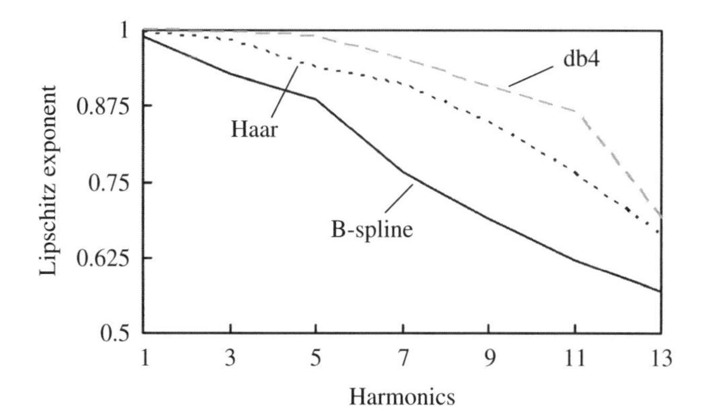

Figure 7.1 Lipschitz exponents of each current harmonic

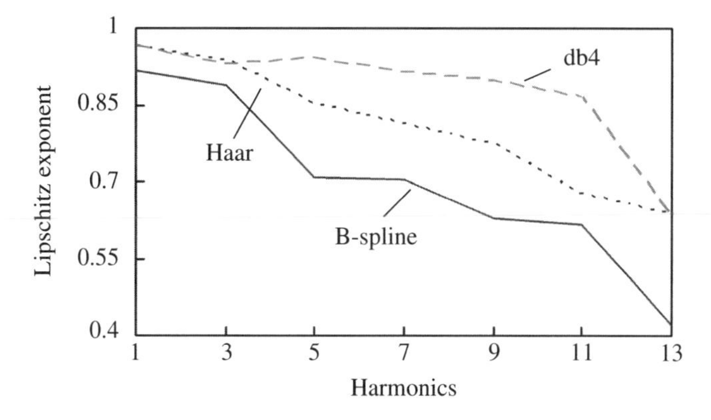

Figure 7.2 Lipschitz exponents of each current harmonic with stochastic interference

2. The Lipschitz exponents of step signal, impulse signal, and white noise are equal to or smaller than zero (≤0). That is to say, the modulus maxima created by these signals have an amplitude that decreases when the scale increases. This conclusion verifies the fact that there is an inverse relationship between the behavior of wavelet maxima dominated by signals and white noise across scales, as well as the fact that wavelet transform can remove white noise from useful signals by analyzing the evolution of the wavelet transform maxima across scales.

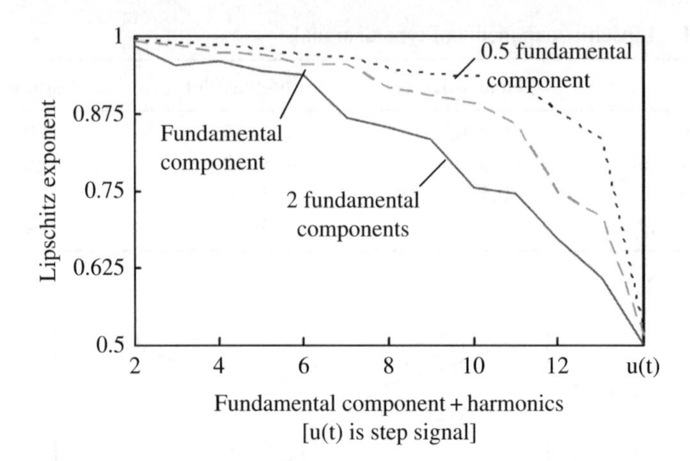

Figure 7.3 Lipschitz exponents of a fundamental component mixed with harmonics

3. After mixing the fundamental component with white noise or harmonics, the Lipschitz exponent of mixed signal decreases and is inversely proportional to the ratio of high-frequency components in the signal.
4. The wavelet decomposition has a band-pass filter property. When decomposing the signal at large scales, the filtered wavelet decomposition contains the fundamental component if we ignore the frequency aliasing. Therefore, the selection of the largest scale should ensure that the frequency band corresponding to the largest scale is higher than that of the fundamental frequency, or the calculation of Lipschitz exponents will be influenced due to the aliasing of the fundamental frequency and the frequency corresponding to the largest scale.

7.1.3 Analysis of fault transient signals

Figure A.3 in the Appendix depicts a one-line diagram of the power system under study, which is a 500 kV transmission system. The transient simulation is carried out by using the power system simulation program PSCAD/EMTDC. The sampling frequency is 20 kHz. Single-phase-to-ground fault is simulated without consideration of the fault arc. The data window is selected as 3.2 ms. The zero sequence current is analyzed using wavelet analysis. Figures 7.4 and 7.5 show the representative simulation results. It can be seen in Figures 7.4 and 7.5 that the proposed window singularity exponent is not only capable of detecting the fault and fault inception time, but also capable of discriminating among different transient signals, such as internal fault signals and external fault signals.

Figure 7.4 depicts that the phase A–to-ground fault occurred when the voltage magnitude is high. The fault-generated current has more high-frequency components for faults at F1 and F2. The window Lipschitz exponents change abruptly after one data window delay, which means that a fault occurred. As seen in Figures 7.4 and 7.5, the calculated window Lipschitz exponents corresponding to internal faults are small for different fault inception times and different fault locations. However, the calculated window Lipschitz

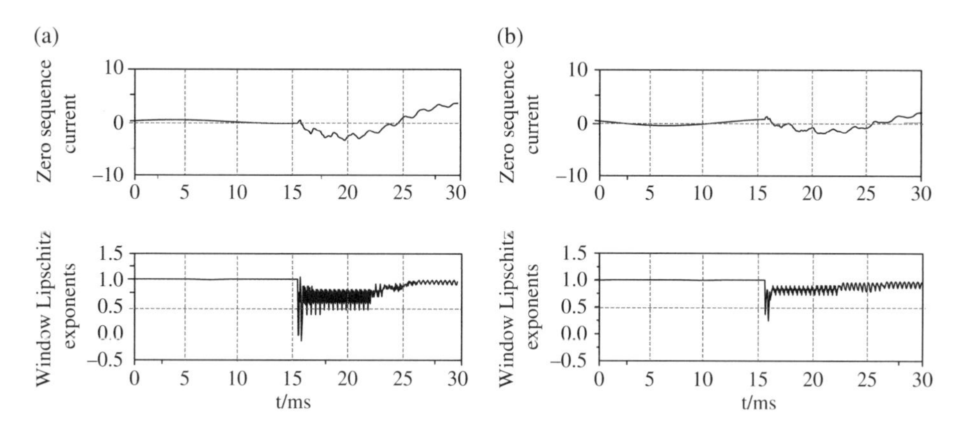

Figure 7.4 Lipschitz exponents for internal and external faults (F1 and F2 are 1 km away from the bus M, and the fault resistance is 0.5 ohm): (a) internal fault occurred at F1 when the voltage magnitude is high, (b) external fault occurred at F2 when the voltage magnitude is high

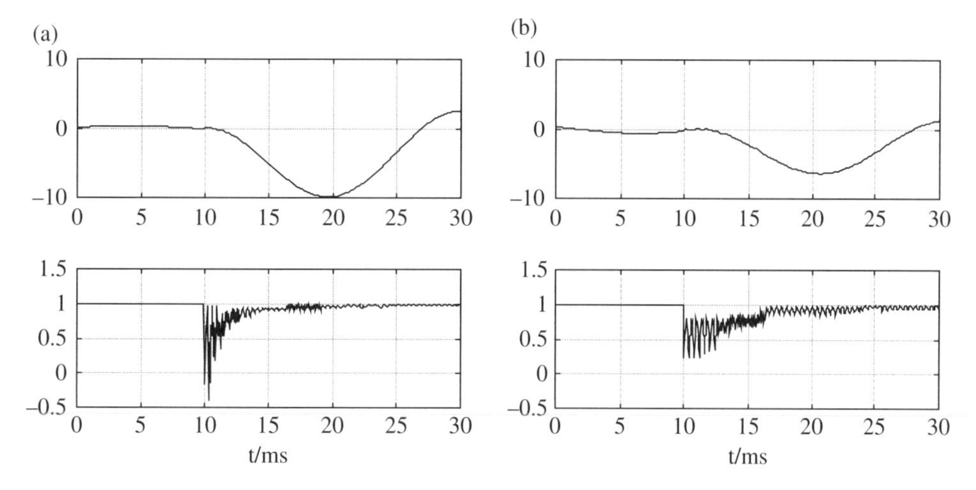

Figure 7.5 Lipschitz exponents for internal and external faults (F1 and F2 are at the middle of the line, and the fault resistance is 0.5 ohm): (a) internal fault occurred at F1 when the voltage magnitude is high, (b) external fault occurred at F2 when the voltage magnitude is high

exponents are greater for external faults. This is because for external faults, bus stray capacitance attenuates the high-frequency components. Then CT (total capacitance) detects less high-frequency components. Therefore, the Lipschitz exponents for external faults are larger than those for internal faults, which is in accordance with the analysis of theoretical signals. Thus, we could distinguish the internal faults from external faults using the Lipschitz exponents.

Actually, the nonlinear arcing faults occur most frequently in extra-high-voltage (EHV) transmission lines. The nonlinear arcing faults will generate high frequency. The window Lipschitz exponents are still valid to detect and discriminate the fault when the fault occurs at a voltage zero point based on the fault arc model. The proposed window Lipschitz exponent presents potential that is applied when distinguishing the internal faults from external faults for EHV transmission lines, with good real-time effectiveness. Therefore, the window Lipschitz exponent based on wavelet transform can be used to develop new detection or protection criteria that are novel and effective.

7.2 Energy distribution analysis method

Since power system transient signals, especially fault transient signals, contain abundant high-frequency components, the detection method based on window energy is widely adopted in power quality analysis, fault detection, relay protection, and so on. However, the fault-generated transient signals are complex because the fault conditions are so different. Consequently, it is difficult to discriminate accurately different transient signals if only window energy is used. For example, it is hard to distinguish the internal faults between external faults with non-unit transient protection for EHV transmission lines. On this basis, referred to the *window energy*, we could extract the signal characteristic from the energy distribution after wavelet decomposition. Hence, we propose the one-wavelet transform postprocessing method that is named energy distribution analysis.

7.2.1 Energy distribution coefficient and energy fluctuation coefficient

7.2.1.1 Basic theory

For an orthogonal wavelet transform, the energy in each scale is the square of its wavelet coefficients. For a biorthogonal wavelet transform, the energy in each scale is the product of its wavelet coefficients and the wavelet coefficients of its dual wavelet. Moreover, according to the wavelet transform representation of transient signals based on multiresolution analysis, the energy in each scale is the energy corresponding to each frequency band. Let $D_j(n)$ be the component of transient signal $x(n)$ at scale j. The energy distribution coefficient B_j and energy fluctuation coefficient S_j, which can be used as characteristics for transient signals' feature extraction or classification, are defined by

$$B_j = E\left[\left\|D_j(k)\right\|^2\right] = \frac{1}{N}\sum_{k=1}^{N}\left\|D_j(k)\right\|^2 = \frac{E_j}{N} \tag{7.6}$$

$$S_j = \frac{B_j}{B_{j+1}} = \frac{E\left[\left\|D_j(k)\right\|^2\right]}{E\left[\left\|D_{j+1}(k)\right\|^2\right]} = \frac{E_j}{E_{j+1}}, j = 1,2,\cdots J \tag{7.7}$$

The definition in Equations (7.6) and (7.7) substantially inspects the transient signal energy and its distribution across scales. We can obtain the detection and classification algorithm based on the energy distribution feature and energy fluctuation feature. Signal energy distribution coefficient B_j and energy fluctuation coefficient S_j have the following features:

1. The energy distribution coefficient reflects the evolution of a transient signal wavelet transform across eigen-subspace. The energy fluctuation coefficient describes the local fluctuation pattern of a transient signal wavelet transform.
2. The wavelet coefficients at fine scales are disturbed by noises, whereas the wavelet coefficients at large scales are unable to describe the fluctuation edge. In practical application, the range of scales corresponding to the dominated frequency band of the signal is chosen to be $j_{\min} \le j_l \le j_{\max}$ ($j_{\min} > 1$, $j_{\max} < J$).
3. Since the dyadic wavelet transform of white noise is a stationary random process, its energy distribution has no relation to scales. Thus, at the dominated frequency band, the noises have a minimum effect on the energy distribution feature and wave shape feature.
4. The signal classification can be achieved by induction or other classification methods with the energy distribution feature and wave shape feature over a range of scales $j_{\min} \le j_l \le j_{\max}$.

7.2.1.2 Simulation analysis

Figures 7.6 and 7.7 display the variation of energy fluctuation coefficients computed with the current in Figures 7.4 and 7.5 in the moving window according to time. As seen in

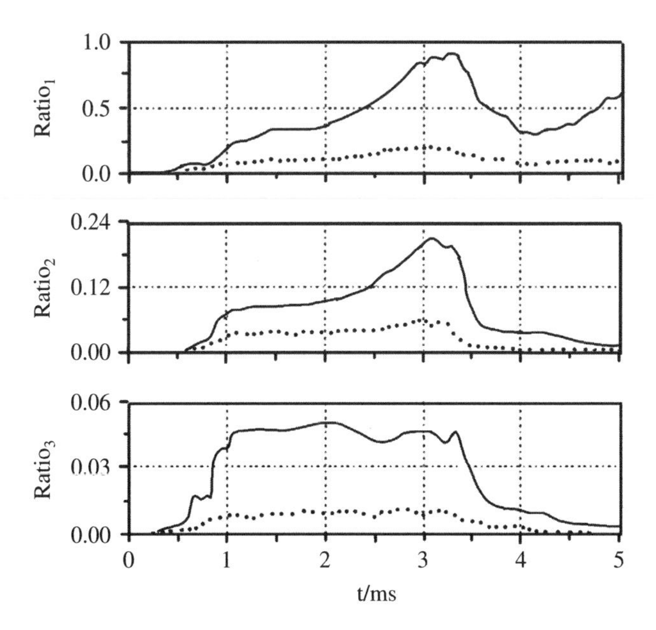

Figure 7.6 The variation of energy fluctuation coefficients computed with the current in Figure 7.4. Ratio_1, $S_1*S_2*S_3$; Ratio_2, $S_1*S_2*S_3*S_4$; Ratio_3, $S_2*S_3*S_4*S_5$

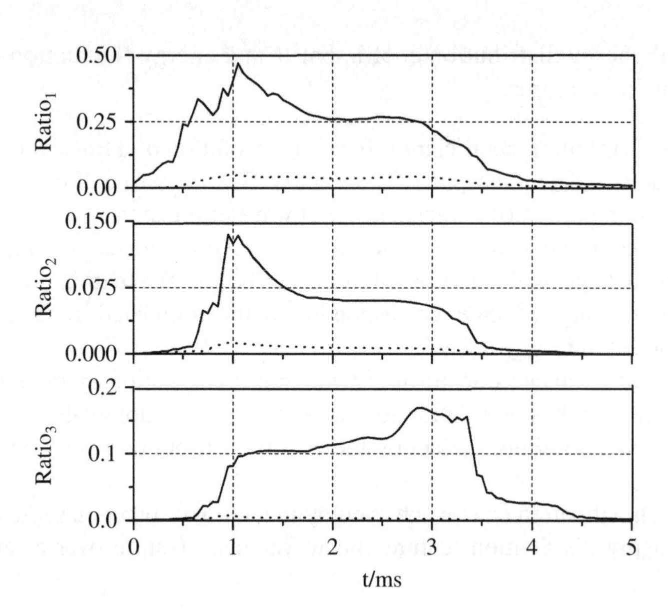

Figure 7.7 The variation of energy fluctuation coefficients computed with the current in Figure 7.5. $Ratio_1$, $S_1*S_2*S_3$; $Ratio_2$, $S_1*S_2*S_3*S_4$; $Ratio_3$, $S_2*S_3*S_4*S_5$

Figures 7.6 and 7.7, the cross-scale energy fluctuation coefficients for internal faults and external faults are significantly different. A large amount of simulation studies indicate that for external faults, the higher frequency components are attenuated more severely due to the existence of bus stray capacitance. According to the different current features of external and internal faults, one can detect and identify the fault signals based on the energy distribution coefficient and energy fluctuation coefficient or cross-scale fluctuation coefficient.

7.2.2 Energy moment

In order to better describe the energy at each decomposition frequency band and its distribution characteristics in a time axis, we introduce the energy moment M_j.

Definition 7.2 Decompose the signal by wavelet transform, and reconstruct the decomposed signals in a single branch. Then the energy moment M_j of signal E_{jk} at each frequency band is given by

$$M_j = \sum_{k=1}^{n} (k \bullet \Delta t) \left| E_j (k \bullet \Delta t) \right|^2 \tag{7.8}$$

where Δt is the sampling time interval and n is the total number of samples.

The energy moment M_j not only considers the energy amplitude but also describes the energy distribution across time. In comparison with the energy spectrum, the energy moment could reveal the energy distribution features better, which is beneficial when extracting signal features.

The basic steps to extract the signal features using the wavelet energy moment are as follows:

1. Select the proper wavelet base and decomposition levels to decompose the sampled signal by wavelet transform. Let x be the original signal, and let D_{jk} be the detail coefficient at level j and instant k of wavelet decomposition. $j = 1,2,\cdots J, J+1$, and J is the largest decomposition level.
2. Reconstruct the detail coefficients to obtain the signal component E_{jk} at each frequency band.
3. Calculate the wavelet energy moment M_j of E_{jk} by Definition 7.2.
4. Construct the feature vectors. The wavelet energy moment of each frequency band is affected greatly when the system is under abnormal operating conditions. Thus, we could construct a feature vector $\mathbf{T}$:

$$\mathbf{T} = \left[M_1, M_2, \ldots, M_{J+1}\right] \Big/ \left[\sum_{j=1}^{J+1}\left(M_j^2\right)\right]^{1/2} \tag{7.9}$$

We calculate the energy moments of two signals as examples; this can help readers understand the energy moment concept more intuitively. The mathematically expressions of signal 1 and signal 2 are shown in Equations (7.10) and (7.11). The sampling frequency is 1 kHz. Figure 7.8 shows the time domain waveform.

$$f_1(t) = \begin{cases} \sin(2\pi \times 50 \times t) & t < 0.1 \\ \sin(2\pi \times 100 \times t) & 0.1 \le t < 0.2 \\ \sin(2\pi \times 150 \times t) & 0.2 \le t < 0.3 \end{cases} \tag{7.10}$$

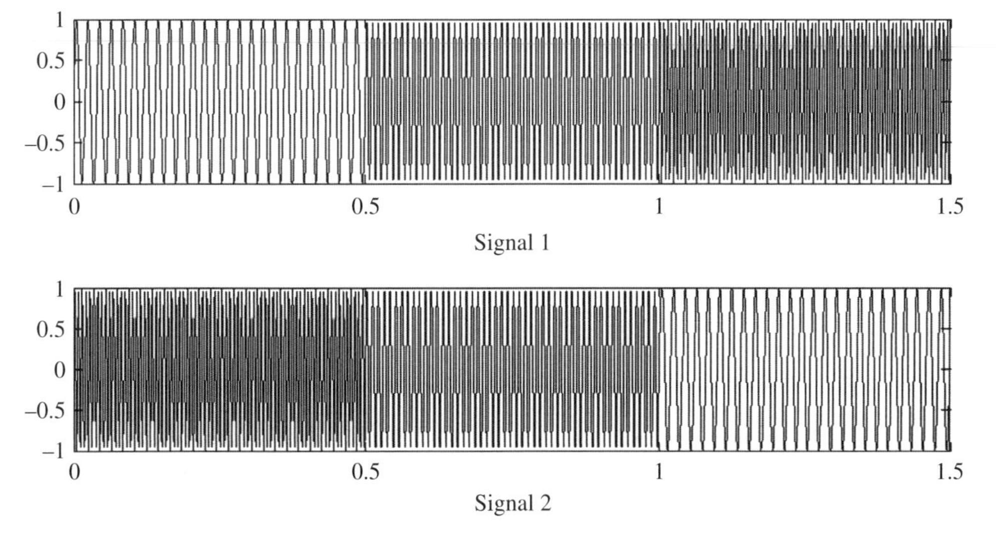

Figure 7.8 The time domain waveform of simulation signals

$$f_2(t)=\begin{cases}\sin(2\pi\times150\times t) & t<0.1\\ \sin(2\pi\times100\times t) & 0.1\le t<0.2\\ \sin(2\pi\times50\times t) & 0.2\le t<0.3\end{cases} \tag{7.11}$$

We decompose signal 1 and signal 2 in five levels. This procedure is used to obtain six reconstructed signals at different frequency bands. The reconstructed signals have the same length as the original signals. The reconstructed signals are analyzed to extract features using a wavelet energy spectrum. A statistical chart of normalized signal energy is shown in Figure 7.9. Although signal 1 is different from signal 2, these two signals could not be identified from the energy spectrum features as shown in Figure 7.9. Figure 7.10 is a statistical chart of a wavelet energy moment for signal 1 and signal 2. It can be seen in Figure 7.10 that the wavelet energy moment is capable of distinguishing these two signals.

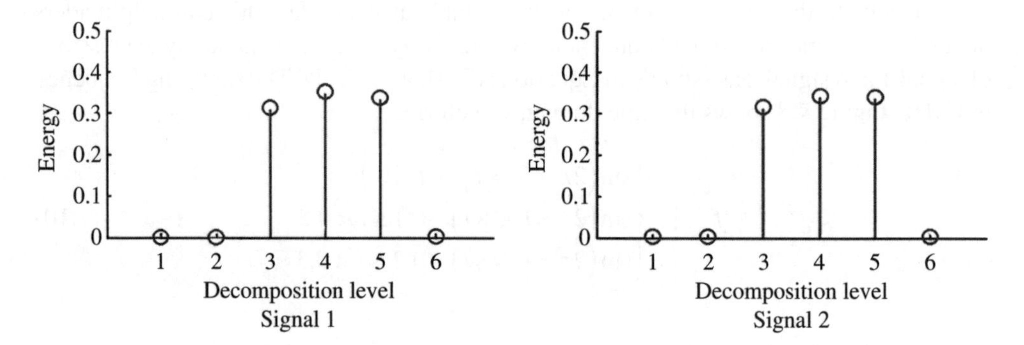

Figure 7.9 Wavelet energy statistical diagram of signal 1 and signal 2

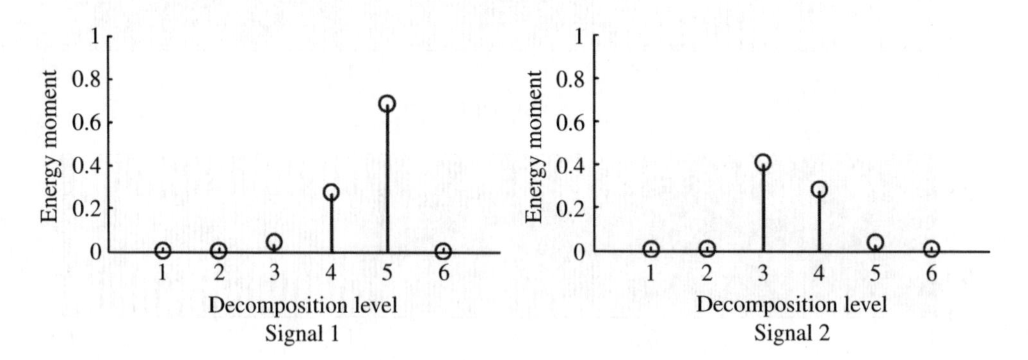

Figure 7.10 Wavelet energy moment statistical diagram of signal 1 and signal 2

7.3 Wavelet neural networks

Artificial neural networks (ANNs) are widely used due to their learning ability, ability to process multiple inputs in parallel, and nonlinear mapping and fault-tolerant abilities, as well as their self-adaptive ability obtained from learning new behavior. However, in practical application, the limitations of ANNs are their lack of definite rules for determining the number of hidden layer nodes, the initialization of various network parameters, and the difficulty of choosing an optimal network structure. As a powerful signal-processing tool, wavelet transform can replace Fourier transform in all applications in principle because it has stronger processing abilities for analyzing the nonstationary signals, but wavelet theory is limited in a small area because its large-scale application costs much (to construct and store the wavelet bases). An ANN is a powerful tool to process large-scale problems. Therefore, it is possible to combine ANNs with wavelet transform.

Currently, there are two main ways to combine wavelets with ANNs, namely, the loose combination and the tight combination. Loose combination means that a wavelet transform performs as a preprocessing tool and provides input eigenvectors for an ANN. Tight combination means that a wavelet transform combines with an ANN directly, and the wavelet function or scaling function is the activation functions of neurons. In general, the wavelet neural network means the latter combination form, so this book pays attention to the tight combination of wavelet transforms and ANNs.

7.3.1 Introduction of a wavelet network

The activation functions in neural networks are one of their most important factors, and they affect the functions of neural networks. The commonly used activation functions include the step function, sigmoid function, and radial basis function (RBF). The sigmoid function is a global function, and its support set is the entire Euclidean space, which results in serious overlapping in space. The radial basis network is one of the methods to represent functions of elements of a compactly supported basis. The locality of the basis functions makes the RBF network more suitable for learning functions with local variations and discontinuities. However, the activation functions in radial basis networks are generally both nonorthogonal and redundant. This means that, for a given function, its RBF network approximation is not unique. Wavelet neural networks have some particular advantages compared with RBF networks. First, the wavelet function can be orthogonal, which can ensure that the approximation function is unique. Second, there are more extensive choices of wavelet functions for different approximation functions; thus, different wavelet functions can be selected based on the characteristics of the approximation function. Finally, the wavelet functions can describe the discontinued functions finely and gradually, and this can obtain approximation that is more valuable.

In 1988, Daugman put forward the idea of constructing a neural network by Gabor transform. So, can wavelet functions be the activation functions of neural networks? The answer is yes. Q.H. Zhang and A. Benveniste had proposed the definition of wavelet neural networks clearly. In fact, the wavelet neural network is an alternative to a feedforward neural network for approximating arbitrary functions. The basic principle is to establish a link between the neural network and wavelet transform by replacing the neurons with wavelons. Zhang and Benveniste provided the parameters initialization method and learning algorithm of a wavelet neural network. To assess the approximation results, the authors approximated the one-dimensional function and two-dimensional function with wavelet decomposition, neural networks, and wavelet neural networks. The uniform wavelet neural network structure for the one-dimensional case and multidimensional case is represented as

$$x(t)=\sum_{i=1}^{N} w_i\psi\left[D_i R_i\left(t-t_i\right)\right]+\overline{x} \tag{7.12}$$

where t_i are the translation vectors; D_i are the diagonal dilation matrices built from dilation matrices; R_i are rotation matrices; w_i are the weights; and $\overline{x}$ is the mean value of x, which is introduced to make it easier to approximate functions with nonzero average.

The activation functions in this wavelet network are wavelet functions, which are not limited to the orthogonal wavelet functions.

In the aspect of follow-up wavelet network construction, some researchers have proposed several wavelet networks, such as the orthogonal wavelet network whose activation functions are scaling functions, the adaptive wavelet network constructed by superposition wavelets, and the orthogonal wavelet network whose activation functions are orthogonal wavelets. Generally speaking, proposing a wavelet network has tight relations to wavelet theory. Wavelet theory provides a reliable theoretical basis for analysis and fusion of a wavelet network. The orthogonal wavelet network is usually derived based on multiresolution analysis, whereas the discrete wavelet network is derived based on frame theory in wavelet analysis. An adaptive wavelet network, which is very similar to the radial basis network, has distinct advantages compared with an RBF network. The selection of center parameters and width parameters in the RBF network depends on experience. However, the selection of dilation parameters and translation parameters is set by time–frequency localization of the wavelet.

7.3.2 *Construction of wavelet neural networks*

The construction of wavelet networks is various and complex due to the variety and complexity of wavelet functions. The reported research indicates that wavelet networks are mainly constructed from continuous wavelet transform, orthogonal wavelet transform, wavelet frame, and wavelet bases fitting. The basis functions of the wavelet network constructed by continuous wavelet transform are not limited to finite discrete values. Moreover, the network representation is not unique and does not confirm the corresponding relationship between wavelet parameters and functions. Therefore, the nonlinear optimization problem of the wavelet network based on continuous wavelet transform is similar to that of a backpropagation network. Nevertheless, wavelet theory is helpful for initializing network parameters and construction. Wavelet theory can also give the network good approximation ability and convergence rates. At present, the application of a wavelet network constructed by continuous wavelet transform occurs infrequently because the continuity makes the network construction inconvenient.

This section will introduce wavelet network construction using orthogonal wavelet transform, wavelet frame, and wavelet basis fitting, respectively.

7.3.2.1 Orthogonal wavelet transform

We know that the wavelet functions $\psi_{m,n}(t) = 2^{m/2}\psi(2^m t - n), n \in (-\infty, +\infty)$ form an orthonormal basis of $L^2(\mathbb{R})$, the space of all square integrable functions on $\mathbb{R}$. $L^2(\mathbb{R}) = \underset{m}{\oplus} W_m$, where W_m is a subspace spanned by $\psi_{m,n}(t)$. The wavelet $\psi(t)$ is often generated from a companion $\phi(t)$, known as the scaling function. The dilations and translations of the scaling function induce a multiresolution analysis (MRA) of $L^2(\mathbb{R})$ (i.e., a series of closed subspaces $\cdots \subset V_{-1} \subset V_0 \subset V_1 \subset V_2 \subset \cdots$. V_m is the subspace spanned by $\{2^{m/2}\phi(2^m t - n)\}_{n=-\infty}^{n=+\infty}$). Moreover, $V_{m+1} = V_m \oplus W_m$. A L^2 function $x(t)$ is decomposed as

$$x(t) = \sum_{m,nk} \langle x, \psi_{m,nk} \rangle \psi_{m,nk}(t) \tag{7.13}$$

Then, for an arbitrary integer m_0,

$$x(t) = \sum_{n} \langle x, \phi_{m_0,n} \rangle \phi_{m_0,n}(t) + \sum_{m \geq m_0} \langle x, \psi_{m,n} \rangle \psi_{m,n}(t) \tag{7.14}$$

For a given set of M and K, the wavelet network described above can approximate the function $x(t)$:

$$\hat{x}(t) = \sum_{k=-K}^{K} c_k \phi_{M,k}(t) \tag{7.15}$$

The approximation in Equation (7.15) can be implemented by a three-layer network, as shown in Figure 7.11.

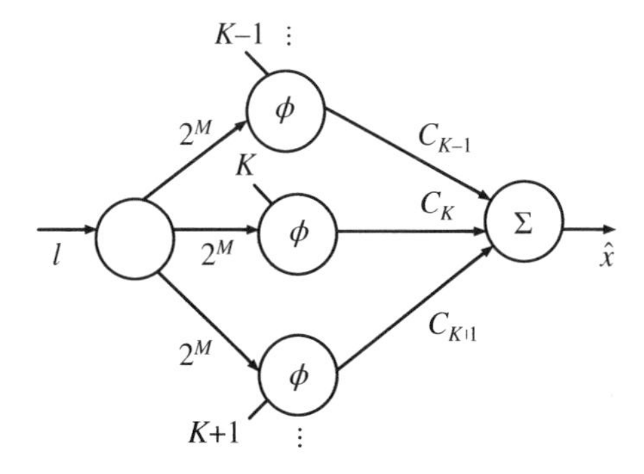

Figure 7.11 Wavelet network induced by orthogonal wavelet transform

The wavelet network induced by the orthogonal wavelet transform substantially selects proper dilation parameters and translation parameters to approximate the function closely. This procedure is achieved based on orthogonal wavelets with compact support and multiresolution analysis.

7.3.2.2 Wavelet frame

The discrete wavelet transform of function $x(t)$ is defined as follows in $L^2(\mathbb{R})$ space:

$$d_{j,k} = \langle x, \psi_{j,k} \rangle = |a|^{-1/2} \int x(t) \bar{\psi}_{j,k} dt \tag{7.16}$$

Given a $L^2(\mathbb{R})$ space and a sequence $\{\psi_{j,k}\} \subset L^2(\mathbb{R})$, $\{\psi_{j,k}\}$ is a frame if there exist two constants $A > 0$ and $B > 0$, such that

$$A\|x\|_2^2 \leq \sum_n |\langle x, \psi_{j,k} \rangle|^2 \leq B\|x\|_2^2 \tag{7.17}$$

where A and B are frame bounds. Every function $x \in L^2(\mathbb{R})$ can be written as

$$x = \sum_{j,k}^{\infty} \langle x, S^{-1}\psi_{j,k} \rangle \psi_{j,k} \tag{7.18}$$

where S is a frame operator. In the actual computation of wavelet transform, function x can be approximated effectively by the sum of limited items in polynomials (Equation (7.18)). The approximation of x is described as

$$\hat{x}(t) = \sum_{k=1}^{K} w_k \psi\left(\frac{t - b_k}{a_k}\right) \tag{7.19}$$

Then, the corresponding wavelet network can be constructed based on Equation (7.19), as shown in Figure 7.12. r is the total number of the wavelet basis after dilations and

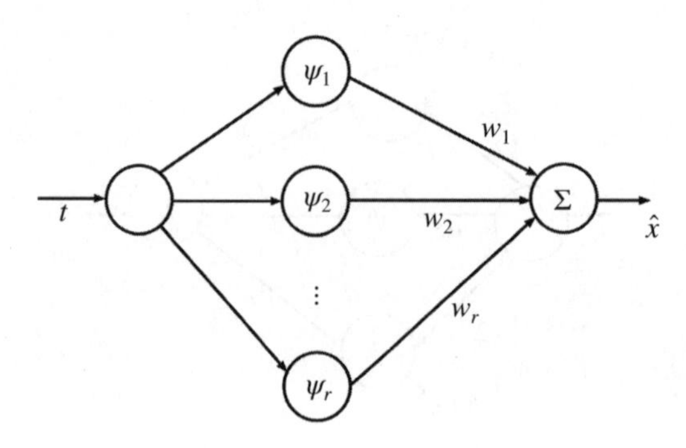

Figure 7.12 Wavelet network induced by a wavelet frame

translations. The essence of the wavelet network constructed by a wavelet frame is to provide optimal approximation by selecting proper dilation parameters and translation parameters, and then adjusting the wavelet coefficients based on wavelet frame theory and the time–frequency localization feature.

7.3.2.3 Wavelet basis fitting

As we know, function $x(t)$ can be fitted through the linear combination of the selected wavelet basis.

$$\hat{x} = \sum_{j,k}^{N} w_{j,k}\psi_{j,k}(t) = \sum_{j,k}^{N} w_{j,k}\psi\left(\frac{t-b_k}{a_j}\right) \tag{7.20}$$

If we select proper weights, dilations, and translations of the network, this wavelet network is able to approximate $x(t)$. Figure 7.13 shows the structure of the wavelet network.

This kind of wavelet network computes the inner product of time domain signals with the wavelet basis whose dilations and translations are fixed. That is to say, when the dilations change, there is only one corresponding translation. This is different from the real wavelet transform to some degree. Strictly speaking, this wavelet network approximates functions based on a form of wavelet combination rather than wavelet transform.

If the wavelet network of $L^2(\mathbb{R})$ is used to learn the energy-finite signals over the interval [0, 1], the approximation space may not match the signal space. One solution to this problem is to use an interval wavelet neural network by redefining the multiscale analysis over the interval [0, 1]. The theory and simulation examples prove that the interval wavelet network has better approximation ability for energy-finite signals over the interval [0, 1] than the aforementioned wavelet networks.

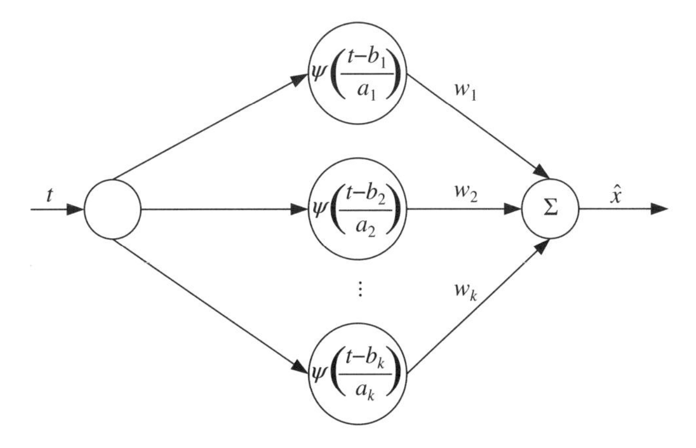

Figure 7.13 Wavelet network induced by a wavelet basis

The discussion here is about one-dimensional wavelets. Multidimensional wavelets can be defined using the direct product based on one-dimensional wavelets. Then, the tensor products of one-dimensional wavelets can construct a multidimensional orthogonal multiresolution analysis.

7.3.3 Learning process of wavelet networks

The learning process of wavelet networks mainly contains the selection of wavelet functions, the initialization of network parameters, the number of hidden-layer nodes, an algorithm for adjusting the parameters, and so on.

1. *Selection of wavelet functions.* There are no unified theories or methods to determine which wavelet function or scaling function would be the best activation function of neurons in different networks. The selection of wavelet functions usually depends on the operator's experience and the practical situation. The selection experience in wavelet analysis also can be used for reference. For example, image compression requires a wavelet function that has compact support, symmetry, orthogonality, and vanishing moments. However, Daubechies has proved that the orthogonal wavelet function does not possess all these properties at the same time. The commonly used wavelet function is the 9/7 biorthogonal wavelet function. In the application of signal approximation and estimation, the selection of wavelet functions should consider the wave shape of the wavelet, its support size, and the number of vanishing moments. In the application of signal detection, an antisymmetric wavelet, which is the first derivative of a smoothing function, is used for edge detection. The symmetrical wavelet, which is the second derivative of a smoothing function, is used for impulse detection.
2. *Initialization of wavelet network parameters.* The initialization of wavelet network parameters includes the initialization of dilations, translations, and weights. The initialization of weights in a wavelet network is the same as the initialization of weights in a neural network. Therefore, we do not introduce weights initialization again. The initialization of dilations and translations is categorized into three groups according to the wavelet network types: parameters initialization of orthogonal wavelet networks, parameters initialization of frame wavelet networks, and parameters initialization of adaptive wavelet networks. The concrete implementation methods can be found in relative references.
3. *Determining the number of hidden-layer nodes in wavelet networks.* There are different methods for determining the number of hidden-layer nodes in different literatures about wavelet networks. For example, for a given scale parameter M, the number of hidden-layer nodes is determined by $2^M + p$, where $p \geq 1$ is a small integer.
4. *Regulating algorithm of network parameters.* Similar to neural network, the regulating algorithm of wavelet network parameters is mainly a backpropagation algorithm and its improved algorithms. Reference [7] trains the network through training the weights c_k. When a training data set is available, c_k can be obtained by minimizing the mean square error.

$$
\begin{aligned}
e_N(x,\hat{x}) &= \frac{1}{N}\sum_{i=1}^{N}\left(x\left(t_1 - \hat{x}(t_i)\right)\right)^2 \\
\{\hat{c}_{-k},\cdots,\hat{c}_k\} &= \arg\min_{\{c_{-k},\cdots,c_k\}} e_N(x,\hat{x})
\end{aligned}
\tag{7.21}
$$

Equation (7.21) can be solved by taking the partial derivatives of the mean square error. Because the scaling functions are orthogonal, for sufficiently large N, c_k has a unique solution. Certainly, other references have also proposed many regulating algorithms. The readers could search them by themselves.

7.3.4 Approximation property of wavelet networks

The approximation property is one of the basic properties in wavelet networks' effective applications. It mainly appears in convergence rates. A few references discuss the approximation property of wavelet networks. Reference [7] proved that the wavelet network has the properties of universal approximation and L^2 approximation. For the wavelet network, the rates of convergence in the universal and L^2 approximation properties can be made arbitrarily rapid in the following sense: for any $\alpha > 0$, there is a Sobolev space H_β such that for any $x \in H_\beta$, there exists a sequence of wavelet networks x_n, where $n = 2^M$, such that

$$
\begin{aligned}
\|x - x_n\|_u &= o\left(n^{-\alpha}\right) \\
\|x - x_n\|_{L^2} &= o\left(n^{-\alpha}\right)
\end{aligned}
\tag{7.22}
$$

where $\|\cdot\|_u$ and $\|\cdot\|_{L^2}$ are the sup and L^2 norms, respectively. The degree of smoothness, β, generally depends on α and dimension d. The larger α and d are, the larger β is required to be. As long as the function is smooth enough, the convergence rates above are independent of d.

7.3.5 Application of wavelet networks

The wavelet networks have been widely applied to many fields such as function or signal approximation, speech identification, spectrum data compression and classification, sonar signal classification, machine fault signal classification, and power system fault signal classification and processing. The applications of wavelet networks in power systems concentrate on power system fault signal analysis, including de-noising, detection, compression, fault location, relay protection, and power equipment fault diagnosis.

7.4 Statistical and clustering analysis methods

7.4.1 Wavelet coefficient statistical analysis methods

It is complex to extract features by directly taking wavelet coefficients as the inputs of a classifier. This process requires large computations, and the classification effect may not be very good. Therefore, in order to simplify computations and reduce dimensions of vector and storage, the statistic characteristics of wavelet coefficients $D_j(k)$ at scale j are used in engineering applications, such as the following

Average value:

$$AVG_j = \sum_k D_j(k) \tag{7.23}$$

Absolute average value:

$$AVG'_j = \sum_k \left| D_j(k) \right| \tag{7.24}$$

Covariance:

$$CONV_j = CONV\left[D_j(k), D_j(k) \right] \tag{7.25}$$

Willison magnitude:

$$WAMP_j = \sum_k \operatorname{sgn}\left(\left| D_j(k) - D_j(k+1) \right| \right) \tag{7.26}$$

Average frequency of a divided frequency band:

$$\bar{f}_j = \frac{\sum_k f_j(k) P_j(k)}{\sum_k f_j(k)} \tag{7.27}$$

where $P_j(k)$ and $f_j(k)$ are the power spectrum and its corresponding frequency at each scale, respectively.

These wavelet postprocessing methods provide direct statistical analysis of wavelet coefficients, and they have advantages such as easy calculation, low dimensions of feature vectors, small storage and computation needs, and rapid computation speed.

7.4.2 Wavelet coefficients clustering analysis method

If we only make statistical analysis of wavelet coefficients at each scale, it may not be enough to describe the differences between different categories. One should increase the dimensions of feature vectors under this circumstance, and one way is to cluster wavelet coefficients at each frequency band. The center or energy of each cluster is the feature. The number of clusters could be different for different scales. Another way is via the technology of vector quantization with a divided frequency band; as the frequency band cluster, it can get the time–frequency information of signals the same as the frequency band clustering. Of course, the computations also increase.

This subsection introduces the wavelet coefficients clustering analysis method with the application of a fuzzy C-means clustering (FCM) algorithm on fault classification for transmission lines.

7.4.2.1 Fuzzy C-means clustering algorithm

The FCM algorithm is a clustering method that allows each data point to belong to a cluster by membership grade. Its basic principle is to partition a given set of data into different subgroups, such that data points belonging to the same group are as similar to each other as possible, whereas data points from two different groups have the maximum difference. The FCM algorithm is modified from the common C-means clustering algorithm. The common C-means clustering algorithm divides data points rigidly, whereas the FCM algorithm is a flexible fuzzy partitioning method.

Let $\mathbf{X} = (x_1, x_2 \cdots x_n)$ be a set of n data points, each having m features. Thus, these data points can construct an $m \times n$ feature matrix $\mathbf{X}_{ij}$. A clustering algorithm partitions n data points into c fuzzy clusters. Similar data points are divided into the same cluster, and the clustering center is calculated to make the cost function minimum. FCM utilizes the fuzzy membership over interval [0, 1] to represent the probability that a data point belongs to a specific cluster. The appropriate partition matrix is represented as

$$\mathbf{U} = \begin{bmatrix} u_{11} & u_{12} & \cdots & u_{1n} \\ u_{21} & u_{22} & \cdots & u_{2n} \\ \vdots & \vdots & \vdots & \vdots \\ u_{c1} & u_{c2} & \cdots & u_{cn} \end{bmatrix} = u_{ij} \tag{7.28}$$

where u_{ij} denotes the grade of membership of the jth element to the ith cluster. $i = 1, 2, \cdots c$. u_{ij} subjects to

$$\sum_{i=1}^{c} u_{ij} = 1, 1 \le j \le n \tag{7.29}$$

$$\sum_{j=1}^{n} u_{ij} > 0, 1 \le i \le c \tag{7.30}$$

Let the vector of clustering centers be

$$\mathbf{S} = s_1, s_2, \cdots s_c = \begin{bmatrix} s_{11} & s_{12} & \cdots & s_{1c} \\ s_{21} & s_{22} & \cdots & s_{2c} \\ \vdots & \vdots & \vdots & \vdots \\ s_{m1} & s_{m2} & \cdots & s_{mc} \end{bmatrix} = s_i \tag{7.31}$$

Then, the object function of the FCM algorithm is defined as follows:

$$J(\mathbf{U},\mathbf{S}) = \sum_{j=1}^{n}\sum_{i=1}^{c}(u_{ij})^q \left\| x_{ij} - s_i \right\|^2 \tag{7.32}$$

where $J(\mathbf{U},\mathbf{S})$ is the weighted sum of squared distances between each data point to the cluster centers. The weight is the membership to the power of q. q is the fuzzy exponent that controls the fuzziness, $q \in (1,\infty)$. The greater the weight is, the fuzzier the partition is. $\| x_{ij} - s_i \|$ represents the distance between the data point x_{ij} and the center of the ith cluster, and usually it is the Euclidean distance.

In FCM algorithms, the clustering rule is to find $\mathbf{U}$ and $\mathbf{S}$ to make the object function minimum. Approximate optimization of $J(\mathbf{U},\mathbf{S})$ by the FCM algorithm is based on iteration through the following necessary conditions for its local extrema:

$$u_{ij} = 1/\sum_{k=1}^{c}\left[\frac{\sum_{i=1}^{m}(x_{ij}-s_i)^2}{\sum_{i=1}^{m}(x_{ij}-s_k)^2}\right]^{\frac{1}{q-1}} \tag{7.33}$$

$$s_i = \sum_{j=1}^{n}(u_{ij})^q x_{ij} / \sum_{j=1}^{n}(u_{ij})^q \tag{7.34}$$

To sum up, the FCM algorithm firstly selects the initial clustering centers to set the area of each class in advance. These initial clustering centers are mostly inaccurate. In addition, it assigns a membership to each data point. The clustering center and membership are updated by iterations to make the clustering centers approach the right positions in the data set. The FCM algorithm can be summarized in the following steps:

1. Select the number of cluster c.
2. Select the weighted exponent q.
3. Set the termination threshold ε.
4. Initialize the fuzzy membership matrix $\mathbf{U}$ and the cluster centers.
5. Let the iteration number t be 1.
6. Update the current cluster centers using Equation (7.34).
7. Update the membership matrix $\mathbf{U}$ by Equation (7.33).
8. If $|\mathbf{U}^{t+1} - \mathbf{U}^{t}| \leq \varepsilon$, then stop; otherwise, go to step 6.

7.4.2.2 Application of the FCM algorithm on fault phase selection for transmission lines based on the wavelet energy ratio

Fault phase selection plays an important role in the protection of a transmission line. For example, distance relay can select different algorithm elements to deal with different fault situations after identifying the type of fault. Moreover, identifying the faulty phase is essential for single-pole tripping and auto-reclosing requirements. The fault types include single-phase grounding fault (AG, BG, and CG), phase-to-phase fault (AB, BC, and AC), phase-to-phase grounding fault (ABG, BCG, and ACG), and three-phase grounding fault (ABCG). That is to say, there are a total of 10 fault types to identify.

7.4.2.3 Basic principle and algorithm flow

When a fault occurs on a transmission line, both high-frequency voltage and current signals are generated and superimposed on the power frequency component. The signal of a faulted phase is different in amplitude and frequency components from the signal of a nonfaulted phase. If we select an appropriate wavelet function to detect the specified frequency band and analyze the differences between the wavelet transform of faulted phase signals and the wavelet transform of nonfaulted phase signals, we can obtain the effective criteria by extracting the fault features to discriminate the faulted phase. This book decomposes the phase currents (i_a, i_b, i_c) in a fixed time window by wavelet transform and extracts the wavelet coefficients $D_{\rho 1}(n)$ at scale 1. ρ represents phase A, phase B, and phase C. The wavelet energy in the fixed time window of wavelet coefficients at scale 1 is calculated as follows:

$$E_{\rho 1}(n) = \sum_{j=n-N}^{n} D_{\rho 1}^{2}(j) \tag{7.35}$$

where $E_{\rho 1}(n)$ is the wavelet energy in the fixed time window of wavelet coefficients at scale 1 corresponding to the phase ρ current at time n. $D_{\rho 1}$ is the wavelet coefficients at scale 1 of the phase ρ current.

Although the amplitudes of energies may vary greatly in different fault situations, the ratio between them can still reflect the fault characteristics. For sake of simplicity, we define

$$g_1 = \frac{E_{a1}}{E_{b1}},\ g_2 = \frac{E_{a1}}{E_{c1}},\ g_3 = \frac{E_{b1}}{E_{c1}} \tag{7.36}$$

For n cases, each case has three features (g_1, g_2, and g_3), which consist of the $n \times 3$ feature matrix g_{ij}. g_{ij} is the input of the FCM algorithm. Thus, Equations (7.33) and (7.34) yield

$$u_{ij} = 1 \Bigg/ \sum_{k=1}^{c} \left[\frac{\sum_{i=1}^{m} (g_{ij} - s_i)^2}{\sum_{i=1}^{m} (g_{ij} - s_k)^2} \right]^{\frac{1}{q-1}} \tag{7.37}$$

$$s_i = \sum_{j=1}^{n} (u_{ij})^q g_{ij} \Bigg/ \sum_{j=1}^{n} (u_{ij})^q, (i \le m, h \le c) \tag{7.38}$$

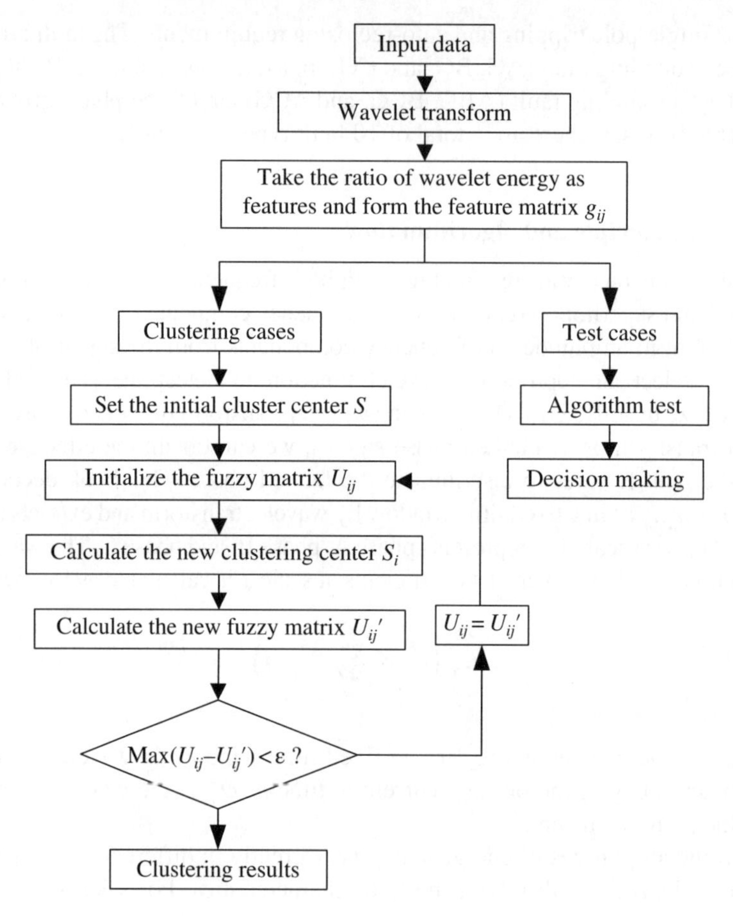

Figure 7.14 Flow chart of the algorithm

7.4.2.4 Simulation study

The simulation model is shown in the Appendix. In power systems, transmission lines are exposed to faults most frequently. According to the operation experience, in high-voltage overhead lines of 110 kV and higher, more than 70% of the faults are single-phase grounding faults. Therefore, this section focuses on the discrimination of AG, BG, and CG faults by FCM.

The fault data are generated from the simulation model under various fault resistances and fault distances. After wavelet transform, the new processed data are shown in Tables 7.2 and 7.3. The fault resistances are considered as 0 Ω, 10 Ω, 20 Ω, 50 Ω, and 100 Ω. The fault distances are 30 km, 50 km, 100 km, 150 km, and 200 km.

Tables 7.2 and 7.3 show the data generated under different fault resistances and fault locations, respectively. It can be seen in Tables 7.2 and 7.3 that the extracted features are able to reveal the features of single-phase grounding fault clearly. The generated data are processed to construct two 15×3 feature matrices g_{ij}. There are 30 cases, of which 24 are clustering cases and the rest are the training cases. The number of training cases is

Table 7.2 Processed data under different fault resistances

Fault resistances (Ω)		0 Ω	10 Ω	20 Ω	50 Ω	100 Ω
Cases		1	2	3	4	5
AG	g_1	129.8	139.1	147.5	166.5	235.5
	g_2	123.6	133.6	143.5	171.3	208.9
	g_3	5.3	5.3	5.3	5.9	7.7
Cases		6	7	8	9	10
BG	g_1	0.0077	0.0075	0.007	0.0058	0.0048
	g_2	5.3	5.3	5.3	5.9	7.7
	g_3	129.8	139.1	147.5	166.5	235.5
Cases		11	12	13	14	15
CG	g_1	5.3	5.3	5.3	5.9	7.7
	g_2	0.0077	0.0075	0.007	0.0058	0.0048
	g_3	0.0081	0.0072	0.0068	0.006	0.0042

Table 7.3 Processed data under different fault distances

Fault distances (km)		30 km	50 km	100 km	150 km	200 km
Cases		16	17	18	19	20
AG	g_1	158.3	166.5	167.5	169.3	167.3
	g_2	163.5	171.3	170.3	175.2	170.1
	g_3	5.5	5.9	6.1	6.0	6.1
Cases		21	22	23	24	25
BG	g_1	0.0061	0.0058	0.0059	0.0057	0.0059
	g_2	5.5	5.9	6.1	6.0	6.1
	g_3	158.3	166.5	167.5	169.3	167.3
Cases		26	27	28	29	30
CG	g_1	5.5	5.9	6.1	6.0	6.1
	g_2	0.0061	0.0058	0.0059	0.0057	0.006
	g_3	0.006	0.006	0.006	0.0059	0.006

[4 9 14 18 23 28]. We performed the FCM algorithm according to the clustering cases to update the cluster center. After iterations, we obtained $n = 15$, $c = 3$, $q = 2$, $\varepsilon = 0.0001$. The optimal target is to minimize the weighted sum of membership and distance between the cluster centers and all data points. The value of the object function changes during iterations until the clusters are converged.

After performing the FCM algorithm, the object functions of the two group training cases are converged after 11 and 9 iterations, respectively. The fuzzy matrices U_1 and U_2 and the final clustering centers S_1 and S_2 also can be obtained. U_1 and U_2 are, respectively, the fuzzy matrices corresponding to different fault resistances and different fault locations. S_1 and S_2 are, respectively, the clustering centers corresponding to different fault resistances and different fault locations.

Each element of the fuzzy matrices represents the membership of a data point belonging to a specific cluster. Thus, the discrimination results can be seen clearly from fuzzy matrices U_1 and U_2. The number of cases belonging to a phase A grounding fault is [1 2 3 5 16 17 19 20]. The number of cases belonging to a phase B grounding fault is [6 7 8 10 21 22 24 25]. The number of cases belonging to a phase C grounding fault is [11 12 13 15 26 27 29 30].

$$\mathbf{U}_1=\begin{bmatrix} 0.9999 & 0.9999 & 0.9999 & 0.9999 & 0.0537 & 0.0239 & 0.0082 & 0.0870 & 0.0586 & 0.0255 & 0.0078 & 0.0758 \\ 0.00004 & 0.00005 & 0.00005 & 0.00009 & 0.9321 & 0.9691 & 0.9892 & 0.8650 & 0.0325 & 0.0151 & 0.0049 & 0.0607 \\ 0.00002 & 0.00002 & 0.00002 & 0.00000 & 0.0142 & 0.0070 & 0.0026 & 0.0480 & 0.90879 & 0.95934 & 0.9874 & 0.8635 \end{bmatrix}$$

$$\mathbf{U}_2=\begin{bmatrix} 0.9967 & 0.99994 & 0.9989 & 0.99993 & 0.000003 & 0.00000 & 0.00000 & 0.00000 & 0.0007 & 0.0000 & 0.0002 & 0.00002 \\ 0.00199 & 0.00003 & 0.00067 & 0.00004 & 0.99999 & 1.0000 & 1.0000 & 1.0000 & 0.0022 & 0.00002 & 0.0004 & 0.00008 \\ 0.00131 & 0.00002 & 0.00046 & 0.00003 & 0.000006 & 0.00000 & 0.00000 & 0.00000 & 0.9971 & 0.99998 & 0.9994 & 0.9999 \end{bmatrix}$$

$$\mathbf{S}_1=\begin{bmatrix} 161.33 & 154.77 & 5.9608 \\ 0.23661 & 6.0285 & 160.9 \\ 6.2544 & 0.36563 & 0.46753 \end{bmatrix} \quad \mathbf{S}_2=\begin{bmatrix} 165.79 & 170.09 & 5.9205 \\ 0.0059 & 5.9205 & 165.79 \\ 5.9201 & 0.00605 & 0.00615 \end{bmatrix}$$

The training cases construct a 6×3 feature matrix. Select the same weight exponent q and threshold ε to perform the FCM algorithm according to the training cases. The fuzzy clustering results indicate that the number of cases belonging to a phase A grounding fault is [4 18]. [9 23] and [14 28] are, respectively, the cases belonging to a phase B grounding fault and phase C grounding fault. The clustering results indicate that the FCM algorithm is feasible and accurate to discriminate the fault types. The discrimination results are immune to the fault resistances and fault locations.

This chapter introduces several wavelet postprocessing methods: modulus maxima and the singularity analysis method, the energy analysis method, the wavelet neural network method, the wavelet coefficients statistic method, and the wavelet coefficients clustering method. The relative theories about all kinds of postprocessing methods are introduced briefly. Simulation studies of theory signals and fault transient signals based on the singularity exponent analysis method are carried out. The energy analysis method defines the energy distribution coefficients, energy fluctuation coefficients, and energy moment. Regarding its combination with a neural network, the wavelet neural network is discussed emphatically. The constructions of wavelet networks, learning processes, and approximation properties are also presented. In statistic and clustering methods, we take the FCM algorithm as an example with which to introduce basic principles, algorithm flow, and simple case studies.

References

[1] He Z.Y., Qian Q.Q., Wavelet transform based window singularity exponent computation method and its application in fault detection. *RELAY*, vol. 29, no. 3, pp. 23–26, 2001.

[2] Mallat S., Hwang W.L., Singularity detection and processing with wavelets. *IEEE transactions on information theory*, vol. 38, no. 2, part 2, pp. 617–643, 1992.

[3] He Z.Y., Wang X.R., Qian Q.Q., A study of EHV transmission lines non-unit transient protection based on wavelet analysis. *Proceedings of the CSEE*, vol. 21, no. 10, pp. 10–14, 19, 2001.

[4] Liu Z.G., Wang X.R., Qian Q.Q., A review of wavelet networks and their applications. *Automation of Electric Power Systems*, vol. 27, no. 6, pp. 73–79, 85, 2003.

[5] Gao X.P., Zhang B., Interval-wavelets neural networks(I) – theory and implements. *Journal of Software*, vol. 9, no. 3, pp. 217–221, 1998.

[6] Gao X.P., Zhang B., Interval-wavelets neural networks(II) – properties and experiment. *Journal of Software*, vol. 9, no. 4, pp. 246–250, 1998.

[7] Zhang J., Walter G.G., Miao Y., et al., Wavelet neural networks for function learning. *IEEE Transactions on Signal Processing*, vol. 43, no. 6, pp. 1485–1497, 1995.

[8] Yang J.W., He Z.Y., Zhao J., Zhang H.P., Wavelet transform-based fuzzy clustering algorithm and its application. *Journal of Dalian Maritime University*, vol. 34, no. 3, pp. 67–71, 2008.

8

Application of Wavelet Analysis in High-voltage Transmission Line Fault Location

High-voltage transmission lines are used to transmit electric power to distant large load centers. However, the transmission lines are exposed to faults because of lighting, short circuits, faulty equipment, misoperation, human errors overload, and aging. A rapid and precise fault location scheme plays an important role in the reliability of fault protection of modern power systems since it enable reducing costs and time for energy supply restoration. The two most common fault location methods are based on impedance measurement and traveling waves. The impedance-based technique is based on the linear relation between the reactance, estimated from the voltage and current of the fault, and the fault location. Most of the impedance-based schemes require pre-fault load measurements or remote end information. The impedance-based method is simple but is not accurate if the systems connected to the two line terminals include generators.

The traveling wave–based fault location methods are other feasible alternatives, and they are widely used in power systems. The essential idea behind this method is based on the correlation between the forward and backward traveling waves along a line or direct detection of the arrival time of waves at terminals. These methods have fast responses and high accuracy, and this kind of method has proved to be immune to power swings and current transformer (CT) saturation, and almost insensitive to fault types, fault resistances, fault inception angles, and source parameters of the system.

In the 1950s, the traveling wave theory was first employed to identify the transmission line fault. Due to the limitations in detecting the high-frequency waves, the earlier investigation of traveling wave–based protection for transmission lines progressed at a slow pace.

In the early 1970s, researchers found that the fault distance was related to the traveling-wave spectrum of fault voltage signals. In 1979, Swift studied the relationship between fault distance and natural frequencies when the system impedance was infinite or zero.

Wavelet Analysis and Transient Signal Processing Applications for Power Systems, First Edition. Zhengyou He.

In 1983, Crossley and McLaren presented the cross-correlation technique to calculate the time delay introduced in the reflected signal, which can be used to obtain the fault distance. If the obtained fault distance is more than the lien length, the distance protection based on the traveling wave operates to isolate the fault line. Recently, modern electronic technology and new mathematical tools have been employed to traveling wave–based fault location, which improves the location accuracy greatly. With the fast development of modern electronics technology, the application of high-speed data acquisition and processing provides a good guarantee for traveling wave–based distance protection. As an emerging signal-processing tool, wavelet transform is applied to many engineering fields and has excellent performance in traveling wave–based fault location. Wavelet transform has the outstanding abilities of time–frequency localization and de-noising. Wavelet transform is not only capable of extracting the traveling waves from transient signals, but also able to distinguish the traveling wave from the disturbance signal such as white noise. Research has indicated that it can improve the sensitivity and anti-disturbance of protection by using wavelet transform to extract traveling waves, and so makes the protection immune to power–frequency components and line parameters, and so on.

8.1 Basic principle of traveling wave–based fault location for transmission lines

8.1.1 Traveling waves generated by faults

When a fault occurs on a transmission line (as shown in Figure 8.1a), the voltage at the point of fault suddenly reduces to a low value. This sudden change produces a high-frequency electromagnetic impulse called the *traveling wave*. These traveling waves propagate away from the fault in both directions. This transient process can be resolved by the law of superposition. The law of superposition in the linear network theory separates a faulted network in Figure 8.1b into pre-fault and pure-fault ones, which are given in

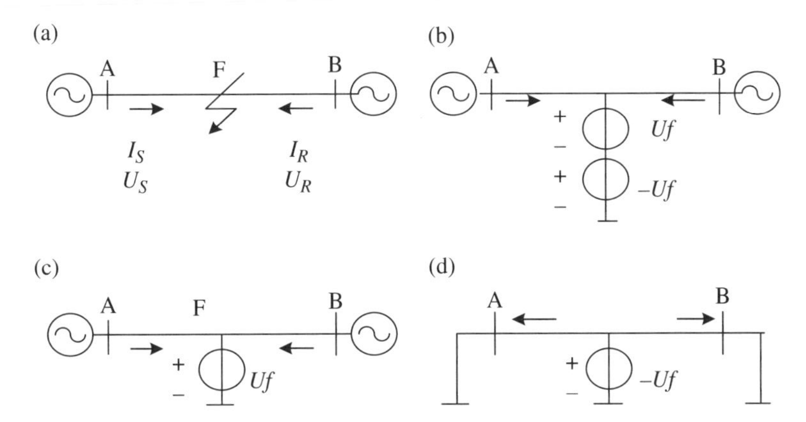

Figure 8.1 The superposition principle for traveling wave analysis: (a) a fault occurs at point F, (b) the equivalent faulted network, (c) a pre-fault network, (d) a pure-fault network

Figure 8.1c and Figure 8.1d. Figure 8.1 shows that the occurrence of a fault is equivalent to the injection of a voltage at the point of fault opposite to the pre-fault steady-state voltage.

8.1.2 Telegrapher's equations of traveling waves

Power transmission lines are of the three-phase type. However, it is much simpler to understand traveling wave concepts by considering wave propagation in single-phase lines. Consider a transmission line with distributed parameters. In the case of a lossless line, the series resistance R and the parallel conductance G are zero. The equations for the voltage and the current distributions are

$$\frac{-\partial u}{\partial x} = L\frac{\partial i}{\partial t} \tag{8.1}$$

$$\frac{-\partial i}{\partial x} = C\frac{\partial u}{\partial t} \tag{8.2}$$

where L and C are the inductance and capacitance per unit length; and x is the distance.

Differentiating the above equations with respect to x provides

$$\frac{\partial^2 u}{\partial x^2} = LC\frac{\partial^2 u}{\partial t^2} \tag{8.3}$$

$$\frac{\partial^2 i}{\partial x^2} = LC\frac{\partial^2 i}{\partial t^2} \tag{8.4}$$

According to J. d'Alembert, the solution to the general wave equation can be expressed as

$$u = u_1\left(t - \frac{x}{v}\right) + u_2\left(t + \frac{x}{v}\right) \tag{8.5}$$

$$i = \frac{1}{Z_c}\left[u_1\left(t - \frac{x}{v}\right) - u_2\left(t + \frac{x}{v}\right)\right] \tag{8.6}$$

where $u_1\left(t - \frac{x}{v}\right)$ represents a function describing a wave propagating in the positive x direction, usually called the forward wave; and $u_2\left(t + \frac{x}{v}\right)$ represents a function describing a wave propagating in the negative x direction, called the backward wave. $v = \frac{1}{\sqrt{LC}}$ is the traveling wave propagation speed, and $Z_c = \sqrt{\frac{L}{C}}$ is the characteristic impedance of the line.

The transmission line equation for three-phase lossless lines with constant parameters is in the time domain:

$$-\begin{bmatrix} \frac{\partial u_a}{\partial x} \\ \frac{\partial u_b}{\partial x} \\ \frac{\partial u_c}{\partial x} \end{bmatrix} = \begin{bmatrix} L_s & L_m & L_m \\ L_m & L_s & L_m \\ L_m & L_m & L_s \end{bmatrix} \begin{bmatrix} \frac{\partial i_a}{\partial t} \\ \frac{\partial i_b}{\partial t} \\ \frac{\partial i_c}{\partial t} \end{bmatrix} \tag{8.7}$$

$$-\begin{bmatrix} \frac{\partial i_a}{\partial x} \\ \frac{\partial i_b}{\partial x} \\ \frac{\partial i_c}{\partial x} \end{bmatrix} = \begin{bmatrix} K_s & K_m & K_m \\ K_m & K_s & K_m \\ K_m & K_m & K_s \end{bmatrix} \begin{bmatrix} \frac{\partial u_a}{\partial t} \\ \frac{\partial u_b}{\partial t} \\ \frac{\partial u_c}{\partial t} \end{bmatrix} \tag{8.8}$$

where $K_s = C_s + 2C_m$, $K_m = -C_m$, L_s is the per-unit-length static self-inductance of the loop defined by the earth and reference conductor; L_m is the per-unit-length mutual inductance between one phase and another phase; C_s is the per-unit-length phase-to-earth capacitance; C_m is the per-unit-length phase-to-phase capacitance of the line; and x is the distance.

The equation above can be rewritten as

$$-\frac{\partial[u]}{\partial x} = [L]\frac{\partial[i]}{\partial t} \tag{8.9}$$

$$-\frac{\partial[i]}{\partial x} = [C]\frac{\partial[u]}{\partial t} \tag{8.10}$$

The greatest difficulty in dealing with three-phase lines derives from the fact that it is necessary to solve systems of coupled partial differential equations. This difficulty is solved by using the phase-mode transformation matrix, which diagonalizes the matrices *LC* and *CL*. In this way, Equations (8.9) and (8.10) are transformed into two new systems, each consisting of three uncoupled scalar equations of the same type as Equations (8.9) and (8.10). Let us consider the change of variables:

$$[u] = [S][u_m] \tag{8.11}$$

$$[i] = [Q][i_m] \tag{8.12}$$

where [*S*] and [*Q*] are the transformation matrices. Substituting Equation (8.11) into (8.9), and Equation (8.12) into (8.10), we obtain

$$-\frac{\partial[u_m]}{\partial x}=[S]^{-1}[L][Q]\frac{\partial[i_m]}{\partial t} \tag{8.13}$$

$$-\frac{\partial[i_m]}{\partial x}=[Q]^{-1}[C][S]\frac{\partial[u_m]}{\partial t} \tag{8.14}$$

where $[u_m]=\begin{bmatrix} u_o \\ u_\alpha \\ u_\beta \end{bmatrix}$, $[i_m]=\begin{bmatrix} i_o \\ i_\alpha \\ i_\beta \end{bmatrix}$.

By derivation and substitution of Equations (8.13) and (8.14), we obtain the two uncoupled equation systems:

$$\frac{\partial^2[u_m]}{\partial x^2}=[S]^{-1}[L][C][S]\frac{\partial^2[u_m]}{\partial t^2} \tag{8.15}$$

$$\frac{\partial^2[i_m]}{\partial x^2}=[Q]^{-1}[L][C][Q]\frac{\partial^2[i_m]}{\partial t^2} \tag{8.16}$$

Since the coupled equations in the phase domain are transformed to decoupled equations in the modal domain, the off-diagonal elements in coefficients matrices of Equations (8.15) and (8.16) are constantly zero, and thus we obtain

$$\left.\begin{array}{l}[S]^{-1}[L][C][S]=[D_u] \\ [Q]^{-1}[L][C][Q]=[D_i]\end{array}\right\} \tag{8.17}$$

where $[D_u]$ and $[D_i]$ are diagonal matrices.

The transformation matrices [S] and [Q] are dependent on the eigenvalues and eigenvectors of $[L][C]$. The common transformation matrices are symmetrical components transformation, Clarke transformation, and Karenbauer transformation.

8.1.3 Reflection and refraction of traveling waves

Traveling waves travel along the transmission line and encounter discontinuities, such as buses and transformers. When traveling waves reach a discontinuity, part of it is reflected back, and the remaining part passes through. The magnitude of the reflected and refracted waves depends on the characteristic impedance of the transmission line and the impedance beyond the discontinuity. At each discontinuity, the total energy of the incident wave is distributed among the reflected and refracted waves.

Consider a transmission line with characteristic impedance Z_{1C} connected with another transmission line with characteristic impedance Z_{2C}, as shown in Figure 8.2. u_1 and i_1 represent the voltage and current in line 1; u_2 and i_2 represent the voltage and current in

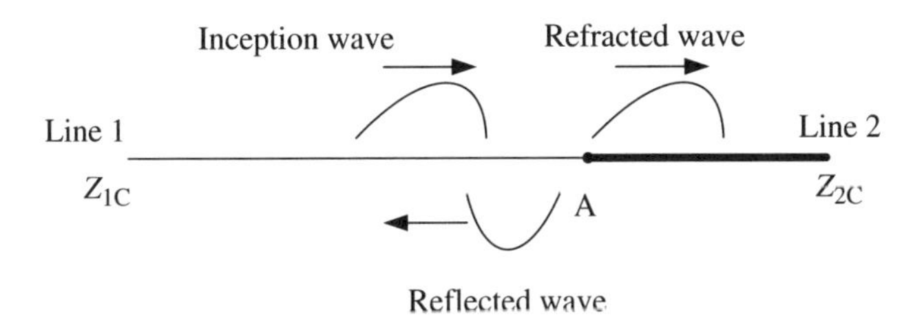

Figure 8.2 Refraction and reflection of a traveling wave

line 2. At the junction A, it is known that $u_1 = u_2$ and $i_1 = i_2$. Assume the subscripts *E*, *R*, and *T* represent the incident waves, reflected waves, and refracted waves, respectively. At the junction A:

$$\left.\begin{aligned} u_T &= u_E + u_R \\ i_T &= i_E + i_R \end{aligned}\right\} \tag{8.18}$$

Meanwhile:

$$\left.\begin{aligned} u_T &= i_T Z_{2c} \\ u_r &= -i_R Z_{1c} \\ u_E &= i_E Z_{1c} \end{aligned}\right\} \tag{8.19}$$

Substituting Equation (8.19) in Equation (8.18) provides

$$\left.\begin{aligned} u_T &= \frac{2Z_{2c}}{Z_{2c} + Z_{1c}} \cdot u_E = \gamma_u \cdot u_E \\ u_R &= \frac{Z_{2c} - Z_{1c}}{Z_{2c} + Z_{1c}} \cdot u_E = \rho_u \cdot u_E \end{aligned}\right\} \tag{8.20}$$

where γ_u is the voltage refraction coefficient; and ρ_u is the voltage reflection coefficient.
Substituting for u_T and u_R from Equation (8.19) in this equation gives

$$\left.\begin{aligned} i_T &= \frac{2Z_{1c}}{Z_{2c} + Z_{1c}} \cdot i_E = \gamma_i \cdot i_E \\ i_R &= \frac{Z_{1c} - Z_{2c}}{Z_{2c} + Z_{1c}} \cdot i_E = \rho_i \cdot i_E \end{aligned}\right\} \tag{8.21}$$

where γ_i is the current refraction coefficient; and ρ_i is the current reflection coefficient.

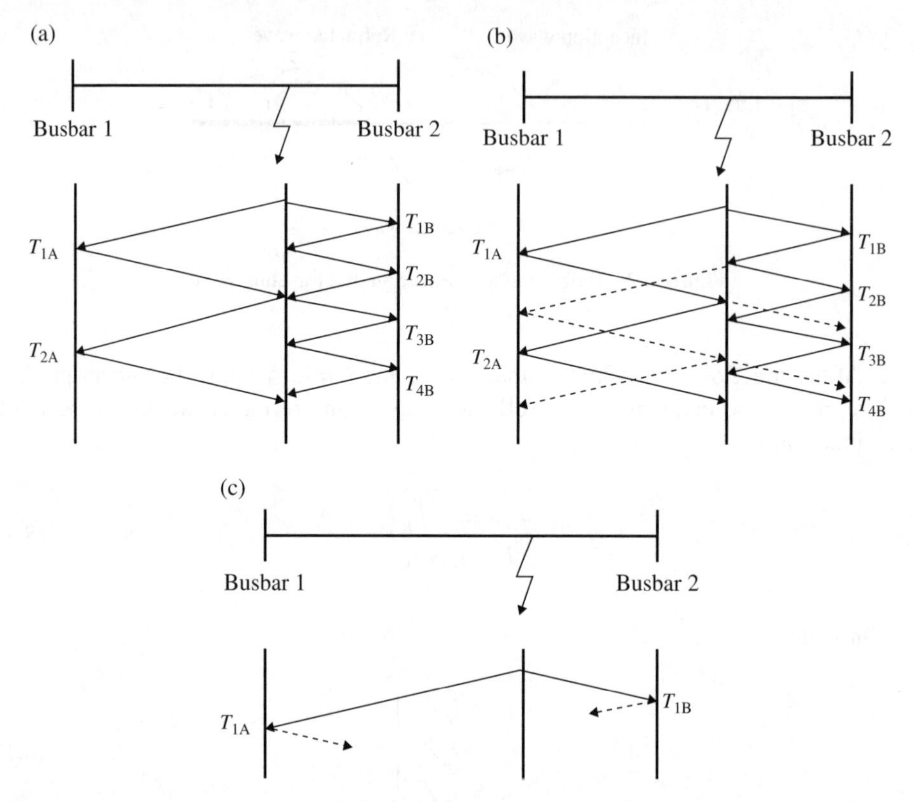

Figure 8.3 Traveling wave fault location principle: (a) single-ended fault location (total reflection), (b) single-ended fault location (partial reflection), (c) double-ended fault location

8.1.4 Basic principle of traveling wave–based fault location

A great number of fault location methods based on traveling waves have been developed in recent years, which can be classified as single-ended methods and double-ended methods. These methods mainly rely on the following information. When a fault occurs on a transmission line, both voltage and current traveling waves are generated and propagate toward the two sides of the line until they meet a discontinuity point, such as the fault point or bus. At this point, part of the waves pass into the adjacent section, called refraction, and the rest are reflected backward, called reflection. This generates additional waves that will propagate through the power systems. Thus, the travel time is proportional to the distance from the fault point to the measurement point. That is to say, the fault distance can be obtained by measuring the travel time of traveling waves.

The single-ended traveling wave fault location principle is shown in Figure 8.3a. The distances from terminal A and B are, respectively,

$$l_A = v \cdot \left(T_{2A} - T_{1A}\right)/2 \tag{8.22}$$

$$l_B = v \cdot (T_{2B} - T_{1B})/2 \tag{8.23}$$

where v is the velocity of propagation of the traveling wave; and T is the arrival time of the traveling wave front. Single-ended traveling wave fault location uses the transient captured at only one end of the line. However, it needs the fault arc to be ionized for the second reflection from the fault. Meanwhile, it assumes that all the energy in any pulses that arrive at the fault point is totally reflected.

In practice the fault resistance, which consists of the nonlinear resistance of the arc path plus any series linear resistance, may not cause total reflection of the pulse energy arriving at the fault, and some fraction will pass through to appear at the opposite end of the line, as illustrated in Figure 8.3b. Under these conditions, the transients produced at the line terminals will be more complex and may require careful analysis to identify the correct pulses. This is because the pulses not only contain the second reflected wave from the fault point but also contain the reflected wave from the opposite end of the line. Certainly, further complications arise if the arc at the fault point is extinguished prematurely. In addition, the amplitudes, polarities, and shapes of the second reflected wave from the fault point will differ depending on whether current or voltage transformers are used, as well as on busbar configuration of the line terminals, which makes the reflected wave identification more difficult. Though the single-ended methods are less reliable or accurate, they do not need time synchronization or communication devices, which makes the single-ended method more economical and one of the hottest subjects of fault location research.

The principle of the double-ended traveling wave fault location method is shown in Figure 8.3c. Only the first arrivals of the wave front at each end of the line have to be detected using synchronized devices. As can be seen in Figure 8.3c, T_{1A} and T_{1B} correspond to the absolute time when the initial traveling wave and the reflected wave from the fault point arrive at terminals A and B, respectively. Having the travel time as well as the velocity of propagation on the transmission line, the fault location from terminals A and B can be calculated:

$$l_A = l/2 + v \cdot (T_{1A} - T_{1B})/2 \tag{8.24}$$

$$l_B = l/2 + v \cdot (T_{1B} - T_{1A})/2 \tag{8.25}$$

where l is the total length of the line.

The double-ended traveling wave fault location principle only employs the arrival time of the first traveling wave to calculate the fault distance. Therefore, the reflection and refraction of the traveling wave are no longer important. Furthermore, the amplitude of the first traveling wave is relatively great, which is helpful for identification. However, the double-ended traveling wave fault location needs time synchronization to keep the two end measurements in sync. Fortunately, synchronization of the measurement is easiest achieved by using the Global Positioning System (GPS).

8.2 Traveling wave fault location method based on wavelet analysis for transmission lines

8.2.1 Traveling wave fault location principle based on wavelet analysis

Double-ended traveling wave fault location principles do not depend on multiple reflections between the station bus and the fault point, but rather calculate the fault distance using the time interval between the absolute arrival time of fault initial surges measured at both terminals of the transmission line, which is more reliable than the single-ended traveling wave method. However, the double-ended traveling wave method needs a communication channel and synchronizing clock. The single-ended traveling wave fault location method identifies the fault location by observing the time delay between successive reflections in the traveling wave signal observed at one terminal. The single-ended traveling wave method needs no communication link, but it encounters the problems in distinguishing between traveling waves reflected from the fault point and from the remote end of the line. The location precision of the single-ended traveling wave method is mainly affected by the arriving time of a wave front. How to accurately acquire the arriving time of a wave front is very important and difficult. The commonly used traveling wave identification methods are the derivative method, correlation analysis method, and matched filter method. The derivative method estimates the arriving time of the wave front by the moment where the first-order derivative and second-order derivative go beyond a given threshold, but this method has low precision and may be influenced by noises. The correlation analysis method estimates the time difference between the arriving instant of the first fault-induced traveling wave and the instant of its subsequent reflected wave. Then, the fault distance can be obtained with the time difference. The accuracy of the correlation analysis method depends on the proper choice of base signal window position and length. It is hard to implement this method in practice. The matched filter method is based on the correlation analysis method. It utilizes the high-pass filter to reflect the wave front, which could improve the reliability. Though this method is applied in practice, it will be influenced by the outlines connected on the busbar. In brief, the methods mentioned above have limitations in practical application.

Since the 1990s, the modern time–frequency analysis technique with wavelet transform as the representative is widely used in power systems. Wavelet transform has both good time localization and frequency localization, which makes it a powerful mathematical tool for analysis of nonstationary signals such as traveling waves.

The modulus maxima theory proves that if the wavelet function is the first derivative of a smoothing function, the modulus maxima of the wavelet transform correspond to the inflection points of the signal. If the wavelet function is the second derivative of a smoothing function, the zero crossings of the wavelet transform correspond to the inflection point of the signal. That is to say, there is a one-to-one relationship between the modulus maxima and the inflection points. The polarity of wavelet modulus maxima identical to the polarity of sudden change of the signal and its amplitude depends on the amplitude and gradient of the sudden change of the signal. Wavelet modulus maxima can detect the singularity of the nonstationary signals.

The high-frequency voltage and current generated by fault are essentially nonstationary signals. The wave front of the initial traveling wave, the reflected wave from the fault point, and the reflected wave from the remote end detected by the fault locator correspond to the sudden changes of traveling waves. The wave front is of great singularity. Modulus maxima of wavelet transform are the local maxima of wavelet transform, which represent the singularity of traveling waves. Because the modulus maxima points have one-to-one correspondence relations to the sudden changes of the traveling wave, one can obtain the moments when the traveling waves arrive at the fault locator through the modulus maxima points. The postfault transient signals are processed with wavelet transform to find the modulus maxima. Then, the time difference between the arrival time of the initial wave and the reflected wave can be used to calculate the fault distance by Equation (8.22) or (8.23).

According to the wavelet transform modulus maxima theory, the wavelet transform modulus maxima of noise decrease when the scale increases, whereas the modulus maxima of the traveling wave increase when the scale increases. The above conclusions are very important to traveling wave analysis because they can help us to isolate a traveling wave signal from a noise-contaminated environment.

8.2.2 *Case study*

As mentioned in Section 8.2.1, the single-ended traveling wave fault location method encounters the problems of distinguishing between traveling waves reflected from the fault point and those from the remote end of the line. There are many different methods to distinguish these two waves, such as using the amplitude and polarity of modulus maxima, or using the relationship between polarities of zero mode and aerial mode modulus maxima. Here we provide a case study that discriminates between the reflected wave from a fault point and that from the remote end employing the polarity information of modulus maxima. When the traveling wave arrives at the measuring end, the modulus maxima of wavelet transform of a current signal appear accordingly. If the fault occurs on the first half of the line, the fault reflection will arrive earlier than the remote end reflection, and the first two wavelet modulus maxima have the same polarity; else, the fault reflection will arrive later than the remote end reflection, and the first two wavelet modulus maxima have opposite polarity.

The simulation system is shown in Figure A.4 in the Appendix. The length of the transmission line is 300 km. Suppose that a single phase-to-ground fault happened at the place 100 km away from the measuring end. The sampling rate is 100 kHz. Transform the current traveling wave into the modal domain using $I_m = I_a - 2I_b + I_c$. After decomposing the mode current with wavelet transform, the wavelet coefficient in high frequency decomposed at scale 2 and the corresponding modulus maxima are shown in Figure 8.4.

As can be seen in Figure 8.4, the first two wavelet modulus maxima have the same polarity, which means that the fault occurred on the first half of the line. The arrival time of the initial wave, the reflected wave from the fault point, and the reflected wave from the remote end are, respectively, 340 μs, 1010 μs, and 1710 μs. Referring to Equation (8.22), the fault location can be calculated using:

$$l_A = v \cdot (T_{2A} - T_{1A})/2 = 3 \times 10^5 \times (1010 - 340) \times 10^{-6}/2 = 100.5 \text{ km}$$

where v is the wave velocity approximate to the velocity of light. The absolute error in the fault location estimation is 500 m.

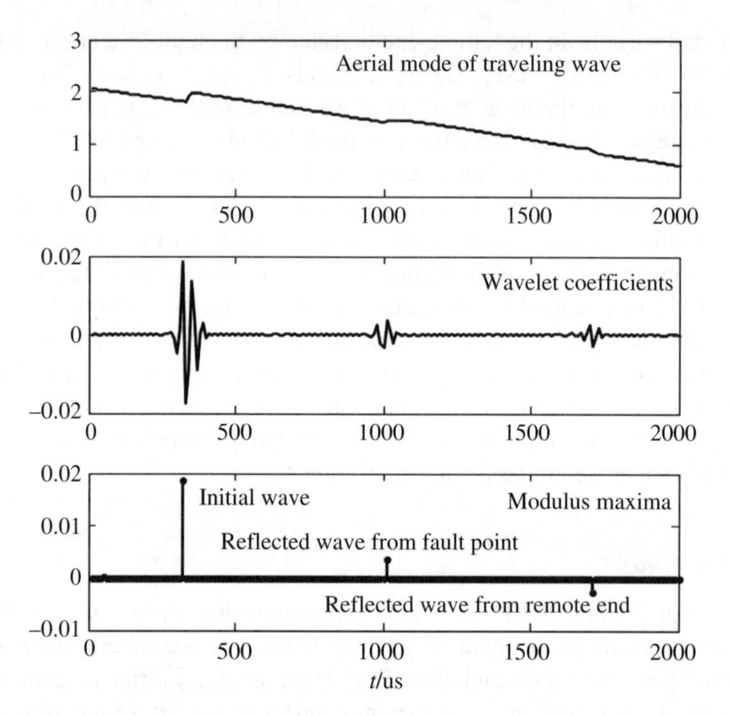

Figure 8.4 Wavelet transform of an aerial mode current when a single phase-to-ground fault occurs 100 km from the measuring end

The case study shows that the modulus maxima of wavelet transform can effectively identify the fault location. Because the estimation error in the fault location depends on the noise and sampling rate largely, there are some limitations in practical application. This limitation is one of the reasons why researchers estimate fault location based on traveling wave natural frequencies rather than wave front identification.

8.3 Fault location method based on traveling wave natural frequencies for transmission lines

Single-ended traveling-wave fault location methods have the advantages of avoiding the complexities and costs of the remote end synchronization. However, the identification of the initial traveling wave and its corresponding reflection from the fault point is a difficult problem, which is apt to limit the algorithms' accuracy. A novel single-ended fault location method based on traveling wave natural frequencies for transmission lines is proposed in References [1, 2]. The relationship among the fault distance, the natural frequencies of traveling waves, and the terminal conditions of transmission lines are described. The fault distance can be derived without any wave front identification efforts from the dominant natural frequency of the traveling wave spectrum, the line parameters, and the boundary conditions. One of the key points of this method is extracting the dominant frequency accurately. This section shows the principle of fault location based on traveling wave natural frequencies and the extraction of natural frequencies using complex wavelets.

8.3.1 Traveling wave natural frequencies

Any sudden change in the power system such as fault occurrence generates voltage and current traveling waves, which propagate in two directions from the point over the line. The spectra of traveling waves are related to fault distance and boundary conditions at line ends. The traveling wave lasts for 1–2 power–frequency cycles for ungrounded faults or 0.5–0.7 power–frequency cycles for grounded faults. Therefore, it is capable to estimate the fault distance employing the spectra of traveling waves.

The frequency spectra of fault traveling waves, which are called natural frequencies, consist of a series of harmonic form frequencies. The lowest order frequency has the largest proportion, and it is the dominant component; other components decrease as the frequencies increase. For example, a three-phase fault happens at the place 100 km away from the measuring end, as shown in Figure A.4 in the Appendix. Fourier transform on the α mode current of one power–frequency period after the fault occurred is conducted. The spectrum appears in Figure 8.5.

In 1979, Swift found that the traveling wave spectra have characteristics as follows.

When the system impedance is large, the dominant natural frequency will be

$$f = \frac{1}{4\tau} \tag{8.26}$$

When the system impedance is very small, the dominant natural frequency is

$$f = \frac{1}{2\tau} \tag{8.27}$$

where f is the dominant natural frequency; and τ is the travel time from the measuring end to the fault point.

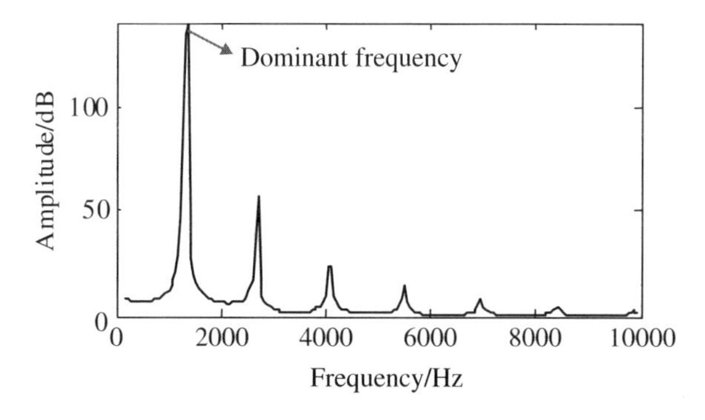

Figure 8.5 Spectrum of the α mode current for a three-phase fault

Actually, if we substitute $\tau = d/v$, we obtain

$$\begin{cases} d = \dfrac{v}{4f}\left(\text{the system impedance is large}\right) \\ d = \dfrac{v}{2f}\left(\text{the system impedance is small}\right) \end{cases} \tag{8.28}$$

The equations above for fault location using traveling wave natural frequencies only cater to special system conditions. It is necessary to discuss the relationship between fault-induced traveling wave natural frequencies and fault distance in all system boundary conditions. First, we need to establish the equivalent circuit of transmission lines and boundary conditions.

8.3.2 The input–state–output description of transmission lines

Since the traveling wave natural frequencies are related to fault distance and transmission line boundary conditions, it is essential to consider boundary conditions when modeling the equivalent circuit of transmission lines. During fault location calculation, the information we can get are voltages and currents at line terminals. Therefore, we represent an entire single-phase transmission line of length d as a Thevenin equivalent two-port circuits. The power system at both line ends can be represented as lumped circuits. Thus, the input–state–output description of a single-phase transmission line and the equivalent circuits of Thevenin type are shown in Figure 8.6.

From Figure 8.6, we obtain

$$u_1(t) = \frac{Z_1}{Z_1 + Z_C} w_1(t) + \frac{Z_C}{Z_1 + Z_C} e_1(t) \tag{8.29}$$

$$u_2(t) = \frac{Z_2}{Z_2 + Z_C} w_2(t) + \frac{Z_C}{Z_2 + Z_C} e_2(t) \tag{8.30}$$

where u_1 and u_2 are terminal voltages; e_1 and e_2 are equivalent Thevenin voltage sources

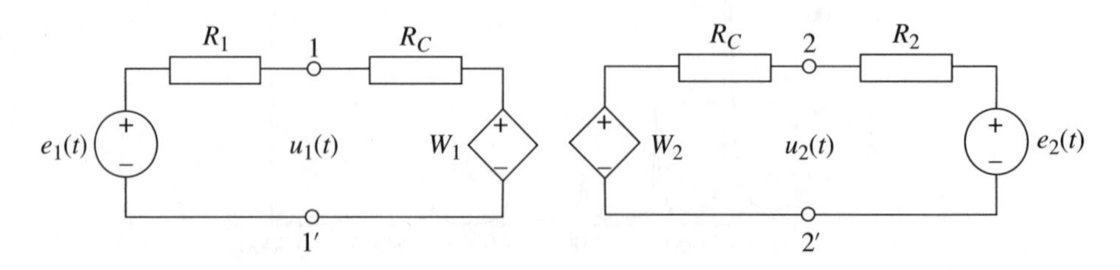

Figure 8.6 Equivalent circuit of a line represented as two ports connecting to power systems represented as lamped circuits

of systems 1 and 2; Z_1, Z_2, and Z_C are, respectively, the equivalent impedance of system 1, that of system 2, and the characteristic impedance of the line; and $w_1(t)$ and $w_2(t)$ are the controlled voltage sources in the time domain and defined as

$$t > T \; w_1(t) = 2u_2(t-T) - w_2(t-T) \tag{8.31}$$

$$t > T \; w_2(t) = 2u_1(t-T) - w_1(t-T) \tag{8.32}$$

Because the parameters of the transmission line are dependent on frequency, we shall use the method of the Laplace transform to study the equivalent circuit.

Substituting Equations (8.29) and (8.30) into Equations (8.31) and (8.32), in the Laplace domain we obtain

$$W_1(s) = \frac{P(s)}{1-\Gamma_1(s)\Gamma_2(s)P^2(s)}\left[\Gamma_2(s)P(s)\frac{2Z_C(s)}{Z_1(s)+Z_C(s)}\cdot E_1(s) + \frac{2Z_C(s)}{Z_2(s)+Z_C(s)}E_2(s)\right] \tag{8.33}$$

$$W_2(s) = \frac{P(s)}{1-\Gamma_1(s)\Gamma_2(s)P^2(s)}\left[\frac{2Z_C(s)}{Z_1(s)+Z_C(s)}E_1(s) + \Gamma_1(s)P(s)\frac{2Z_C(s)}{Z_2(s)+Z_C(s)}E_2(s)\right] \tag{8.34}$$

where $\Gamma_1(s)$ and $\Gamma_2(s)$ are the Laplace transform of the voltage reflection coefficients at line terminals.

$$\Gamma_1(s) = \frac{Z_1(s)-Z_C(s)}{Z_1(s)+Z_C(s)} \tag{8.35}$$

$$\Gamma_2(s) = \frac{Z_2(s)-Z_C(s)}{Z_2(s)+Z_C(s)} \tag{8.36}$$

$P(s)$ is a delay operator of the line in the Laplace domain

$$P(s) = e^{-sT} = e^{-\frac{s\cdot d}{v}} \tag{8.37}$$

For lossy lines, the wave is no longer traveling at light speed and is frequency dependent. The propagating velocity v under a certain frequency is

$$v = \frac{\omega}{\beta} = \frac{\omega}{imag(\sqrt{Z(j\omega)Y(j\omega)})} \tag{8.38}$$

where *imag* denotes an operator, which returns to the value of the imaginary of the complex number. $Z(j\omega)$ and $Y(j\omega)$ are, respectively, the per-unit-length impedance and admittance of the line.

8.3.3 The relation of fault distance and natural frequencies

According to Equations (8.29) and (8.30), the terminal voltages contain frequency components of source voltage E and state variables W. Let us assume E has only 50 Hz power frequency (fundamental frequency). The frequencies of W (i.e., the natural frequencies of the line) are the roots of Equation (8.39):

$$1 - P(s)\Gamma_1(s)P(s)\Gamma_2(s) = 0 \tag{8.39}$$

Substituting the Laplace operator s with $s = \sigma + j\omega$, and using Euler's equation, we obtain

$$e^{2(\sigma_n + j\omega_n)T} = \Gamma_1\Gamma_2 e^{j2k\pi} \quad (k = 0, \pm 1, \pm 2, \cdots) \tag{8.40}$$

where σ_n and ω_n are the damping constant and angular frequency of the nth natural frequency component, respectively. Since the dominant frequency, which is used for the fault distance calculation, has the largest amplitude, it can be identified easily.

When a fault occurs on the transmission line, the reflection coefficient at the fault point in the Laplace domain is given by

$$\Gamma_f(s) = |\Gamma_f| e^{j\theta_f} = -\left[2Z_C(s)^{-1} + Y_f(s)\right]^{-1} Y_f(s) \tag{8.41}$$

where $Y_f(s)$ is the fault admittance in the Laplace domain.

It can be derived from Equation (8.40) that

$$e^{2(\sigma_n + j\omega_n)T} = |\Gamma_1| e^{j\theta_1} |\Gamma_f| e^{j\theta_f} e^{j2k\pi} \tag{8.42}$$

Substituting $T = d/v$ into Equation (8.42), we obtain

$$d = \frac{(\theta_1 + \theta_f + 2k\pi)v}{4\pi f_n} \tag{8.43}$$

where f_n (n = 1, 2, 3…) is the nth natural frequency component.

For the dominant natural frequency:

$$d = \frac{\left(\theta_1 + \theta_f\right)v}{4\pi f_1} \tag{8.44}$$

In a transmission system, $Z_1(s)$ is the equivalent impedance of the transmission boundary as known in advance. The reflection angle of the terminal can be calculated from Equation (8.35). However, the fault admittance $Y_f(s)$ is indefinite. Therefore, the reflection angle of the fault point is unknown. From Equation (8.41), the reflection coefficient of the fault point can also be represented as

$$\Gamma_f(s) = \left|\Gamma_f\right| e^{j\theta_f} = \frac{-Z_C(s)}{2Z_f(s) + Z_C(s)} \tag{8.45}$$

where $Z_f(s) = 1/Y_f(s)$, which is usually much smaller than the characteristic impedance of the line. Therefore, the reflection angle θ_f is approximately equal to π. The fault location equation can be written as

$$d = \frac{\left(\theta_1 + \pi\right)v}{4\pi f_1} \tag{8.46}$$

For multiconductor transmission lines, it is necessary to, first of all, transform phase components into modal components. The phase-mode transformation matrices are normally different for current and voltage depending on frequency. Here, we only consider a balanced three-phase line (i.e., ideally transposed lines), or a delta configured line very high in the air. Then, the two transformation matrices are the same and constant with frequency. Under those conditions, we use the Clarke transformation matrix to decouple the three-phase transmission lines. For three-phase transmission lines, the traveling wave natural frequency is determined as

$$\mathbf{det}\left[\mathbf{I} - \mathbf{P}(s)\Gamma_1(s)\mathbf{P}(s)\Gamma_2(s)\right] = \mathbf{0} \tag{8.47}$$

where $\mathbf{I}$ is a 3 × 3 identity matrix; $\mathbf{P}(s)$ is a 3 × 3 modal time delay operator matrix; $\boldsymbol{\Gamma}_1(s)$ and $\boldsymbol{\Gamma}_2(s)$ are modal reflection coefficient matrices of system terminal and fault point, respectively; and det is an operator that calculates the determinant of the matrix.

After phase-mode transformation, the Laplace domain operator $\mathbf{P}(s)$ matrix is diagonal. The reflection matrices Γ_1 and Γ_f could be diagonal if the system and the fault are, respectively, balanced. Therefore, for a three-phase short circuit, Equation (8.47) is decoupled. Either mode α current or mode β current is capable of dominant natural frequency extraction and fault distance estimation using Equation (8.46).

For phase-to-phase fault and double phase-to-ground fault, when the proper selection of a phase-mode transformation matrix such as mode β involves the two faulted phases, then mode β is also decoupled. The calculation can be made on mode β as before.

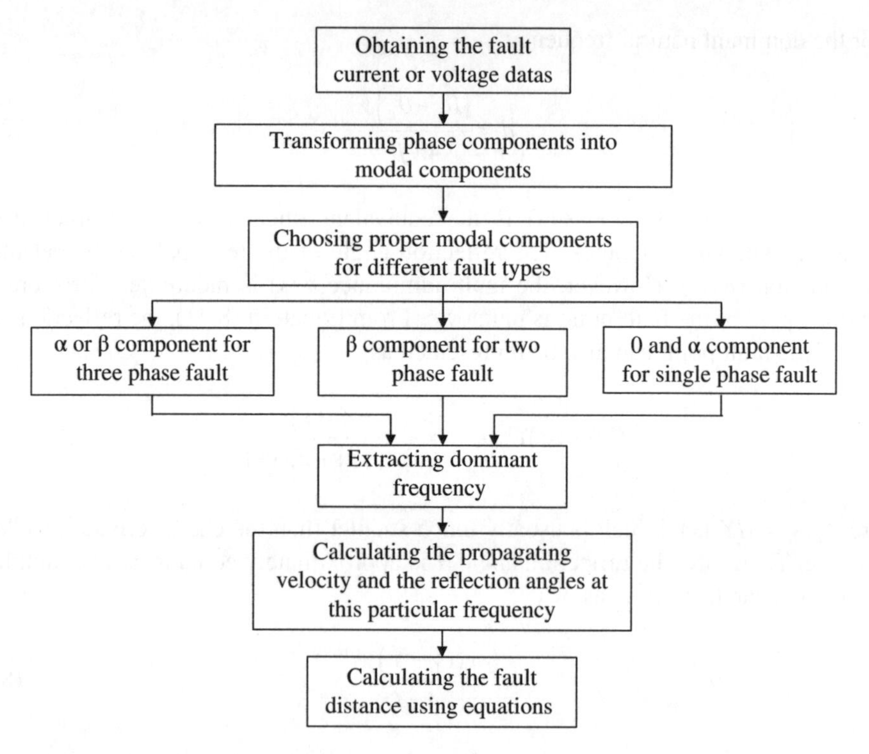

Figure 8.7 Flow chart of a fault location using traveling wave natural frequencies

When there is a single phase-to-ground fault, the module components are coupled with each other (i.e., signals in one mode are mixed of all modes). This is the most general phenomenon of modal mixing, which increases the difficulty for valid fault location. In a natural frequency–based fault location method, the compensation of module mixing error is possible as follows.

If the selection of modal transformation matrix is that mode β involves the two healthy phases, then there are only mode zero and α, which are mixed together. The dominant frequency is extracted at the mode α current. Using this dominant frequency, two different distances can be calculated, respectively, with mode zero and mode α parameters. The final fault distance is obtained by averaging these two distances. The fault location chart flow based on the traveling wave natural frequency is shown in Figure 8.7.

8.3.4 Extraction of natural frequencies based on wavelet transform

One of the keys to natural frequency–based fault location method is extracting the dominant natural frequency accurately. Contrary to real wavelets, which are ideal for detecting sharp signal transitions, the complex wavelet can measure the time evolution of frequency transients in a way similar to that of windowed Fourier transform (WFT). For fault location, it offers extra advantages over WFT because it results in the calculated distance in equal steps by scaling and translating one base function; thus, the algorism is much simpler. Complex wavelets are a group of base functions whose mother wavelets are complex numbers. Therefore, it can separate amplitude and phase components. The most commonly used complex wavelets are complex Gaussian, complex Morlet, complex Shannon, and complex frequency B-spline wavelet. We will use complex Gaussian for our application. The *P*th complex Gaussian wavelet is defined by the *P*th derivative of the complex Gaussian function:

$$\psi(x) = d^P \left(C_p e^{-ix} e^{-x^2}\right) / dx \tag{8.48}$$

where C_p is constant.

The continuous complex wavelet transform of signal x is given by

$$W_x(b,a) = x, \psi_{b,a} = |a|^{-\frac{1}{2}} \int_{-\infty}^{\infty} x(t) \psi^* \left(\frac{t-b}{a}\right) dt \tag{8.49}$$

where a and b are the scale parameter and translation parameter, respectively.

8.3.5 Case study

The simulation system has the same topology of Figure A.4 in the Appendix. The only change is the line length is 600 km. The voltage source at bus M is ideal. Assume a three-phase fault occurred 400 km away from the bus M. The sample rate is 100 kHz. The post-fault phase current signal of the first cycle is transformed into a modal signal, and the mode α current is analyzed employing the 20th complex Gaussian wavelet. The scale is chosen from 1 to 512 with the scale step as 1. The amplitude part of complex wavelet transform coefficients is shown in Figure 8.8. For each scale, transform results in one continuous variable curve, whose color is proportional to coefficients. The brighter color stands for higher amplitude. As can be seen in Figure 8.8, the dominant frequency is at around scale 300. The second and third natural frequencies are also visible at scale 160 and scale 100. Other natural frequencies at smaller scales, corresponding to higher frequencies, decay too fast to be identified.

Because it is hard to identify accurately the scale at which transform results have the highest amplitude from Figure 8.8, the norm entropy at each scale is calculated. One chooses the scale corresponding to the dominant natural frequency. The scale of maximum norm entropy is chosen as the scale corresponding to the dominant natural frequency. The norm entropy at each scale is defined by

$$E_a(s) = \Sigma |W_b|^q \tag{8.50}$$

where a represents scale; W_b is magnitude coefficients at scale a; and q is the norm order and $q = 1$. The distribution of norm entropy for scale 1 to 512 is shown in Figure 8.9. Each peak value in Figure 8.9 represents different-order natural frequencies. The peak value,

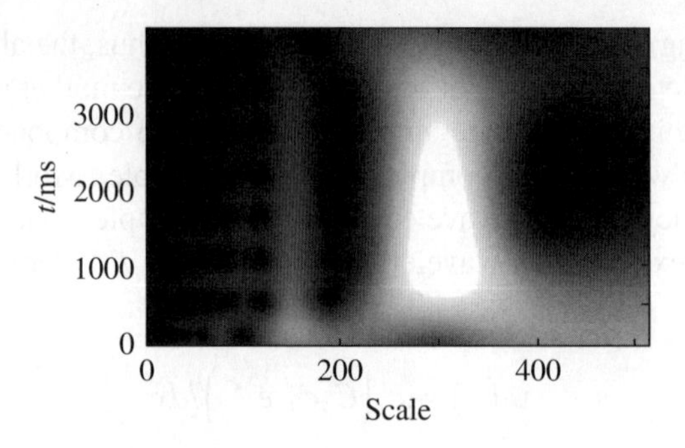

Figure 8.8 The magnitude of complex wavelet transform of an α mode current

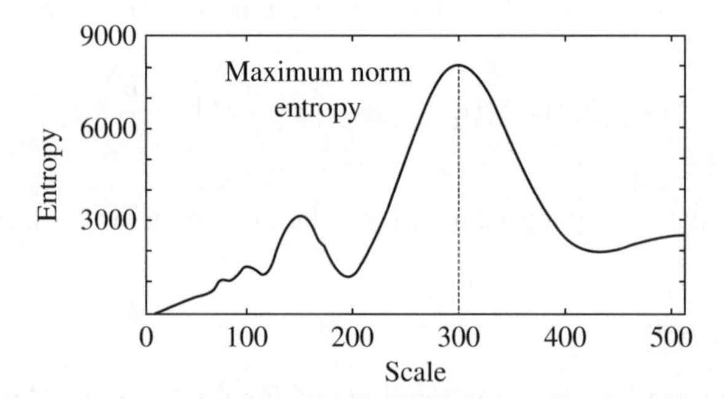

Figure 8.9 Curve of norm entropy for scale 1 to 512

which has the greatest amplitude, corresponds to scale 298, as shown in Figure 8.9. The center frequency of this scale is 369.1 Hz. The mode α current wave velocity under this frequency is 295258 km/s; therefore, the fault distance is calculated as

$$d = 295258/369.1/2 = 399.97 \text{ km}$$

The absolute error is 30 m.

In comparison with the fault location method based on modulus maxima, the fault location method based on natural frequencies is insensitive to noise. This is because it utilizes the dominant natural frequency rather than the highest frequency band.

8.4 Fault location method based on time–frequency characteristics of traveling wave

8.4.1 Continuous wavelet transform

Wavelet transform maintains the correlation between time and frequency of the observed signal, and it has excellent time–frequency localization abilities. This proves to be advantageous in determining the wave front of the initial traveling wave and the subsequent reflections.

The definition of continuous wavelet transform (CWT) for a given signal $x(t)$ is given as

$$CWT(a,b) = \frac{1}{\sqrt{|a|}} \int_{-\infty}^{+\infty} x(t)\phi^{*}\left(\frac{t-b}{a}\right)dt \tag{8.51}$$

where $\phi(t)$ is the mother wavelet; * denotes a complex conjugate; and a and b are the scale factor and the translation factor, respectively.

In the frequency domain, the signal can be decomposed into different scales using the CWT. The pseudo-frequency, which corresponds to each scale coefficient, is calculated as

$$F_a = F_S F_C / a \tag{8.52}$$

in which F_S is the sampling rate of the input signal; and F_C is the center frequency for the mother wavelet.

In the time domain, the modulus maxima point of the wavelet transform coefficient (WTC) corresponds to the singularity of the signal. The mutation strength and direction in the signal can be characterized by the amplitude and polarity of modulus maxima, respectively.

On the other hand, the singularity of the signal can be described using Lipschitz exponent ε. The smaller the number ε represents, the higher the singularity of the signal is. And Mallat has proved that the WTCs satisfy the following formula relationship with the Lipschitz exponent:

$$|CWT(a,t)| \leq Ha^{\varepsilon} \tag{8.53}$$

where $CWT(a, t)$ is the WTCs when the translation factor $b = 1$; a is the decomposition scale; ε is the signal's Lipschitz exponent; and H is a constant.

In Equation (8.53), the equality is taken when the $CWT(a, t)$ occurs as the wavelet modulus maxima (WMM), $W_{mm}(a)$, so it then can be rewritten as follows:

$$\log|Wmm(a)| = \log H + \varepsilon \log a \tag{8.54}$$

Equation (8.54) acts as a first-order function, which uses "log a" as the independent variable, "log$|W_{mm}(a, t)|$" as the dependent variable, and ε as the gradient of the function. Therefore, the Lipschitz exponent can represent the time domain and frequency domain characteristics of the signal conjointly.

8.4.2 Traveling wave time–frequency characteristics analysis

The fault-generated transient traveling waves are wideband signals that cover the entire frequency range, and in the frequency domain, the magnitude of the individual signal components decreases and the traveling velocity increases as the frequency increases. Therefore, in the time domain, the arrival time of different frequency components is not alike. The component of the highest frequency will reach the locator first, and the

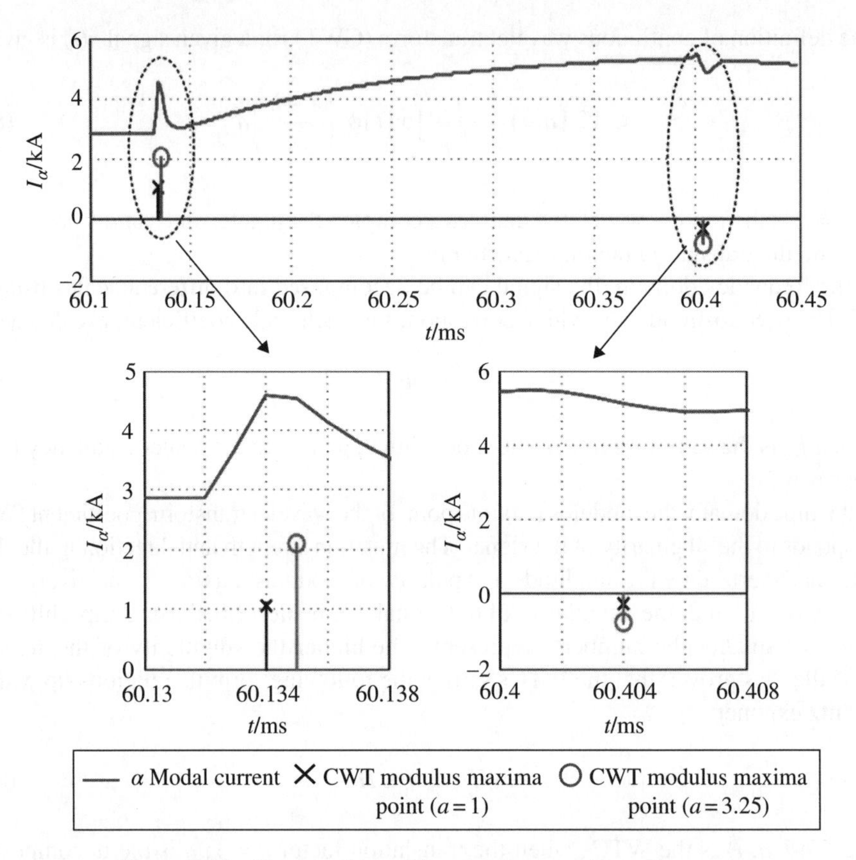

Figure 8.10 The waveform of the α mode current traveling wave

components of other frequencies will be postponed to some time to arrive at the measuring point. So, the transient wave front detected at the locator is not an ideal step signal, but is spread out by the rise time or fall time, as shown in Figure 8.10.

The highest frequency detected at the locator is the highest one that is able to propagate to the locator and is limited by the sampling frequency. Therefore, as shown in Figure 8.10, the arrival time of the initial wave front is that of the component of the detected highest frequency. Meanwhile, the time–frequency characteristics of the second transient wave front are analyzed according to the fault position.

When the fault occurs in the first half section of the transmission line, the second wave front detected by the locator consists of the components of the initial traveling wave that can still spread to the observer after two reflections at the close-in bus and the fault point. When the fault occurs in the second half of the line, the second transient wave front detected at the measurement point consists of the component of the initial traveling wave that can still reach the locator after the reflection at the remote-ended bus and the refraction at the fault point. In the propagation, the component of the traveling wave with higher frequency not only decays faster but also possesses a smaller reflection coefficient and refraction coefficient. Consequently, the component with the highest frequency of the initial traveling wave may not reach the locator again. Thus, the frequency components of the second traveling wave detected are different from those of the initial traveling wave, and the frequency band will move down. Simultaneously, the farther the traveling wave propagation distance is, the larger the movement range of the frequency band is, and the more the traveling velocity of the wave front decreases. Furthermore, the propagation velocity of the traveling wave decreases with the frequency, and the lower the frequency is, the more the influence by the frequency is. All of these variables make the time delay between the arrival time of different frequency components of the second traveling wave become longer, and the rise time or fall time of the wave front becomes longer too, which means the transient signal singularity decreases.

8.4.3 Proposed method based on time–frequency characteristics of traveling waves

Assume that the fault occurs at t_0 and the distance from the fault point to the locator is d. For m modal traveling wave signal, suppose that the frequency component f_1 of the initial traveling wave front reaches the measurement at t_1 and with the corresponding traveling velocity v_{m1}. Similarly, the arrival time and propagation velocity of the frequency component f_2 of the second transient wave front are t_2 and v_{m2}, respectively. Consequently, for the initial traveling wave

$$t_1 - t_0 = \frac{d}{v_{m1}} \tag{8.55}$$

And, when the fault occurs in the first half section of the transmission line, for the second traveling wave

$$t_2 - t_0 = \frac{3d}{v_{m2}} \tag{8.56}$$

So we can get the fault distance by associating Equation (8.55) and Equation (8.56):

$$d = \frac{v_{m1} v_{m2}}{3v_{m1} - v_{m2}} \left(t_2 - t_1 \right) \tag{8.57}$$

If the fault occurs in the last half section, the second transient wave front satisfies

$$t_2 - t_0 = \frac{2L - d}{v_{m2}} \tag{8.58}$$

Thus, combining Equations (8.55) and (8.58), the fault distance d can be obtained as

$$d = \frac{2v_{m1}}{v_{m1} + v_{m2}} L - \frac{v_{m1} v_{m2}}{v_{m1} + v_{m2}} \left(t_2 - t_1 \right) \tag{8.59}$$

The propagation velocity v_{m1} and v_{m2} can be obtained based on the frequency-dependent line parameters. Hence, determining the traveling velocities means determining the frequency of the transient wave fronts. And determining the arrival time of the traveling waves t_1, t_2 means detecting the singularity of traveling wave signals that can be accomplished by using wavelet transform modulus maxima.

8.4.4 *Implementation of the proposed method*

The recorded current traveling wave signals are first decoupled into their modal components, and then the proper modal component is selected to be decomposed into different scales using CWT. Regard the pseudo-frequency of the first scale, which is the highest frequency in the subbands, as the detected frequency of the initial wave front. Therefore, the propagation velocity v_{m1} can be calculated and the arrival time t_1 can be detected corresponding to the first CWT modulus maxima point.

The time domain and frequency domain characteristics of the second wave front have changed compared with the initial one. The Lipschitz exponent of the signals can relate the WMM to its corresponding decomposition scale as represented in Equation (8.54). Namely, the arrival time and the frequency component can be determined jointly by using the Lipschitz exponent. So, the procedures of determining the propagation velocity and arrival time of the second wave front are realized as follows:

1. Search the WMM of the second wave front on each decomposition scale $(a_i, W_{mm}(a_i))$, in which $i = 1, 2, \ldots, n$; n is the largest decomposition level of CWT.
2. According to Equation (8.54), use the least-squares method (LSM) to compute the Lipschitz exponent ε and the constant H. And then, based on the computed ε and H, calculate the new wavelet modulus maxima $W_{mm}(a_i)'$ of each scale using Equation (8.54) again.
3. Select the pseudo-frequency of the best scale, a_{best}, on which the difference between $|W_{mm}(a_i)'|$ and $|W_{mm}(a_i)|$ is smallest, as the determined frequency f_2 of the second transient wave front for fault location. Thus, the propagation velocity v_{m2} is able to be estimated using the frequency-dependent parameters corresponding to f_2.
4. On the selected best scale a_{best}, compare the wavelet modulus maxima $|W_{mm}(a_i)'|$ with the wavelet transform coefficient value $|CWT(a_{best}, t)|$ one by one within the wavelet cone of influence (COI). And choose the point, t_{best}, whose corresponding wavelet transform coefficient value $|CWT(a_{best}, t)|$ is closest to $|W_{mm}(a_i)'|$, as the detected arrival time t_2 of the second wave front of the traveling wave.

Consequently, based on v_{m1}, v_{m2}, t_1, and t_2, the fault distance can be calculated according to Equation (8.57) or (8.59).

As a performance index, the estimation error of the fault location is expressed using the following equation:

$$\%\text{Error} = \frac{|\text{Calculated fault location} - \text{Actual fault location}|}{\text{total line length}} \times 100 \tag{8.60}$$

8.4.5 Numerical example analysis

The simulation platform is the model described in Figure A.4. The short-circuit faults in the transmission lines under various conditions are simulated. Set the sampling frequency to be 1 MHz. And the one-order Gaussian wavelet, whose center frequency is 0.2 Hz, is chosen as the mother wavelet for CWT. Furthermore, the decomposition scale of CWT is in increasing order from 1 to 4 by 0.25; thus, the pseudo-frequency of each scale decreases from 200 kHz to 50 kHz according to Equation (8.52).

Taking the recorded traveling wave shown in Figure 8.10 as an example, the proposed fault location method is implemented and compared with the traditional traveling wave–based fault location method. The fault conditions of the traveling wave are as follows: the fault distance is 40 km, fault type is A–G, fault resistance is 10 Ω, and fault inception angle is 90°. Meanwhile, the transmission line is transposed. Therefore, Clarke transformation matrix is applied to decouple the phase components into modal components where phase A acts as the foundational phase, and the α mode component is selected for fault location. Then, the CWT is employed.

Search the WMM of the initial wave front of the traveling wave on each decomposition scale, and then the Lipschitz exponent ε of the initial transient signal can be obtained by using LSM, that is, 0.4447. The emergence time of the modulus maxima of the initial wave front on the first decomposition scale is 60.134 ms, and its corresponding frequency is 200 kHz. Therefore, the propagation velocity is 2.98002×10^5 km/s, which is obtained under the R_α, L_α, G_α, and C_α corresponding to the frequency.

As shown in Table 8.1, the WMM of the second transient wave front on each decomposition scale is searched. Hence, ε is calculated as 0.7948. It is bigger than that of the initial transient signal, which indicates that the singularity of the second wave front reduces as traveling waves propagate along the transmission line and reflect or refract at the discontinuous point.

Using the Lipschitz exponent, the new WMM of the second transient wave front on each decomposition scale is obtained by Equation (8.54), as shown in Figure 8.11a and Table 8.1. By comparing the actual searched WMM with the calculated one, the pseudo-frequency 61.538 kHz corresponding to the decomposition scale 3.25 is selected as the frequency of the detected second wave front, as shown in Figure 8.11a. Then the propagation velocity is 2.96860×10^5 km/s, calculated based on the frequency-dependent parameters. Furthermore, on this decomposition scale, the arrival time of the wave front is determined, which is 60.404 ms, as shown in Figure 8.11b.

So, the fault distance can be calculated by Equation (8.57). Also see Figure 8.11.

However, in the traditional traveling wave–based fault location system, the propagation velocity of the traveling wave is taken as 2.9975×10^5 km/s, and the arrival times of both the initial and the second wave fronts are assumed as the modulus maxima points on the smallest wavelet decomposition scale. Thus, for the above example, the arrival times of the first and second transient wave fronts are detected as 60.134 ms and 60.404 ms, respectively, and then the fault location is estimated by Equation (8.22).

Also, comparing the results of the proposed method with those of the conventional method shows that the accuracy of the fault location method based on traveling wave time–frequency characteristics is obviously improved.

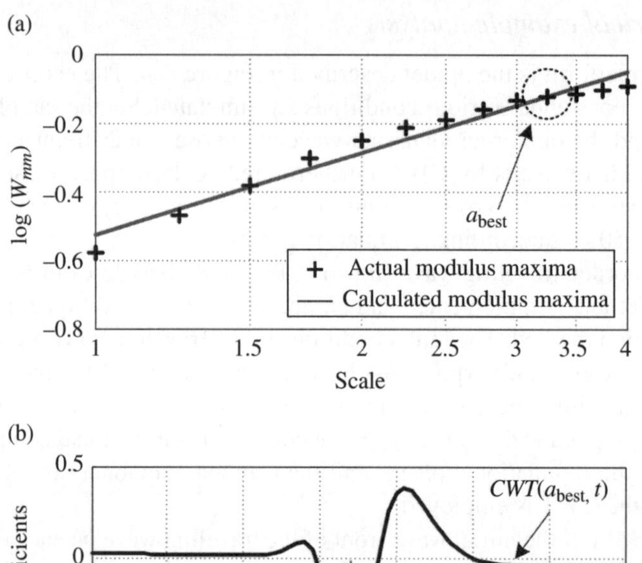

Figure 8.11 The selected best: (a) decomposition scale and (b) arrival time of the second transient wave front

Table 8.1 The WMM of the second transient waterfront on each scale

Scale a_i	Actual modulus maxima W_{mm}	Calculated modulus maxima W_{mm}'	$\|\|W_{mm}\|-\|W_{mm}'\|\|$
1	−0.2627	0.2963	0.0336
1.25	−0.3377	0.3538	0.0161
1.5	−0.4144	0.4089	0.0055
1.75	−0.4953	0.4622	0.0331
2	−0.5631	0.5140	0.0491
2.25	−0.6124	0.5644	0.0479
2.5	−0.6470	0.6137	0.0332
2.75	−0.6928	0.6620	0.0308
3	−0.7339	0.7094	0.0245
3.25	**−0.7569**	**0.7560**	**0.0009**
3.5	−0.7642	0.8019	0.0377
3.75	−0.7891	0.8471	0.0580
4	−0.8049	0.8917	0.0867

In summary, wavelet transform is capable of providing the time and frequency information simultaneously. It can extract the polarity information of the wave front, or pick up the dominant frequency. It is nearly impossible to utilize the traveling wave to estimate fault location until wavelet transform emerges. That is because there are no other mathematical methods to describe and represent traveling waves in both the time domain and frequency domain. Thus, wavelet transform performs meritorious deeds never to be obliterated in fault location.

References

[1] Wu L.Y., He Z.Y., Qian Q.Q., A frequency domain approach to single-ended traveling wave fault location. *Proceedings of the CSEE*, vol. 28, no.25, pp. 99–104, 2008.

[2] Wu L.Y., He Z.Y., Qian Q.Q., A new single-ended fault location method of extracting wave natural frequency. *Proceedings of the CSEE*, vol. 28, no.10, pp. 69–75, 2008.

Further reading

Ge Y. Z., A new type of relay protection and fault location principle and technology. *Xi'an Jiaotong University Press*, 1996.

Li X. B., Research on transmission line faults location based on wavelet analysis. *Tianjin University of Technology*, 2008.

9

Application of Wavelet Transform in Fault Feeder Identification in a Neutral Ineffectively Grounded Power System

The neutral grounded types in China can be categorized into two groups: the neutral effectively grounded type and the neutral ineffectively grounded type. Power systems of 110 kV and higher are neutral effectively grounded systems. In them, the fault current is great when a single phase-to-ground fault occurs. The neutral effectively grounded system therefore is also called a great current grounded system. The 6–66 kV distribution networks are neutral ineffectively grounded systems. In them, the fault current is very low. So the neutral ineffectively grounded power system is also called a small current grounded system. Neutral ineffectively grounded systems include neutral ungrounded systems, neutral grounded via Peterson coil systems (resonant grounded systems), and neutral grounded via resistance systems. The low-fault current of a neutral ineffectively grounded system has an insignificant influence on the power supply equipment in a short time. Therefore, operation of the system can continue without any interruption for 1–2 hours. However, with the growth of system capacity, the feeders, especially the cable feeders, increase evidently, which causes the capacitive current to increase. If the fault cannot be isolated within a reasonable time, there is a risk of double phase-to-ground fault. Moreover, the fault current possibly leads to intermittent arc–earth overvoltage with long duration and wide influence, which will damage the equipment and destroy the safe and stable operation of the system. It is therefore very important to detect and isolate the faulty feeder automatically.

Wavelet Analysis and Transient Signal Processing Applications for Power Systems, First Edition. Zhengyou He.

Currently, the main reasons for a faulty feeder are difficult identify, but they include the following aspects:

1. The fault conditions are complicated. The fault may be a stable fault or intermittent fault, or a resistance fault or arc fault. Fault quantities differ greatly under various fault conditions.
2. For a single phase-to-ground fault, fault currents are predominantly capacitive and much lower than the maximum load current. The zero-sequence current may even be smaller than the lower limit value of total capacitance (CT), which causes great measurement error.
3. The electromagnetic interference in the field and the zero-sequence circuit will amplify the high harmonics and transients, which degrade the signal-to-noise ratio of the measured quantities greatly.
4. A zero-sequence filter, which is used to obtain the zero-sequence current, causes unbalanced current. Imbalance in the power grid also causes zero-sequence current. These unbalanced currents are hard to separate from the weak fault current.

From this analysis, the root cause leading to faulty line selection difficulties is that the fault signal is too weak and is enslaved to the accuracy of the measurement process. Faulty line selection is therefore affected significantly by various unfavorable factors.

Research on faulty line selection for small current grounded systems is mainly based on steady components for quite a long period. Because the amplitude of steady components is relatively small, along with the complexity of the power grid and load, it is difficult to obtain the fault signal effectively in the practical application. When a single phase-to-ground fault occurs, there is a complicated transient process. The amplitude of the characteristic signal is relatively large. In particular, the amplitude of a phase-to-ground capacitance current is several or dozens of times larger than that of the steady component. The transient process contains abundant fault transient information, which can detect the faulty line to overcome the shortages of a steady component–based method. Wavelet analysis is an outstanding time–frequency analysis tool that provides solutions to faulty line selection. As wavelet analysis has good localizing property, it is capable of improving the performance of detecting singular and instantaneous fault signals. It is also helpful to analyze the fault transient signals deeply and extract the favorable features to select the faulty line.

9.1 The basic principle of faulty line selection in neutral ineffectively grounded systems

9.1.1 Steady-state analysis of single phase-to-ground faults

9.1.1.1 Neutral ungrounded systems

Figure 9.1 shows a simplified diagram of a neutral ungrounded system. Under normal operation conditions, the phase capacitance current that leads the phase voltage by 90° is generated by the shunt capacitance. The sum of three-phase capacitance currents is zero. Under fault conditions, such as during a phase A–to-ground fault, the phase voltage at the

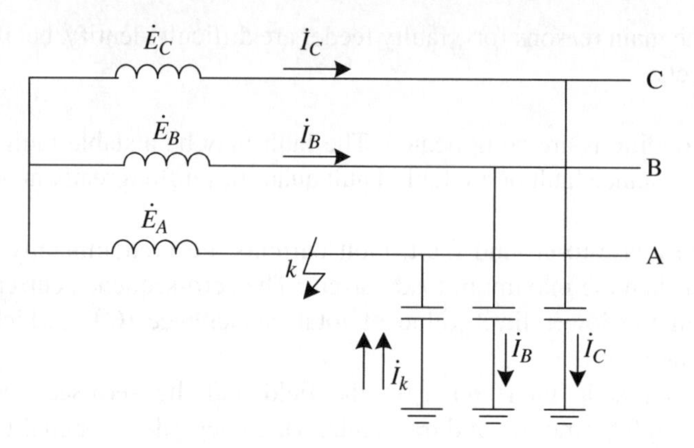

Figure 9.1 Simplified diagram of the neutral ungrounded system

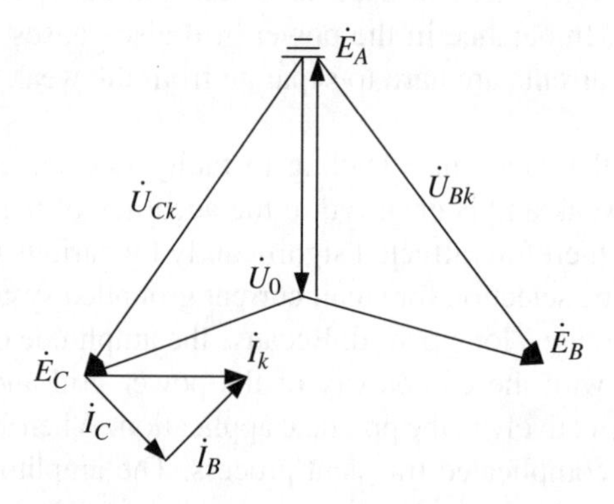

Figure 9.2 Phasor diagram of a phase A–to-ground fault

fault point is brought to zero, whereas the healthy phases are at potentials equal to the phase-to-phase voltages. The capacitance currents of the healthy phases increase $\sqrt{3}$ times correspondingly. The phasor diagram appears in Figure 9.2 [1].

Because the line voltages and load currents are still three-phase symmetrical, we here analyze only the phase-to-ground voltages and currents. After a phase A–to-ground fault occurred, if we neglect the voltage generated on the line impedance by load current and capacitance current, the three-phase-to-ground voltages at fault point are

$$\begin{aligned} \dot{U}_A &= 0 \\ \dot{U}_B &= \dot{E}_B - \dot{E}_A = \sqrt{3}\dot{E}_A e^{-j150^\circ} \\ \dot{U}_C &= \dot{E}_C - \dot{E}_A = \sqrt{3}\dot{E}_A e^{j150^\circ} \end{aligned} \tag{9.1}$$

The zero-sequence voltage at the bus neutral point is

$$\dot{U}_0 = \frac{1}{3}\left(\dot{U}_A + \dot{U}_B + \dot{U}_C\right) = -\dot{E}_A \tag{9.2}$$

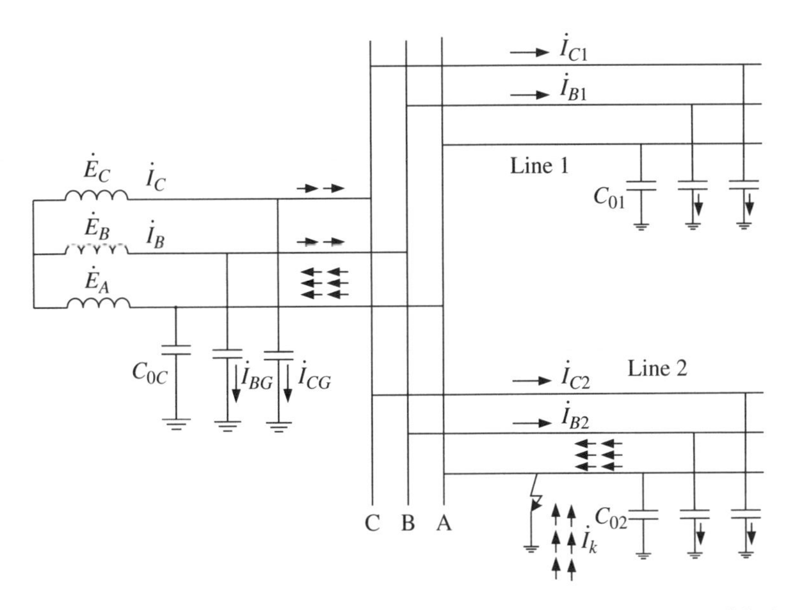

Figure 9.3 The distribution diagram of the current after a single phase-to-ground fault occurred

The capacitance currents of the healthy phases flow to the fault point. The capacitance currents of the healthy phases are

$$\begin{cases} \dot{I}_B = \dot{U}_B j\omega C_0 \\ \dot{I}_C = \dot{U}_C j\omega C_0 \end{cases} \tag{9.3}$$

The effective values are $I_B = I_C = \sqrt{3}U_\varphi \omega C_0$, where U_φ is the effective value of phase voltage.

Because phase A voltage is zero in the whole system, the capacitance currents of phase A of each element are zero. The current flowing from the fault point is the sum of capacitance currents of all the healthy phases (i.e., $\dot{I}_k = \dot{I}_B + \dot{I}_C$). As shown in Figure 9.2, its effective value is $I_k = 3U_\varphi \omega C_0$, which is three times the capacitance current under normal operation.

If the power system has generator G and multiple lines, as shown in Figure 9.3, we use C_{0G} and C_{0n} (n is the line number) to represent the capacitance of generator and each line. When there a phase A–to-ground fault occurs at line 2, the capacitance current distribution is indicated by "→". The current of phase A in the healthy line is zero. The currents flowing in phase B and phase C are capacitance currents. Therefore, the zero-sequence current at the line terminal is

$$3\dot{I}_{01} = \dot{I}_{B1} + \dot{I}_{C1} \tag{9.4}$$

Its effective value is shown as follows according to the phasor diagram in Figure 9.2:

$$3I_{01} = 3U_\varphi \omega C_{01} \tag{9.5}$$

The characteristic of a healthy line is that the current of the healthy line is generated by the capacitance to the ground. The capacitive power flows from the bus to the line. This conclusion applies to all healthy lines.

For generator G, first, it has the current generated by the shunt capacitances of phase B and phase C. Besides, it is the power source of other capacitance currents. Hence, phase A of generator G has all the capacitance currents, whereas phase B and phase C have the capacitance currents of the same-name phases of each line. The zero-sequence current of generator G is still the sum of three phase currents. Figure 9.3 shows that the capacitance currents of each line flow in from phase A and flow out from phase B and phase C. Therefore, the capacitance currents of each line cancel each other out, and only the capacitance currents of the generator remain.

$$3\dot{I}_{0G} = \dot{I}_{BG} + \dot{I}_{CG} \tag{9.6}$$

The effective value is $3I_{0G} = 3U_{\varphi}\omega C_{0G}$. The zero-sequence current of a generator is its capacitance current. The capacitive power flows from the bus to the generator.

For faulty line 2, the current of line 2 is generated by the shunt capacitances of phase B and phase C. Moreover, the capacitance currents of all the B phases and C phases will flow to the fault point. The current at fault point is

$$\dot{I}_k = \left(\dot{I}_{B1} + \dot{I}_{C1}\right) + \left(\dot{I}_{B2} + \dot{I}_{C2}\right) + \left(\dot{I}_{BG} + \dot{I}_{CG}\right) \tag{9.7}$$

Its effective value is

$$I_k = 3U_{\varphi}\omega\left(C_{01} + C_{02} + C_{0G}\right) = 3U_{\varphi}\omega C_{0\Sigma} \tag{9.8}$$

where $C_{0\Sigma}$ is the sum of phase-to-ground capacitances of each line and generator.

The current outflow from phase A is $\dot{I}_{A2} = -\dot{I}_k$. Thus, the zero-sequence current at the beginning terminal of line 2 is

$$3\dot{I}_{02} = \dot{I}_{A2} + \dot{I}_{B2} + \dot{I}_{C2} = -\left(\dot{I}_{B1} + \dot{I}_{C1} + \dot{I}_{BG} + \dot{I}_{CG}\right) \tag{9.9}$$

Its effective value is

$$3I_{02} = 3U_{\varphi}\omega\left(C_{0\Sigma} - C_{02}\right) \tag{9.10}$$

The characteristic of a faulty line is that the zero-sequence current is equal to the sum of capacitance currents generated by the healthy elements of the whole system. The capacitive power flows from the line to the bus, which is inverse to the power flow of healthy lines.

According to this analysis, we can draw the zero-sequence equivalent network of a single phase-to-ground fault, as shown in Figure 9.4a. There is a zero-sequence voltage

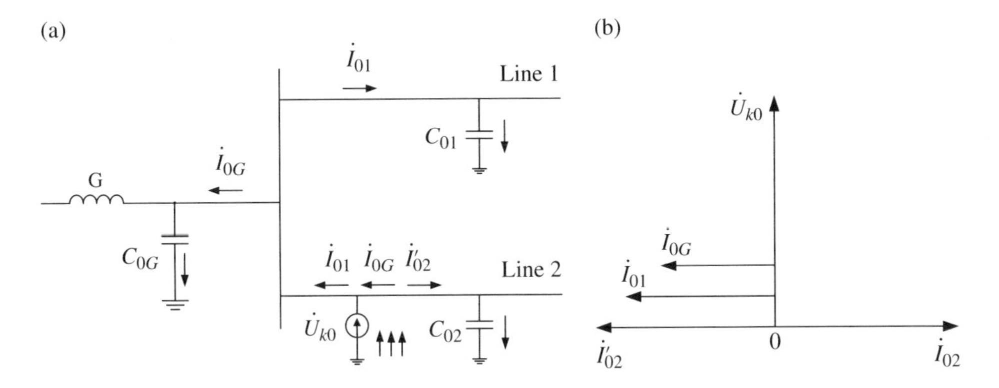

Figure 9.4 The zero-sequence equivalent network of a single phase-to-ground fault and its phasor diagram: (a) Equivalent network, (b) phasor diagram

$\dot{U}_{k0}$ at the fault point. The zero-sequence current loop is constructed by the phase-to-ground capacitances of each element. The phasor diagram appears in Figure 9.4b. I'_{02} in Figure 9.4 is the zero-sequence capacitance current of line 2. We can calculate the zero-sequence current and obtain the zero-sequence current distribution in terms of the zero-sequence equivalent network shown in Figure 9.4.

From this analysis, the characteristics of zero-sequence component distribution in neutral ungrounded systems can be summarized as follows:

1. The loop of a zero-sequence network is constructed by phase-to-ground equivalent capacitances of all elements with the same voltage level, which differ greatly from the neutral directly grounded system. The zero-sequence impedance is very great.
2. When a single phase-to-ground fault occurs, the fault can be modeled by a zero-sequence voltage source, which is equal in magnitude and opposite in sign to the pre-fault voltage at the fault point. The zero-sequence voltage will appear in the whole system.
3. The zero-sequence currents in the healthy elements are generated by their phase-to-ground capacitances. The capacitive reactive power flows from the bus to the line.
4. The zero-sequence current in the faulty element is the sum of capacitance currents generated by phase-to-ground capacitances of all the healthy elements. The capacitive reactive power flows from the line to the bus.

9.1.1.2 Neutral grounded via Peterson coil systems

From the analysis in this chapter, we know that when there is a single phase-to-ground fault in neutral ungrounded systems, the currents of fault points are the sum of capacitance currents of all of the elements. If the current is too great to cause arc at the fault point, the arc overvoltage will further improve the phase voltage of the healthy phases, which may make the insulation damage cause two points (or multiple points) of

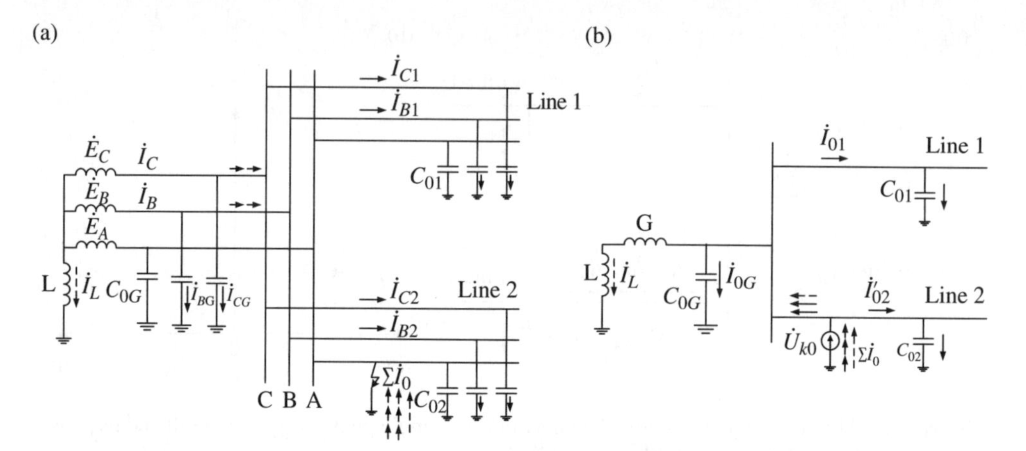

Figure 9.5 The current distribution of a neutral grounded via Peterson coil system after a single phase-to-ground fault occurred, and its zero-sequence equivalent network: (a) Current distribution of neutral grounded via Peterson coil system, (b) the zero-sequence equivalent network

grounded fault. In particular, if there are combustible gas, the arc may cause explosion. In order to solve this problem, the neutral point of the distribution network is generally grounded via an inductance coil, as shown in Figure 9.5a. Thus, when a single fault occurs on a line, the Peterson coil can compensate the capacitive fault current and the arc can hardly appear. The inductance coil is therefore called the Peterson coil or arc suppression coil.

In various voltage-level networks, the system should be grounded via Peterson coil if the capacitance current of the whole system exceeds the following values: 30 A for a 3–6 kV network, 20 A for 10 kV, and 10 A for 22–66 kV.

When the neutral point of the network is grounded via Peterson coil, the current distribution will change greatly. As shown in Figure 9.5, the coil is typically connected to the neutral of the distribution transformer. After phase A–to-ground fault occurred on line 2, the capacitance current distribution differed from that of a neutral grounded without Peterson coil system. The difference is that the current of the fault point consists of not only the capacitance current but also the inductance current flow past the Peterson coil. Thus, the total current flow past the fault point is

$$\dot{I}_k = \dot{I}_L + \dot{I}_{C\Sigma} \tag{9.11}$$

where $\dot{I}_{C\Sigma}$ is the phase-to-ground capacitance current of the whole system, which can be calculated by Equation (9.7); and $\dot{I}_L$ is the inductance current generated by Peterson coil. $\dot{I}_L = \frac{-\dot{E}_A}{j\omega L}$, where $\dot{I}_L$ is the inductance of Peterson coil.

Because the phase relationship between $\dot{I}_{C\Sigma}$ and $\dot{I}_L$ is 180°, $\dot{I}_k$ will be decreased by the compensation of Peterson coil. Figure 9.5b shows the zero-sequence equivalent network of the system.

The arc suppression coil (Peterson coil) is categorized into three compensation types: complete compensation, undercompensation, and overcompensation.

1. *Complete compensation.* A system in which the inductance current is equal to the capacitance current is completely compensated. That is to say, the fault current is zero. Its disadvantage is that the system satisfies the series resonance. The zero voltage will generate great current in the series resonance circuit if the three-phase capacitances are different or the three-phase contactors of the breaker are closed asynchronously. This current will produce a great voltage drop across the arc suppression coil, which increases the voltage of the neutral of the system remarkably. The insulation of equipment will be damaged. In practice, therefore, the complete compensation type is not used.
2. *Undercompensation.* A system in which the inductive current is slightly smaller than the capacitive current is called *undercompensated.* In such a system, the current after compensation is capacitive. The disadvantage of undercompensated systems is that when the operation mode of the system changes (e.g., one element is cut, or the breaker switches because of the fault), the capacitive current will decrease and be equal to that of the inductive current. In that case, the circuit is tuned. Undercompensation is therefore not used in practice.
3. *Overcompensation.* A system in which the inductive current is slightly larger than the capacitive earth fault current is called *overcompensated.* The residual current after compensation is inductive. The overcompensation cannot cause overvoltage of series resonance and is applied widely in practice. The overcompensation degree P is described as

$$P = \frac{\dot{I}_L - \dot{I}_{C\Sigma}}{\dot{I}_{C\Sigma}} \tag{9.12}$$

P is selected as 5~10%. After compensation, the fault current through the protection at the faulty line is inductive current. The phase relationship between this inductive dcurrent and zero-sequence voltage is the same as the phase relationship at the wholesome line. The amplitude of this inductive current is close to that of the capacitive current of the wholesome line. The commonly used faulty line selection principles, such as the zero-sequence current-based principle and zero-sequence power direction-based principle, are not applicable in neutral indirectly grounded systems.

9.1.2 Transient analysis for ground fault

When there is a single phase-to-ground fault in a neutral grounded system via Peterson coil, the transient grounding current at the fault point consists of the transient capacitive current and the transient inductive current. The frequency and amplitude of these two currents are significantly different, so these two currents cannot compensate each other. In that case, the concepts of residual current and the detuning are not applicable. In order

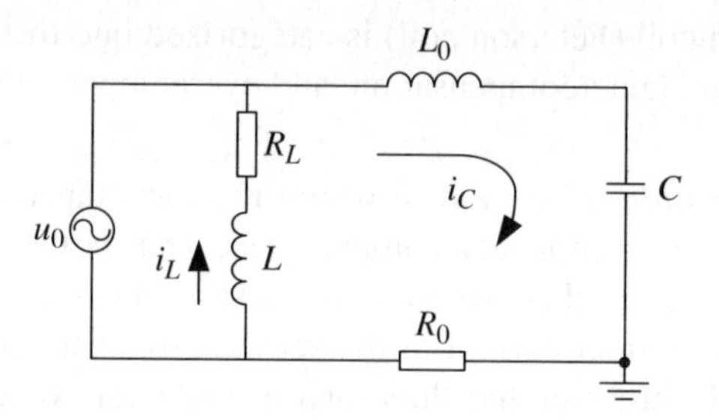

Figure 9.6 The equivalent circuit for calculating the single phase-to-ground transient current

to analyze the transient process of ground faults, we first establish the equivalent circuit of single phase-to-ground faults.

Figure 9.6 shows the equivalent circuit for calculating the single phase-to-ground transient current. C is the three-phase-to-ground capacitance. L_0 is the equivalent inductance of a three-phase line and transformer in a zero-sequence circuit. R_0 is the equivalent resistance in a zero-sequence circuit, which consists of fault resistance and arc resistance. R_L and L are, respectively, the resistance and inductance of the arc suppression coil. u_0 is the zero-sequence source voltage. When a single phase-to-ground fault occurs, the transient current is

$$u_0 = U_m \sin(wt + \varphi) \tag{9.13}$$

$$\begin{aligned} i_d &= i_C + i_L \\ &= (I_{Cm} - I_{Lm})\cos(wt + \varphi) \\ &+ I_{Cm}\left(\frac{\omega_f}{\omega}\sin\varphi\sin\omega t - \cos\varphi\cos\omega_f t\right)e^{-t/\tau_C} + I_{Lm}\cos\varphi e^{-t/\tau_L} \end{aligned} \tag{9.14}$$

where $\begin{cases} I_{Cm} = U_m\omega C \\ I_{Lm} = U_m / \omega L \end{cases}$, ω_f is the free oscillation angular frequency of the free oscillation current; ω is the power–frequency angular frequency; φ is the initial phase of the zero-sequence voltage; and τ_C and τ_L are, respectively, the time constant of the inductance loop and capacitance loop.

The first part of Equation (9.14) is the steady component of the ground current, which is equal to the difference of steady capacitance current and steady inductance current. The other part of Equation (9.14) is the transient component of the ground current, which is the sum of the transient components of capacitance current and that of inductance current. To prevent the oscillation, the distribution network is overcompensated. The distribution network is an aperiodic oscillation loop, in which $R_0 \geq 2\sqrt{L_0 / C}$. The transient capacitance current is

$$i_{\text{C.os}} = \frac{u_0}{L_0\omega_0} e^{-\delta t} \sin\omega_0 t \tag{9.15}$$

where ω_0 is the free oscillation angular frequency.

The transient component of the ground current is

$$i_{\mathrm{d.os}} = i_{\mathrm{C.os}} + i_{\mathrm{L.os}} = \frac{u_0}{L_0\omega_0} e^{-\delta t} \sin \omega_0 t + I_{Lm} \cos\varphi e^{-t/\tau_L} \tag{9.16}$$

In normal conditions, ω_0/ω is great. The frequency difference of $i_{\mathrm{C.os}}$ and $i_{\mathrm{L.os}}$ is also great. Therefore, at the initial stage of ground fault, the main characteristics of transient components are determined by the transient capacitance current.

Because the transient process is very short, the influence of source voltage change on the transient capacitance current is negligible. When $R_0 \approx 2\sqrt{L_0/C}$, the free oscillation component is

$$i_{\mathrm{C.os}} \approx \left(u_0 / L_0\right) t e^{-\delta t} \tag{9.17}$$

After derivation to Equation 9.17, we obtain that the maximum of free oscillation components appears at $t = 1/\delta = 2L_0/R_0$. The maximum is

$$i_{\mathrm{C.os\,max}} \approx 2u_0 e^{-1} / R_0 \tag{9.18}$$

From the analysis given here, we can conclude that once the single phase-to-ground fault occurs, the transient capacitance current with fast decay and the transient inductance current with slow decay will flow through the fault point. Whether the neutral point is resonant or indirectly grounded, the amplitude and frequency of the transient ground current are mainly determined by the transient capacitance current. The maximum of the transient current has a phase relationship with the first half wave of the zero-sequence voltage, which can be used to select the faulty line. For radial distribution, the polarity between the transient current and the zero-sequence voltage is opposite for the faulty line but the same for the healthy line. Although the amplitude of the transient current is great, the transient current exists for a relatively short time. It is about 0.5~1 of a power–frequency cycle. As for the inductance current in a transient process, the initial value of a direct current (DC) component in the inductance current is relative to the initial phase angle and core saturation. In order to balance this DC component, the ground current generates a DC component that is equal in amplitude and opposite in direction to the DC component of the inductance current. The generated DC component increases the amplitude of the transient ground current.

9.2 The application of wavelet analysis in faulty line selection in small current grounded systems

9.2.1 Fault simulation

The fault in small current grounded systems is a single phase-to-ground fault. The single phase-to-ground fault is affected by many factors, such as the fault inception angle, fault location, fault resistance, and fault line length [2].

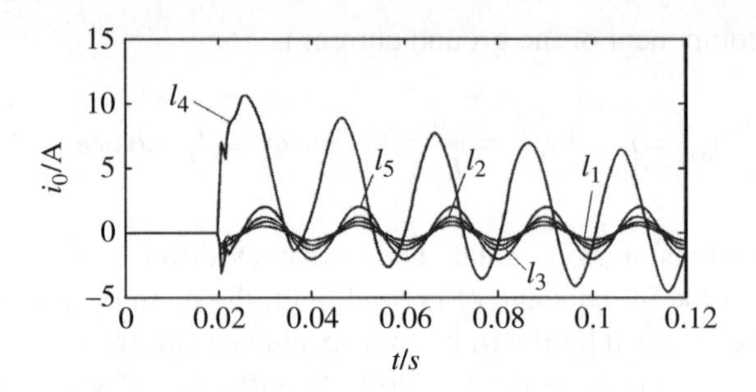

Figure 9.7 Zero-sequence current of each line

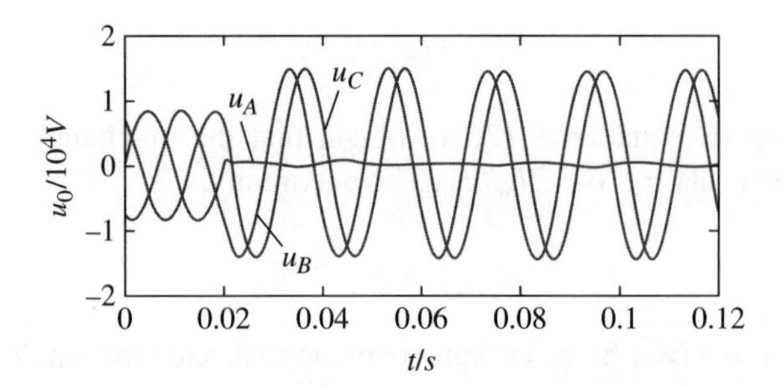

Figure 9.8 Three phase voltages of a busbar

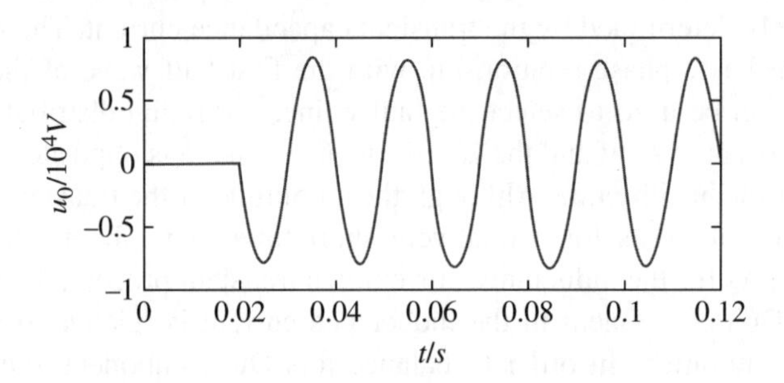

Figure 9.9 Zero-sequence voltage of a busbar

The simulation model appears in Figure A.5 in the Appendix. The single phase-to-ground fault occurred in line L_4. The fault is 60 km away from the busbar with the fault of phase A. The fault resistance is 20 Ω. Figure 9.7 shows the zero-sequence current of each line. Figure 9.8 and Figure 9.9 show, respectively, the three phase voltages of the busbar and the zero-sequence voltage of the busbar.

As can be seen in Figure 9.7, the zero-sequence current of the faulty line is far from the time axis. When there is a voltage zero-crossing fault, the high-frequency oscillation component of the faulty line and the healthy line is small. However, the inductive damped DC component of the faulty line is great. This DC component flows through the faulty line and arc suppression coil. Figures 9.8 and 9.9 indicate that the three phase voltages are symmetrical and the zero-sequence voltage is zero under the normal operating condition. After a single phase-to-ground fault occurred, the phase-to-ground voltage at the fault point drops to a small magnitude, whereas the healthy phases are at potentials equal to those of the phase-to-phase voltages. The zero-sequence voltage increases to the phase voltage.

9.2.2 *Faulty line selection based on wavelet packet decomposition*

9.2.2.1 Faulty line selection principle

The transient current obtained by a zero-sequence current transformer or zero-sequence current filter is decomposed by a wavelet packet. The basic principle of a wavelet packet is to implement high- and low-pass filtering to divide the signal into different frequency bands. The sampling interval increases by double, whereas the data samples decrease 50% after one filtering. The sampling sequence of a transient signal is decomposed with proper bandwidth to obtain different frequency band signals. The energy of each frequency band is calculated by Equation (9.19):

$$\varepsilon = \sum_{n} \left[w_k^{(j)}(n) \right]^2 \tag{9.19}$$

where $w_k^{(j)}(n)$ is the coefficient of subband (j, k). After multiresolution analysis of a zero-sequence current, the lowest frequency band (0, 4) that contains a power–frequency component is removed. The high-frequency band with maximum energy is selected as the characteristic band. The characteristic band contains the main features of transient capacitance current, and then the faulty line can be selected by the phase relationship between $i_{\mathrm{C.os\,max}}$ of the transient capacitance current and the first half wave of the zero-sequence voltage [3].

9.2.2.2 Simulation study

The free oscillation frequency of a transient capacitive current is generally 300–1500 Hz for an overhead line and 1500–3000 Hz for a cable line. Considering this, we adopt 6400 Hz as the sampling frequency. According to a sampling theorem, the accurate frequency that can be extracted accurately is up to 3200 Hz, which satisfies the requirement of free oscillation frequency. In order to eliminate the spectrum leakage and spectrum overlapping, the wavelet function should have a good frequency property. Here, we select a db10 wavelet as the mother wavelet. The decomposition level is 4. Thus, the subband width is 200 Hz.

The faulty line selection steps are described as follows. First, take wavelet packet decomposition to the zero-sequence current of one cycle before fault and four cycles after fault. Second, determine the characteristic frequency band according to the maximum energy. Finally, compare the polarity of the modulus maximum of the zero-sequence current with the polarity of zero-sequence voltage to determine the faulty line.

Example A single phase-to-ground fault occurred at 60 km away from the bus on line L4. The fault occurred near the maximum of A-phase voltage, and the fault resistance is 100 Ω.

Figure 9.10 shows the wavelet packet decomposition results of current and the zero-sequence voltage. As can be seen in Figure 9.10, the modulus maxima in the characteristic band corresponding to lines L_1, L_2, L_3, L_4, and L_5 are, respectively, −1.8, −4.22, −1.44, 10.92, and −5.98. Obviously, line L_4 has the greatest modulus maximum, and its polarity is different from that of other lines. The first half wave of the zero-sequence voltage has the negative polarity, which is the same as the healthy lines L_1, L_2, L_3, and L_5 but different from the faulty line L_4.

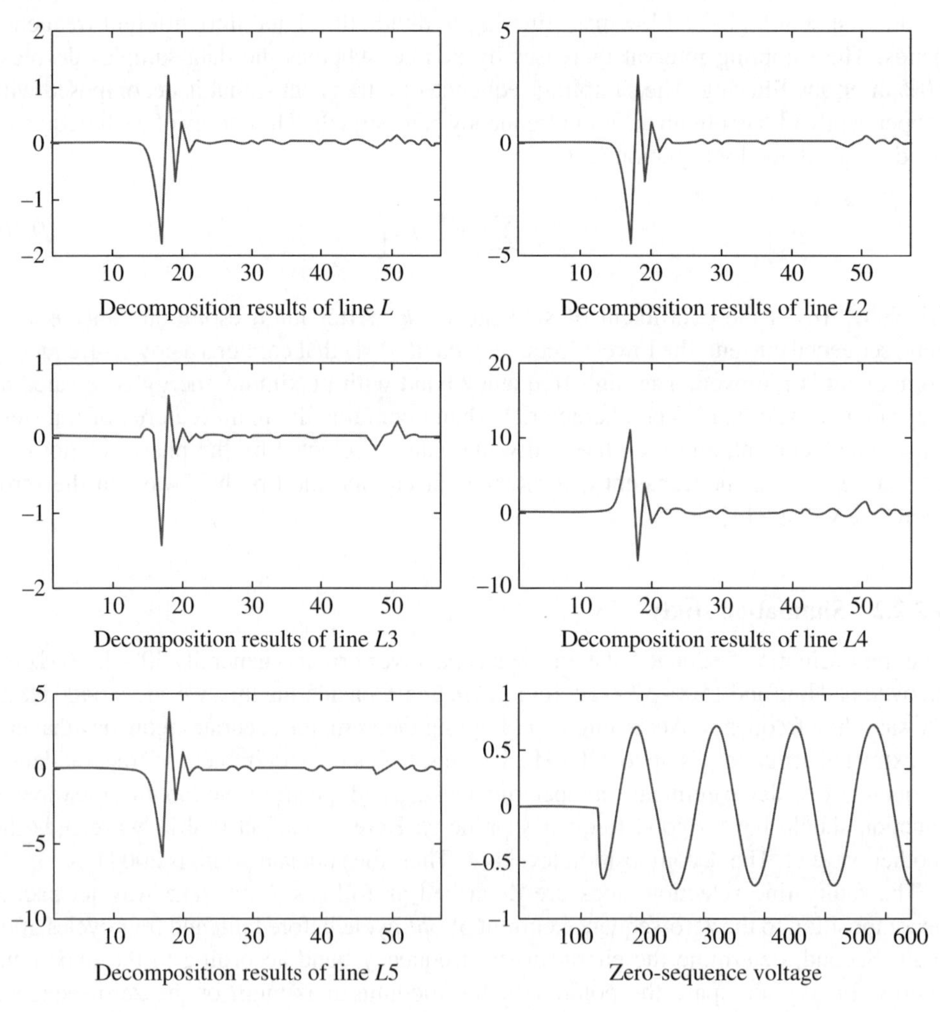

Figure 9.10 Current decomposition results of a wavelet packet and the zero-sequence voltage

From the simulation study, we know that the following remarkable features of the faulty line and the healthy line can be used to achieve the faulty line selection:

1. The modulus maximum of the faulty line is greater than that of the healthy lines.
2. The modulus maximum of the faulty line has the opposite polarity to the first half wave of the zero-sequence voltage. The modulus maximum of the healthy line has the same polarity as the first half wave of the zero-sequence voltage.
3. When a fault occurred at the busbar, the modulus maxima of all the lines have the same polarity as the first half wave of the zero-sequence voltage.

Wavelet packet decomposition based on the faulty line selection method analyzes zero-sequence current and distinguishes the faulty line by the polarity comparison. This method has good anti-interference ability. The selection results are reliable, accurate, and insensitive to grounding type and fault resistance, as well as the voltage zero-crossing fault or any fault occurring near the maximum of the phase voltage.

9.2.3 Faulty line selection method based on the current traveling wave modulus maximum

9.2.3.1 Faulty line selection principle

When a single phase-to-ground fault occurs in one line, the traveling waves generated by the fault will propagate from the fault point to the distribution network. These traveling waves will be reflected and refracted when they arrive at the impedance discontinuities. The refracted wave will propagate along the healthy line. Assume that there are N lines connected to the busbar. If a single phase-to-ground fault occurred at line 4, the propagation path of the fault-generated traveling wave is simply shown in Figure 9.11. The initial

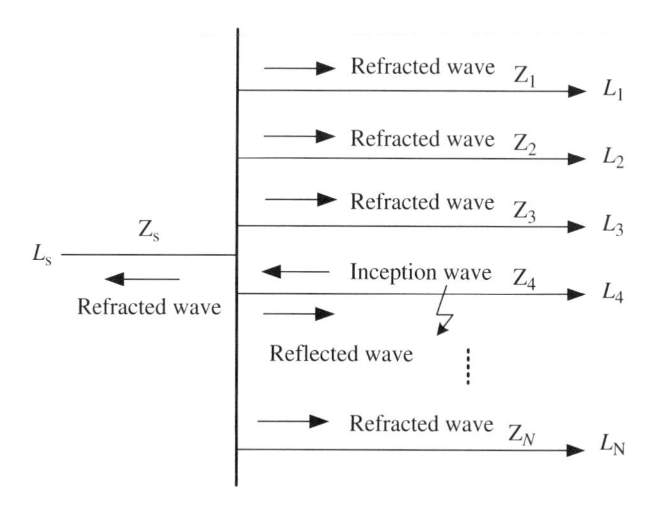

Figure 9.11 The propagation of the initial traveling wave

traveling wave of the faulty phase on the faulty line L_4 is the superimposition of the inception wave and reflected wave. The initial traveling wave of the faulty phase of line L_4 will generate the initial traveling wave of the healthy phase of line L_4 because of coupling. The traveling waves of line L_4 produce the initial traveling waves of other lines. From the analysis given in this chapter, the initial traveling waves of the faulty line and the healthy lines have remarkable differences. The amplitudes and polarities of the initial waves of each line can be represented by wavelet modulus maxima. The faulty line selection criterion is established by the amplitude difference and polarity difference of the initial waves of each line [4].

The three-phase lines have significant electromagnetic coupling between the conductors, which makes it difficult to analyze the fault feature and extract the fault information. Using a modal transformation, the three-phase system can be decoupled into several independent modes. In terms of the characteristics of three-phase lines and the current transformer installation, different faulty line selection methods can be obtained based on the traveling wave modes.

9.2.3.2 Faulty line selection method based on the β mode and γ mode currents

Only the currents of phase A and phase C can be measured when the phase A and phase C current transformer are installed. After a phase-to-modal transformation, we can get β mode and γ mode currents according to Equation (9.20):

$$\begin{cases} I_\beta = \dfrac{I_A - I_C}{3} \\ I_\gamma = \dfrac{I_A + I_C}{3} \end{cases} \tag{9.20}$$

When phase A–to-ground or phase C–to-ground fault occurs, the traveling waves of other phases are generated by coupling with the opposite polarity. The β mode current is therefore always greater than the γ mode current. The traveling waves of the healthy lines originate from the reflected wave or refracted wave of the faulty line. The modulus maxima of the faulty line are always greater than, and opposite in polarity to, those of the healthy lines. The faulty line can be selected by comparing the amplitude and polarity of the β mode current. When phase B–to-ground fault occurs, the traveling waves of phase A and phase C are coupled waves with the same polarity. The γ mode current is always greater than the β mode current. Other features are the same as phase A– or phase C–to-ground fault. When the fault occurs on the busbar, all the mode currents have the same polarity. In conclusion, we find the faulty line selection method based on the amplitudes and polarities of β mode and γ mode currents. The mode currents with greater modulus maxima of each line are employed to identify the faulty line by comparing the polarities.

9.2.3.3 Faulty line selection method based on zero-sequence current

If the distribution network only installs the zero-sequence current transformer, the zero-sequence current can be obtained as

$$I_0 = \frac{I_A + I_B + I_C}{3} \tag{9.21}$$

The zero-sequence traveling wave contains a three-phase current. The traveling wave of the faulty phase has the greatest amplitude. The coupled wave and the refracted wave have smaller amplitudes and opposite polarities. Therefore, the zero-sequence current modulus maxima of the faulty line are greater than those of the healthy line, and their polarities are inverse. For busbar fault, all the mode currents have the same polarity. From the analysis given in this chapter, we find the faulty line selection method based on the amplitudes and polarities of the zero-sequence current. The faulty line can be selected by comparing the polarities of the zero-sequence currents.

9.2.3.4 Simulation analysis

The simulation system is shown as Figure A.5 in the Appendix. The zero-sequence current transformer is installed in the system. The simulation analysis is carried out for the current traveling wave under the single phase-to-ground fault.

The single phase-to-ground fault occurred 6 km away from the busbar in line L4. The fault resistance is 10 Ω. The sampling frequency is 1 MHz. Figure 9.12a shows the zero-sequence currents of each line. Wavelet transforms are processed to the zero-sequence currents using the db3 wavelet. Figure 9.12b shows the modulus maxima of zero-sequence currents at scale 2. As seen in Figure 9.12, the modulus maximum of the zero-sequence currents of line L_4 is greater than that of other lines, and it has the opposite polarity to other modulus maxima. Therefore, line L_4 is identified as the faulty line.

The transient traveling wave can effectively exclude the influence of the unbalanced electrical quantities under the system's normal operation. The faulty line selection using a traveling wave and wavelet transform can improve significantly the faulty line selection performance. It can also provide a new idea or starting point for the study or application of faulty line selection.

This chapter introduced the faulty line selection in small current grounded systems based on wavelet packets and current traveling waves. The faulty line selection method based on wavelet packets takes the frequency band that has the greatest energy as the characteristic frequency band. Then, the faulty line is discriminated by comparing the polarities of the zero-sequence current in the characteristic band and the zero-sequence voltage. The current traveling wave is a transient fault component that contains abundant fault information. The current traveling wave is immune to the unbalanced electrical quantities under the normal operating conditions and the small phase-to-ground capacitance current. The wavelet modulus maxima are utilized to describe the amplitude, polarity, and time property of the traveling wave. The wavelet modulus

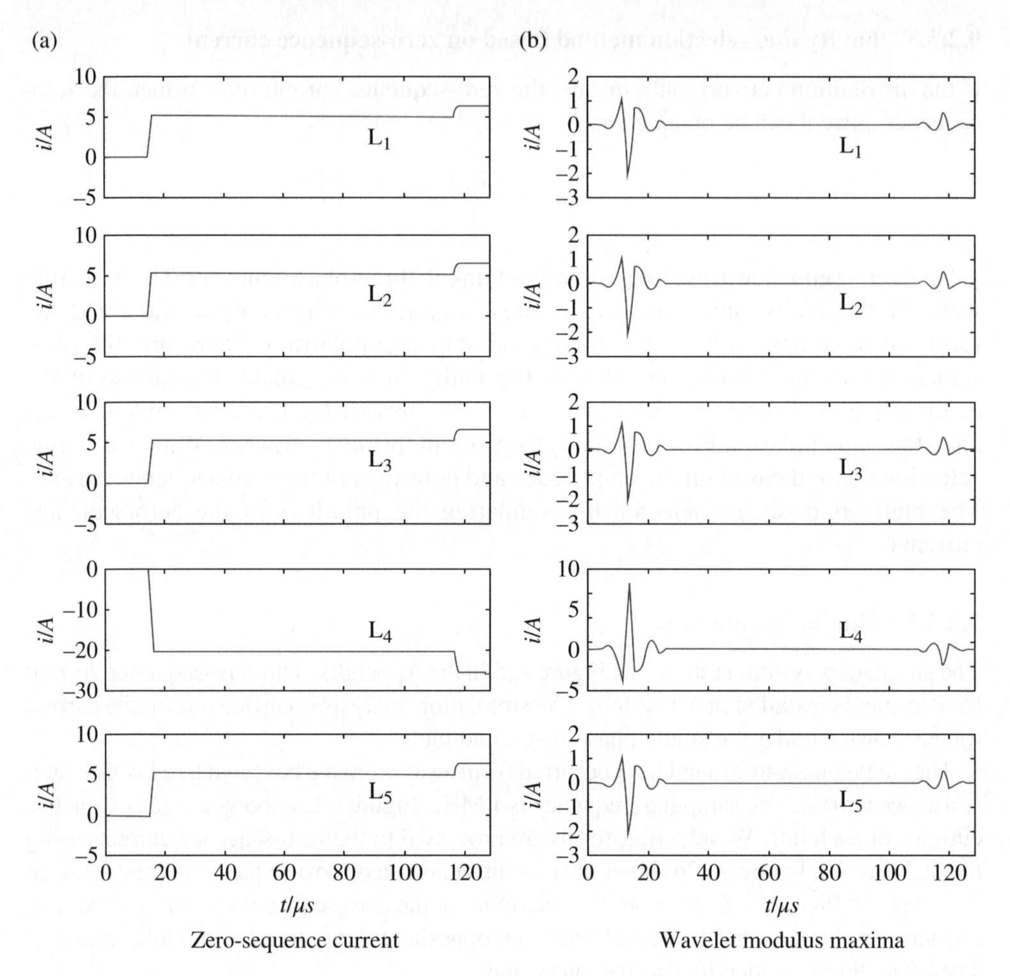

Figure 9.12 The zero-sequence current of each line and their wavelet modulus maxima under the single phase-to-ground fault in line L4

maxima can extract the fault information accurately. The faulty line selection method based on current traveling waves thus can be achieved according to the configuration of current transformers in practice.

References

[1] Shu H.C., The power distribution network fault line selection. *Proceedings of the CSEE*, vol. 24, no. 6, pp. 54–58, 2004.

[2] Yao H.N., Cao M.Y., Power system resonance grounding. *China Electric Power Press*, Beijing, 2008.

[3] Wang Y.N., Huo B.L., Wang H., et al., A new criterion for earth fault line selection based on wavelet packets in small current neutral grounding system. *Proceedings of the CSEE*, vol. 24, no. 6, pp. 54–58, 2004.

[4] Kong R.Z., Dong X.Z., Bi, J.G., Test of fault line selector based on current traveling wave. *Automation of Electric Power Systems*, vol. 30, no. 5, pp. 63–67, 2006.

10

Application of Wavelet Transform to Non-unit Transient Protection

Regular relay protection usually uses power–frequency components to develop the fault detection criterion and blocking criterion. However, the fault-generated power–frequency components are associated with the non power–frequency components. Hence, the relay can only extract the power–frequency components by filtering and delay, which makes it difficult to achieve high-speed operations. The fault-generated high-frequency transient signals have more information than power–frequency signals, such as fault type, fault direction, fault position, and fault duration. This information exists over the whole frequency domain. It can well solve the contradiction by developing a transient protection technique using the high-frequency components. Traveling wave–based protection is a typical transient protection, and its products have already been used in power systems. However, the traveling wave–based protection techniques suffer from faults at zero crossings on the voltage signal due to the fact that, at this point, the generated traveling waves are zero. Considering the arcing faults in high-voltage and extra-high-voltage (EHV) transmission lines, Johns proposed a non-unit high-frequency protection, which utilizes high-frequency voltage signals in two specified bands captured by the line traps and a tuned circuit connected to a conventional capacitor voltage transform (CVT) [1]. This proposed non-unit protection technique overcomes the limitations of the traveling wave–based protection techniques. Thereafter, Z.Q. Bo proposed a new noncommunication protection technique for transmission lines [2]. The technique utilized the high-frequency current signals captured by a multichannel filter unit. In addition to no requirement of a communication link, the technique also avoided the complexity and high costs caused by line traps and the tuned circuit. Therefore, the non-unit protection technique has good application prospects for developing extra-large-scale and extra-high-voltage transmission lines. The non-unit protection technique has also been one of the hot topics in relay protection for quite a long time.

Nevertheless, the key to the implementation of transient protection is the transient features extraction and the establishment of protection mechanisms. The fundamental reason

Wavelet Analysis and Transient Signal Processing Applications for Power Systems, First Edition. Zhengyou He.

for the poor performance of traveling wave–based protection is that the fault information-processing methods fall behind. For example, the digital filter extracts the transient features with the fixed filter bandwidth, which is not adjustable to changes in fault instant, fault position, fault type, and fault resistances in the practical system. The wavelet transform, developed in the 1990s, has a time–frequency window that automatically adapts to give an appropriate resolution. Wavelet transform provides distinctive advantages in processing nonstationary transient signals because it can localize in the time and frequency domains at the same time. Meanwhile, wavelet transform has fast algorithms, which is similar to FFT. Therefore, wavelet transform is expected to apply well in the transient protection field.

10.1 Principle of transient protection for EHV transmission lines

10.1.1 Basic configuration

The basic configuration of transient protection is shown in Figure 10.1. The current transformer (CT) installed on the line near the busbar transfers the current of the line. Then the current is sent to a microprocessor to be analyzed with wavelet transform after filtering and through an analog-to-digital (A/D) converter. In order to cover all fault types in practice, the modal computation circuit receives the current signals from an A/D converter and combines the three-phase currents into modal currents (e.g., $I_m = I_a + I_c - 2I_b$, or zero-sequence current). The wavelet transform results are then processed by the appropriate postprocessing method such as spectrum energy calculation. From here, the protection criterion is established. The tripping signal will be sent if the protection criterion is satisfied.

10.1.2 Basic principle of transient protection for EHV transmission lines

Figure A.3 in the Appendix is used to describe the basic principle of transient protection. When a fault occurs on a transmission line, current traveling waves are generated and propagate toward the two sides of the line until they meet a discontinuity point, such as the fault point or busbar. At this point, part of the waves pass into the adjacent section, called refraction, and the rest are reflected backward, called reflection. Hence, the

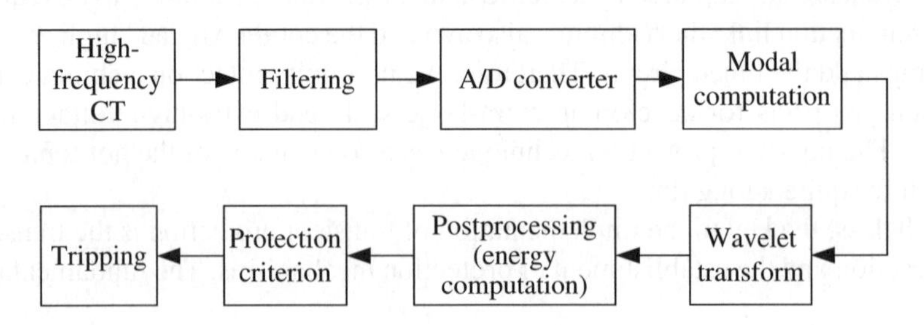

Figure 10.1 Basic configuration of transient protection based on wavelet transform. A/D, analog-to-digital; CT, current transformer

current wave, particularly in the high-frequency range, will be reflected back and forth between the fault point and busbar. Research shows that the transient current will be attenuated due to the effect of busbar capacitance. This characteristic makes it possible to develop a non-unit protection technique utilizing the single-ended transient current wave of the line. As shown in Figure A.3, the protection is installed at busbar P and protects the line PM. Assume the fault point F_1 is in line PM and fault point F_2 is in line MN. When an external fault occurred at F_2, as shown in Figure A.3, the transient current I_1 will be generated toward busbar M. When signal I_1 reaches busbar M, a significant portion of signal I_1 will travel into line PM, while some of the signal I_0 will be shunted to earth through the busbar. As a result, the current detected by CT is $I = I_1 - I_0$. That is to say, the current detected by CT will be attenuated in comparison with the initial current I_1. However, for the fault occurring at F_1, there is no attenuation in the current detected by CT. Therefore, the currents detected by CT have different frequency components for faults at point F_1 and point F_2. Thus, we can establish the protection criterion based on this difference.

10.2 Transient protection based on wavelet transform for EHV transmission lines

The frequency range of transient current is influenced by many factors, such as transmission line parameters, line structures, fault instants, fault positions, fault distances, fault types, and so on. Therefore, it is insufficient to extract the high-frequency components using the filter with the fixed center frequencies. As mentioned in this chapter, wavelet transform has the adaptive ability and is appropriate for extracting various bands of fault current. After wavelet transform, the spectrum energies are obtained by calculating the sum of the square of wavelet coefficients over a data window. Let E_{j_1} and E_{j2} be the energies corresponding to scale 2^{j_1} and 2^{j_2}:

$$\begin{cases} E_{j_1}(n\Delta T) = \sum\limits_{k=n-M}^{n} \left|WI_m(j_1, k\Delta T)\right|^2 * \Delta T \\ E_{j2}(n\Delta T) = \sum\limits_{k=n-M}^{n} \left|WI_m(j_2, k\Delta T)\right|^2 * \Delta T \end{cases} \tag{10.1}$$

$$\text{Ratio} = \frac{E_{j_1}(n\Delta T)}{E_{j_2}(n\Delta T)} K \tag{10.2}$$

where ΔT is the time step length; K is the attenuation factor; and M is the number of samples in the window. If $j_2 > j_1$, E_{j_1} represents the energy of high-frequency components at scale j_1, whereas E_{j_2} represents the energy of low-frequency components at scale j_2. E_{j_1} and E_{j_2} will change abruptly after $M\Delta T$ delay of the fault current wave arriving at CT. From the aforementioned principle, it is obvious that the ratio is greater for internal faults than for external faults. A threshold should be set to distinguish between the internal faults and external faults. To improve the anti-interference ability, the internal fault is identified only if the ratio is greater than the threshold value for a preset time threshold. The fault detection criterion and the protection tripping criterion are as follows [3].

Fault detection criterion:

$$\left\{E_{j_1}\left(n\Delta T\right)\geq E_{1set}\right\}\ and\ \left\{E_{j_2}\left(n\Delta T\right)\geq E_{2set}\right\} \tag{10.3}$$

Protection tripping criterion:

$$\text{delay}\left(\text{Ratio}\geq 1\right)>x \tag{10.4}$$

where delay means the duration time of keeping Ratio ≥ 1.

In comparison with other protection principles, this protection principle makes full use of wavelet transform and the high-frequency current from CT. There is no requirement for communication devices and the delay time is short, which makes it capable to operate fast with high sensitivity and flexibility.

10.2.1 The choice of wavelet basis

Most of the transient signals have a nonlinear phase, so we can extend the requirement of wavelet symmetry. The orthogonal wavelet is chosen because the proposed EHV line protection criterion is based on the energy detection method after wavelet transform. Meanwhile, the wavelet with strict frequency division must be selected in order to reduce the frequency aliasing as far as possible. Moreover, real-time protection requires the wavelet to have a compact support. Taking the factors discussed here into account, the db10 wavelet is chosen as the wavelet base because it is orthogonal, compactly supported, and more regular, and has a good filtering effect.

10.2.2 Relay response of typical internal and external faults

The sampling frequency is $f=200$ kHz. Point F1 and F2 are both 1 km away from the busbar *N*. The arcing fault is considered. The fault arc length is 10 m. Figures 10.2 and 10.3 show the relay responses to typical phase A–to-ground faults that occurred at points F1 and F2. The fault occurred at the maximum of the phase A voltage (t=15 ms).

It can be seen that the currents of the fault phase increase significantly and have abundant high-frequency transients. Due to the coupling effect, the wholesome phases B and C also have transients. However, the fault at F1 cannot be distinguished from the fault at F2, only from the fault currents. Figures 10.2b,d and 10.3b,d show the wavelet coefficients of the zero-sequence current at scale $j=5$ and $j=1$. According to multiresolution analysis, the wavelet coefficients stand for, respectively, the components of zero-sequence current in subbands 50~100 kHz and 3.125~6.25 kHz. The data window is chosen as 2.5 ms. The calculated energy curves are shown in Figures 10.2c,e and 10.3c,e. It can be seen

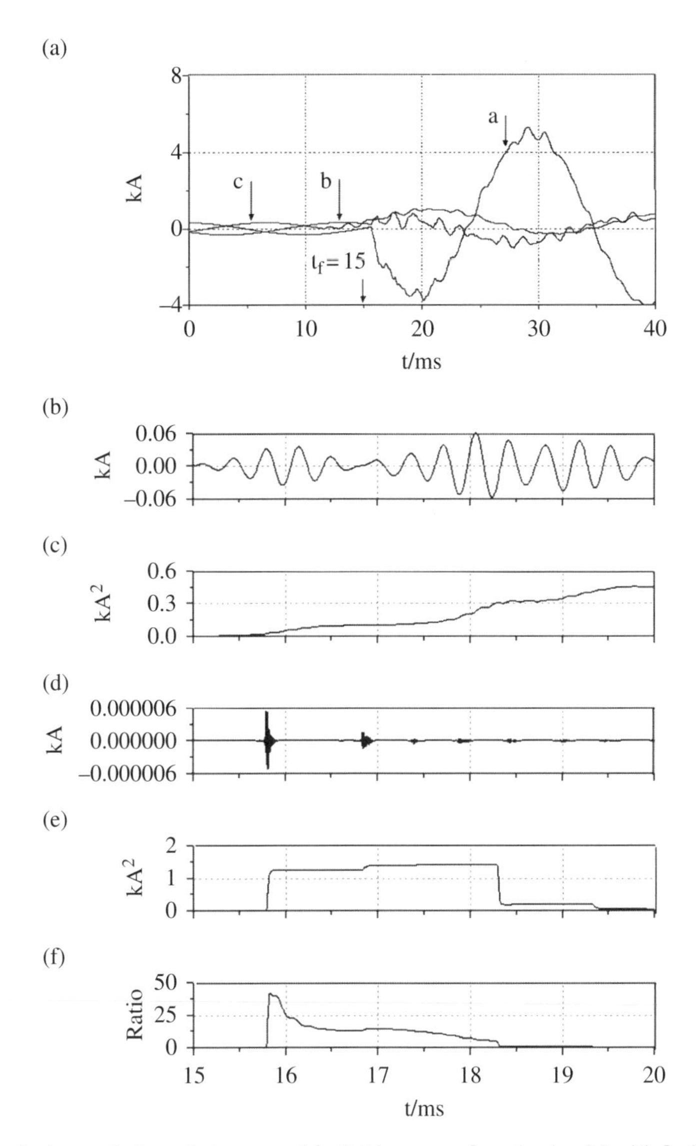

Figure 10.2 An internal phase A–to-ground fault 1 km away from busbar M with fault time t=15 ms

that the energies increase significantly or change abruptly. This is the basis of establishing the protection criterion. There is a delay between the time of abrupt change and the fault instant. This is because the current wave needs time to travel from the fault point to the CT. The travel time can be used to locate the fault point, which displays the superiority of the wavelet in the fault location. The ratio of energies of different bands are shown in Figures 10.2f and 10.3f. When an internal fault occurred at F1, the ratio in Figure 10.2f is substantially greater than 1. When an external fault occurred at F2, the ratio in Figure 10.3f is several orders of magnitude smaller than that of the internal fault and is substantially

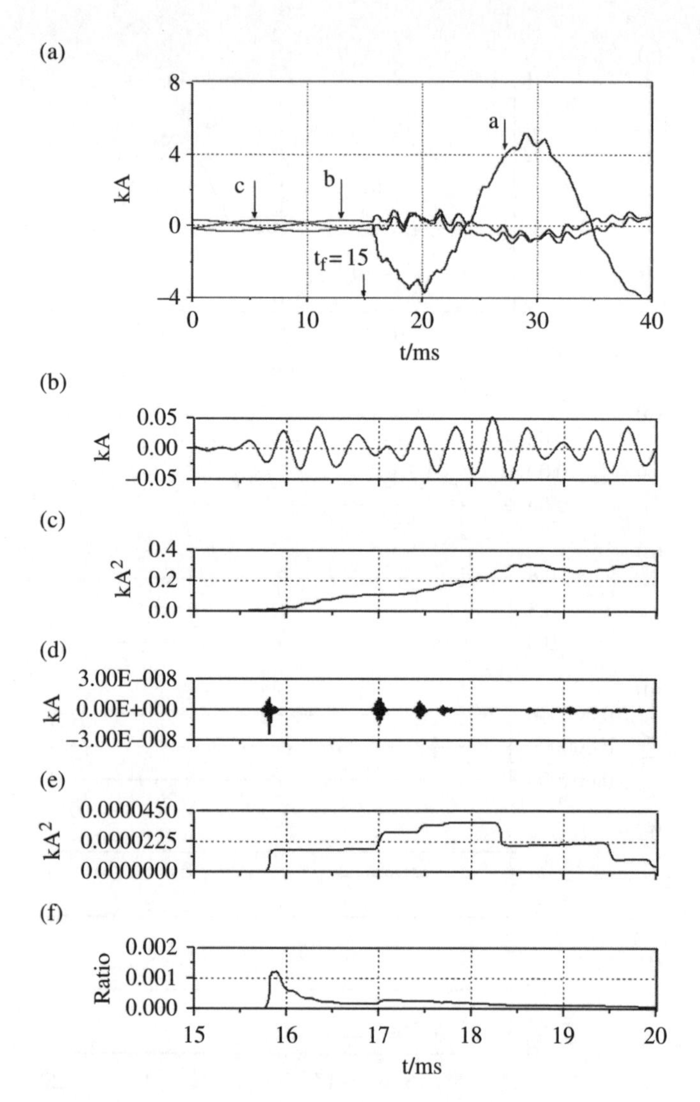

Figure 10.3 An external phase A–to-ground fault 1 km away from busbar M with fault time t=15 ms

smaller than 1. Thus, the criteria in Equations (10.3) and (10.4) can detect the fault and discriminate the internal faults from external faults.

1. *Effect of fault time.* Figures 10.4, 10.5, 10.6, and 10.7 show the simulation and analysis results of internal and external faults with fault inception angles of 0° (t=10ms) and 60° (t=13ms). In comparison with the results in Figures 10.2 and 10.3, the amplitudes of wavelet transform for voltage zero-crossing faults at scales j = 5 and j = 1 are minimum, and the energy is minimum as well. For the fault occurring with high-voltage

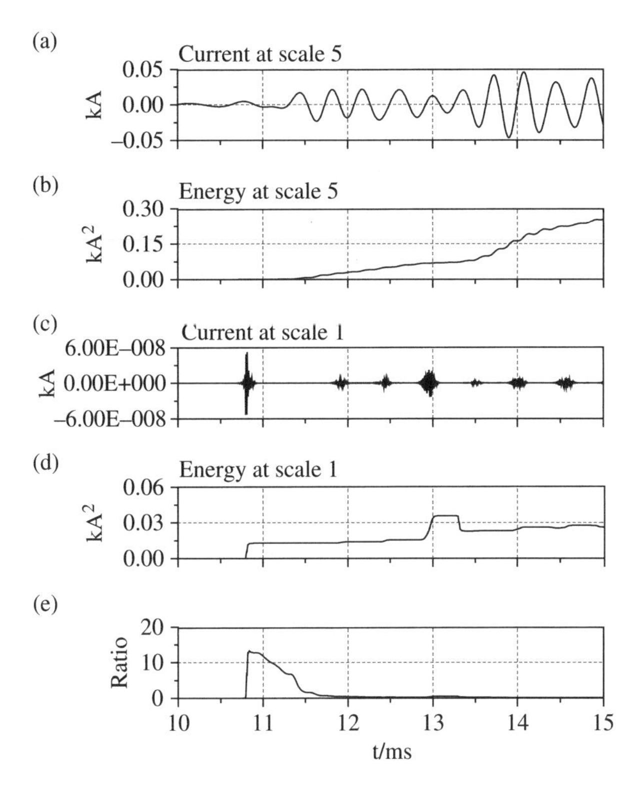

Figure 10.4 An internal phase A–to-ground fault 1 km away from busbar M with fault time t=10 ms

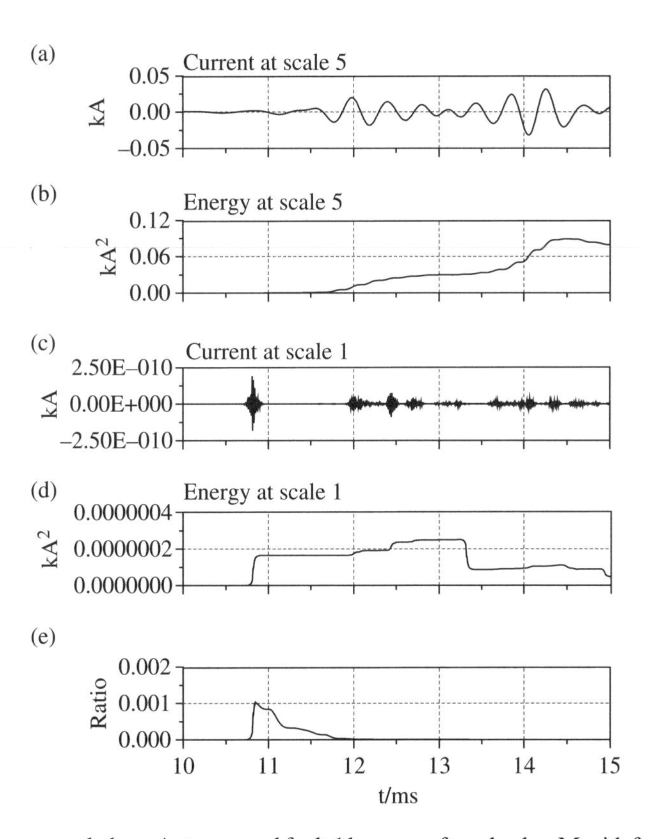

Figure 10.5 An external phase A–to-ground fault 1 km away from busbar M with fault time t=10 ms

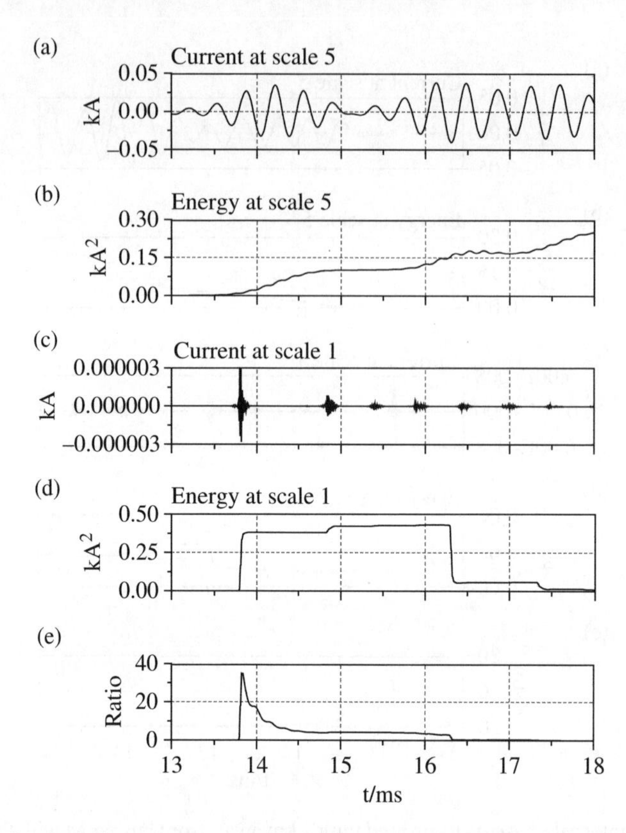

Figure 10.6 An internal phase A–to-ground fault 1 km away from busbar M with fault time t=13 ms

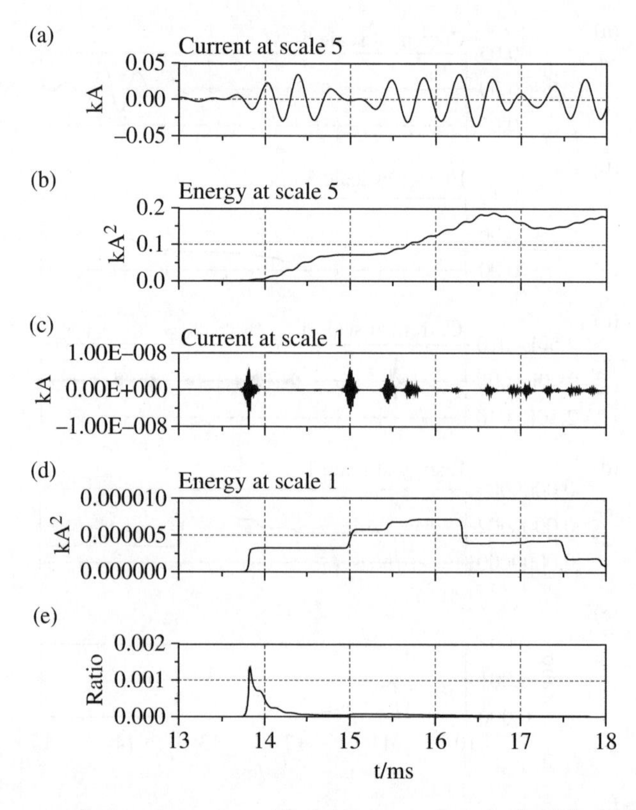

Figure 10.7 An external phase A–to-ground fault 1 km away from busbar M with fault time t=13 ms

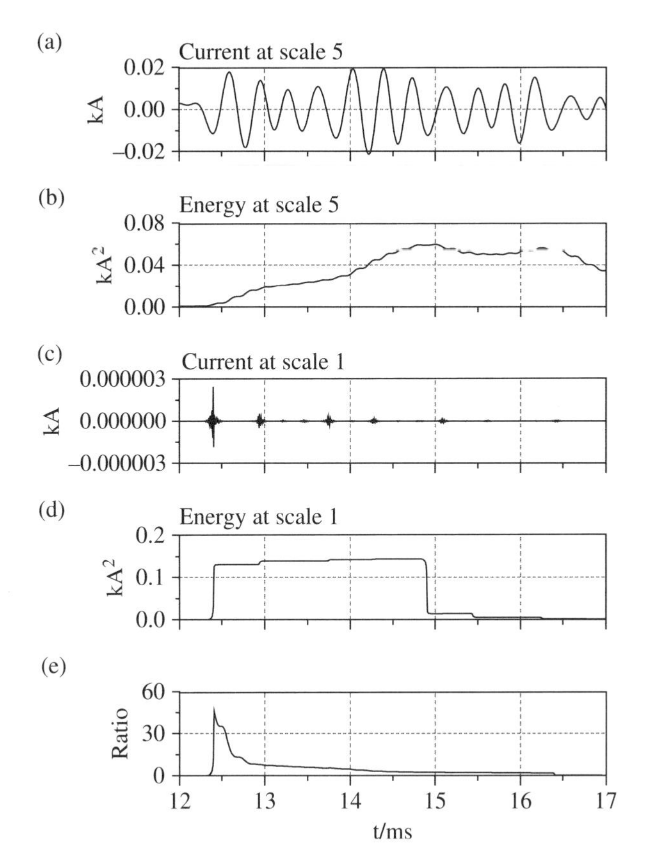

Figure 10.8 An internal phase A–to-ground fault 85 km away from busbar M with fault time t=10 ms

magnitude, the energy of the data window is great. However, the common feature of these faults is the increase or abrupt change of energy, which means the detection criterion is valid. The ratio for internal faults is greater than 1, whereas the ratio for external faults is smaller than 1. A large amount of simulation indicates that the protection does not suffer from the fault time.

2. *Effect of fault position.* In order to investigate the validity of protection criterion under different fault positions, the simulations are carried out under different fault positions. Figures 10.8 and 10.9 show the simulation results of faults occurring at the middle of lines PM and MN. Figure 10.10 shows the simulation results of a fault occurring at F1 500 m away from busbar P. The simulation results indicate that the protection is insensitive to the fault positions.
3. *Effect of fault types.* Figures 10.11 and 10.12 show the phase B–and phase C–to-ground fault with zero-crossing voltage occurring at F1 and F2. When the zero-sequence current is put into use, the protection criterion is effective for all asymmetric faults on transmission lines. For symmetric faults, namely, three-phase-to-ground fault, other modes of current can be employed.

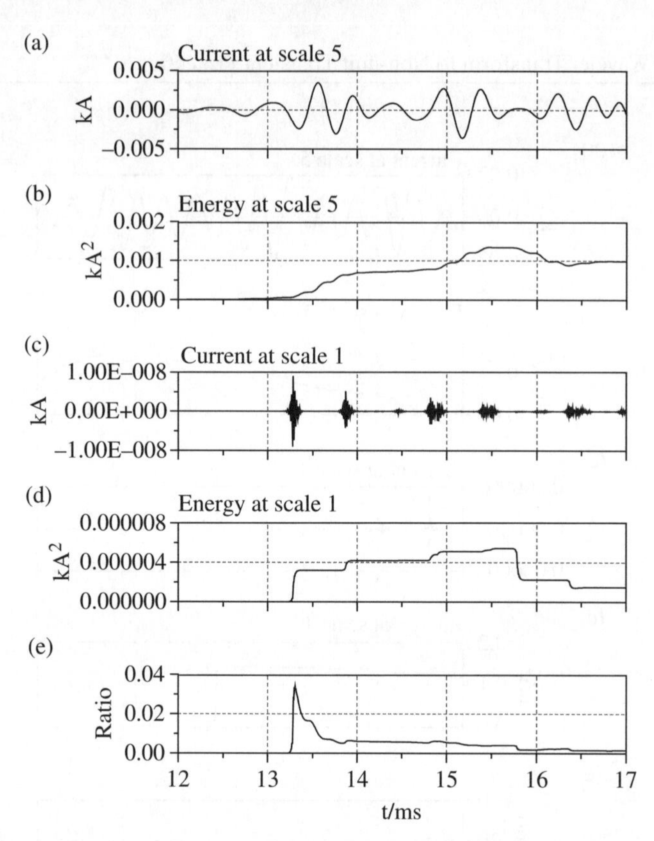

Figure 10.9 An external phase A–to-ground fault 95 km away from busbar M with fault time t=10 ms

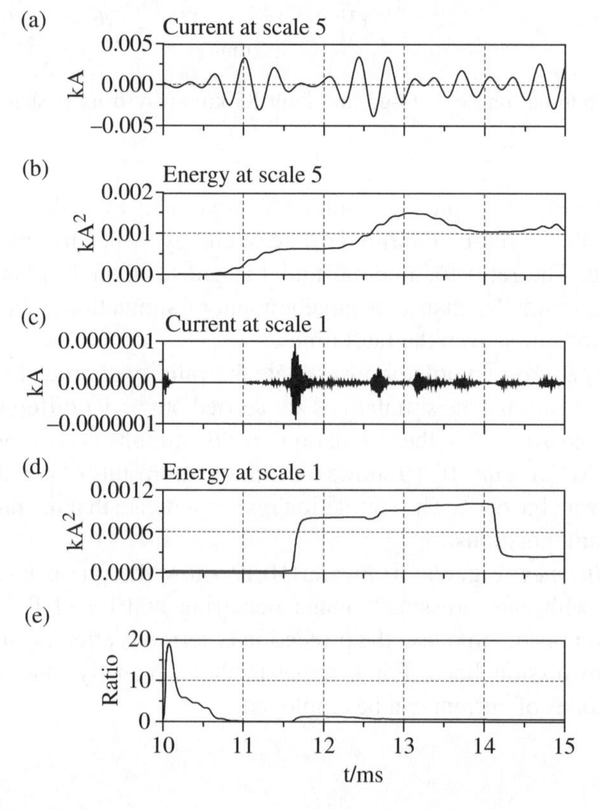

Figure 10.10 An internal phase A–to-ground fault 0.5 km away from busbar M with fault time t=10 ms

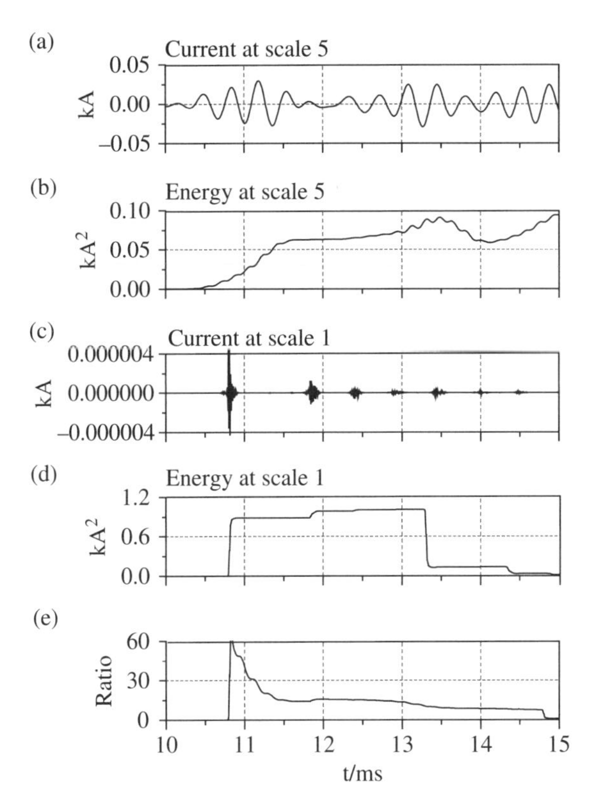

Figure 10.11 An internal phase B– to phase C–to-ground fault 1 km away from busbar M with fault time t=10 ms

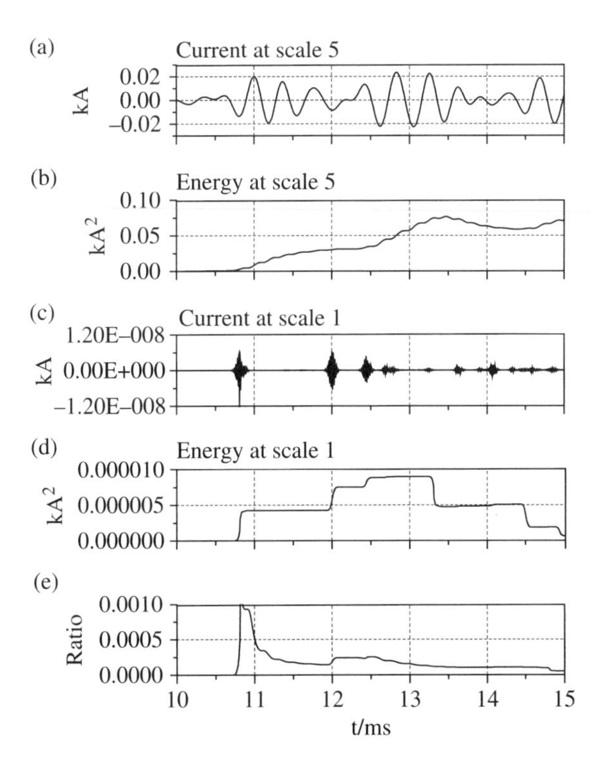

Figure 10.12 An internal phase B– to phase C–to-ground fault 1 km away from busbar M with fault time t=10 ms

4. *Effect of fault resistances.* One disadvantage of the traditional power–frequency protection principle is the difficult detection and identification of high-impedance faults. The transient protection based on wavelet transform is also valid for high-impedance faults. Figures 10.13 and 10.14 show the simulation results with a fault resistance of 200 Ω.
5. *Effect of arc length.* Figures 10.15 and 10.16 display the simulation results with an arc length of 15 m. Due to the increase of fault impedance, the wavelet coefficients and energy of the data window decrease. However, the protection can still operate reliably.
6. *Effect of sampling frequency.* The feasibility of the proposed protection is also discussed with a sampling frequency of 20 kHz, as shown in Figures 10.15 and 10.16. It should be noted that, at this point, the attenuation factor K is different. It can be seen that the protection principle still works with a low sampling frequency if we choose proper frequency bands and attenuation factors.

This chapter introduces a transient protection principle based on wavelet transform for EHV transmission lines by utilizing the transient current and the good properties of

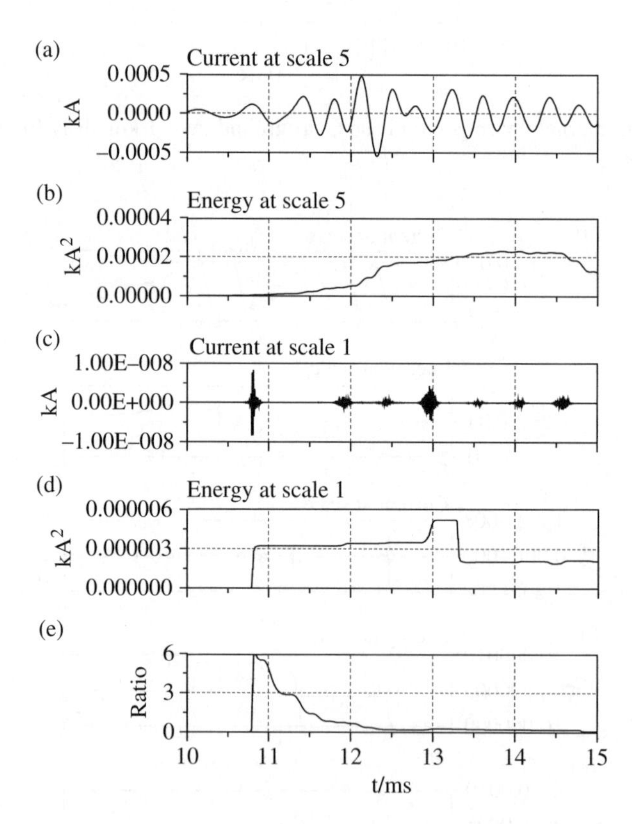

Figure 10.13 An internal phase A–to-ground high-impedance fault 1 km away from busbar M with fault time t=10 ms

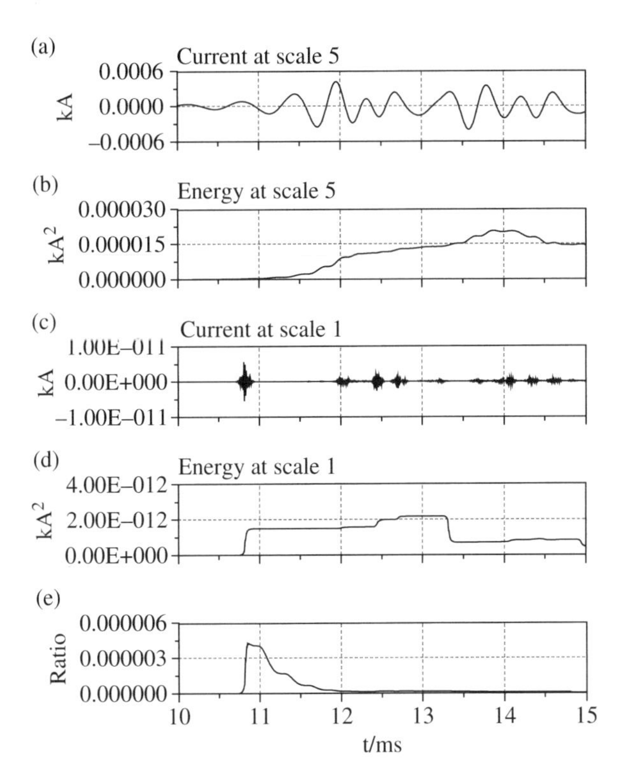

Figure 10.14 An external phase A–to-ground high-impedance fault 1 km away from busbar M with fault time $t=10\,\text{ms}$

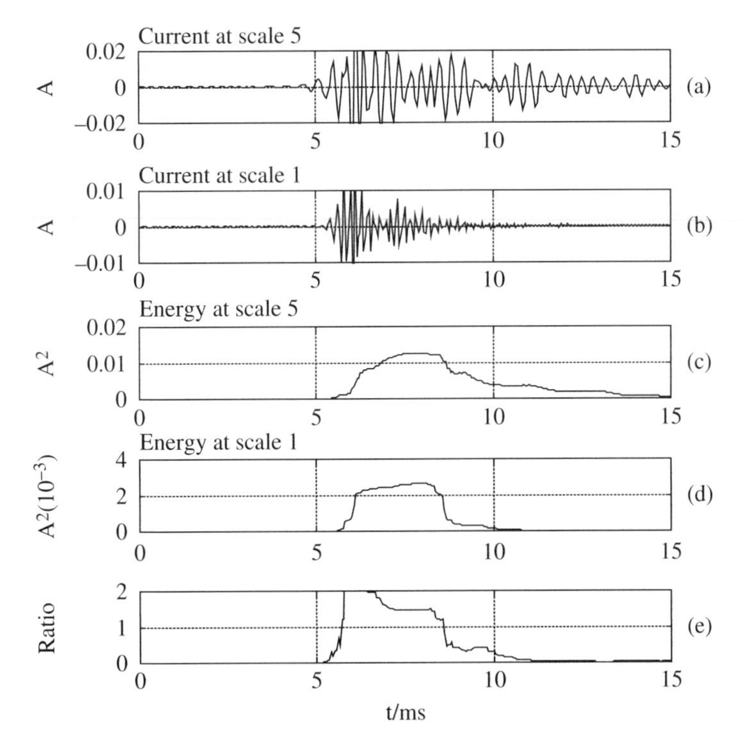

Figure 10.15 An internal phase A–to-ground fault at the middle of line PM with a sampling frequency of 20 kHz

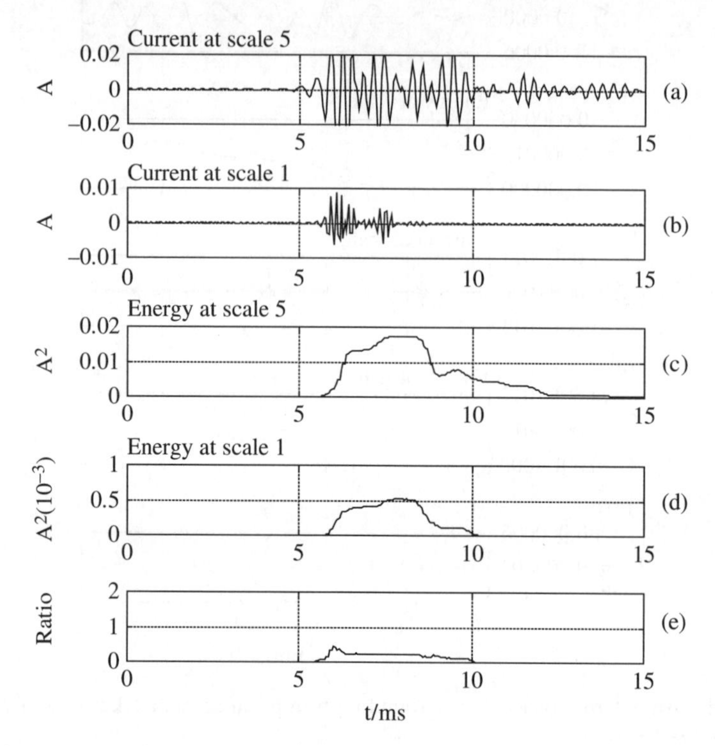

Figure 10.16 An external phase A–to-ground fault at the middle of line MN with a sampling frequency of 20 kHz

wavelet transform such as zooming, filtering, and de-nosing. The wavelet-based transient protection not only can detect the fault but also can distinguish the internal faults from external faults. The principle is simple and has good real-time performance. There is no requirement for communication channels, which saves costs and reduces the complexity of the relay. Thus, this protection is practical. However, there are some issues to be considered and solved, such as the discrimination between fault transients and breaker switching or capacitor switching, the phase selection based on transients, and the flexibility under various noisy environments.

References

[1] Jayasinghe J.A.S.B., Aggarwal R.K., Johns A.T., et al., A novel non-unit protection for series compensated EHV transmission lines based on fault generated high frequency voltage signals. *IEEE Transactions on Power Delivery*, vol. 13, no. 2, pp. 405–413, 1998.

[2] Bo Z.Q., A new non-communication protection technique for transmission lines. *IEEE Transactions on Power Delivery*, vol. 13, no. 4, pp. 1073–1078, 1998.

[3] He Z.Y., Wang X.R., Qian Q.Q., A study of EHV transmission lines non-unit transient protection based on wavelet analysis. *Proceedings of the CSEE*, vol. 21, no. 10, pp. 10–19, 2001.

11

The Application of Wavelet Analysis to Power Quality Disturbances

11.1 Problem description of power quality disturbances

With the development of economy and society, electric power has gained a lot of attention because it is an environmental and efficient secondary energy. However, the expanding scale of power systems and the widespread usage of power electronic equipment and nonlinear loads have caused many power quality disturbances, such as voltage sag, voltage swell, voltage fluctuation, harmonics, voltage notch, impulse transients, and oscillation transients. The power quality disturbances will increase the line loss, cause protection relay maloperation, increase the measurement error of the metering devices, and decrease the life of the compensation capacitor, which will make a huge influence on people's normal lives. Therefore, it is important to provide overall research on power quality disturbances [1].

Power quality disturbances can be divided into steady-state power quality disturbances and transient power quality disturbances. The steady-state power quality disturbances have been studied deeply. Their main indices include frequency, voltage fluctuation, voltage unbalance, harmonics, and voltage flicker. At present, the International Electrotechnical Commission (IEC) makes strict standards for steady-state power quality disturbances. The corresponding detection device, monitoring system, professional analysis and simulation software, and engineering management methods are mature and well known. Research on transient power quality disturbances started late in China. Transient power quality disturbances are an extension of the steady-state power quality disturbances. The main indices of the transient power quality problems include voltage impulse, voltage surge, voltage sag, and voltage interruption. There are no uniform standards for transient power quality disturbances yet. The amplitudes and durations of the indices are usually utilized to describe the transient power quality disturbances.

Wavelet Analysis and Transient Signal Processing Applications for Power Systems, First Edition. Zhengyou He.

The power quality disturbance analysis discussed in this chapter comprises disturbance detection, recognition, and source location. Disturbance detection means detecting the disturbance occurrence and its end time. The difficulty is how to detect the disturbance in a strong noise environment and under conditions in which the disturbance is not immediately evident. The disturbance recognition is to recognize the disturbance type by features. The difficulty is how to extract the proper features. The disturbance source location is similar to the fault location of transmission lines. Its main object is to locate the disturbance.

11.2 Power quality disturbance detection based on wavelet transform

The present wavelet transform–based power quality disturbance detection methods are mainly categorized into a continuous wavelet transform–based detection method and multiresolution analysis–based detection method. The continuous wavelet transform has a great computational burden. Multiresolution analysis makes it difficult to detect the disturbance directly and needs to reconstruct the signal because of the binary sampling. The dyadic wavelet is widely applied to transient signal detections due to its completeness, stability, and translation invariance [2].

Sharp variation detection of the signal is an important aspect of wavelet transform application. The sharp variation points correspond to the extrema or zero crossings of the wavelet coefficients modulus. The signal singularity is also relative to the evolution across scales of the wavelet transform modulus maxima. It is, therefore, very meaningful to describe the instantaneous features of signals by wavelet transform. If the wavelet is the first derivative of a smoothing function, its modulus maxima correspond to the sharp variations. The sharp variations can be located by detecting the wavelet's modulus maxima. Since the Mallat algorithm is achieved by means of a filter bank that decomposes the signal at different resolutions, the problem is that after each decomposition, the signal samples are reduced by half by means of downsampling operations. Hence, the signal samples may be too few to obtain the full view of the signal.

Equivalent exchange is an important property in multirate signal processing. One can make an equivalent exchange to each level of filter banks of the Mallat algorithm, in which the decimation will be moved to the last. One can verify that the dyadic decimation after the convolution of the signal with the filter is equivalent to making convolution of the signal with a filter on which upsampling is performed by inserting zeros. This is the so-called algorithme à trous in French [3]. Algorithme à trous not only has the rapidity of the Mallat algorithm but also keeps the effective information. Algorithme à trous can be seen as a fast dyadic transform. A fast dyadic wavelet transform is calculated by convolution of a signal with the filters h_0 and h_1, which are inserted $2^j - 1$ zeros between each sample. Assume c_0 is the discrete signal; the calculation process is as follows:

$$d_{j+1} = c_j * h_{1j} \tag{11.1}$$

$$c_{j+1} = c_j * h_{0j} \tag{11.2}$$

where h_j is the inserted $2^j - 1$ zeros between each sample of h.

Table 11.1 The filter bank coefficients of a quadratic spline wavelet

n	−2	−1	0	1	2	3
h_0	0	0.125	0.375	0.375	0.125	0
h_1	0.0061	0.0869	0.5798	0.5798	0.0869	0.0061

The selection of wavelet bases is very important when using wavelet transform to process signals. Different results may be obtained from the same original signal if different wavelet bases are selected. Thus, the proper wavelet bases should be selected for practical application with consideration of the signal features, computation burden, and so on. Because quadratic spline wavelets perform well in detecting the sharp variations of signals, this chapter provides an example of transient power quality disturbance detection by quadratic spline wavelet. The coefficients of the filter banks are shown in Table 11.1. Besides, in order to study the influence of wavelet symmetry on sharp variations detection, the db4 wavelet (which is a compactly supported, orthonormal, and nonsymmetric wavelet) is compared to the quadratic spline wavelet. Meanwhile, the biorthogonal spline wavelet bior3.1 (symmetric wavelet) and rbio3.1 (antisymmetric wavelet) are used to study the influence of symmetric form on sharp variations detection.

11.2.1 The detection of disturbance signals with different sampling frequencies based on dyadic wavelets

Figures 11.1 and 11.2 show, respectively, the transform results based on algorithme à trous of two voltage sag disturbances in a quadratic spline wavelet. The sampling frequencies of these two voltage sag disturbances are 1 kHz and 50 kHz. It can be seen from the figures that the sharp variations of voltage sag disturbance at 1 kHz are hard to observe in the detail coefficients. However, the sharp variations of voltage sag disturbance at 50 kHz can be detected in five scales' coefficients. Hence, with the increase of sampling frequency, the sharp variation of signal can be seen in higher scales.

The algorithme à trous is achieved by removing the downsamplers and upsamplers in the discrete wavelet transform, and upsampling the filter coefficients by a factor of 2^{j-1} in the jth level of the algorithm. The output of each level of algorithme à trous contains the same number of samples as the input. The numbers of samples at each cycle at Figures 11.1 and 11.2 are, respectively, 20 and 100. The original length of the filter banks of the dyadic wavelet is 6. After inserting zeros in the filters, if the length of the filters are nearly the same as the number of samples at one cycle, the singularity of the signal will not be revealed, as can be seen in Figure 11.1. With the increase of sampling frequency, the singularity of the signal is incarnated at higher decomposition scales. In practice, the signal suffers noise. From Chapter 3, we know that the Lipschitz exponent of noise is negative, which means that the singularity of noise decreases as the decomposition scale increases. Therefore, we can find the singular point of the signal at higher scales.

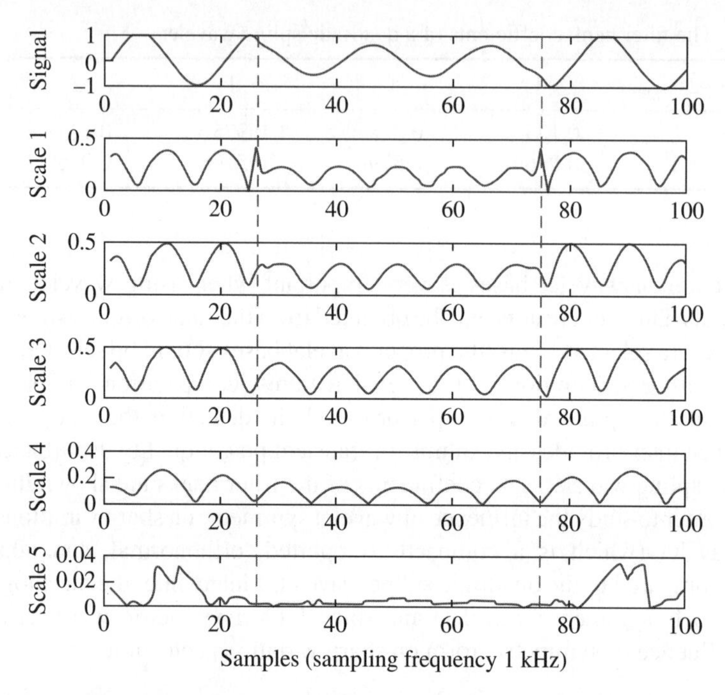

Figure 11.1 Transform results of voltage sag disturbance at sampling frequency of 1 kHz

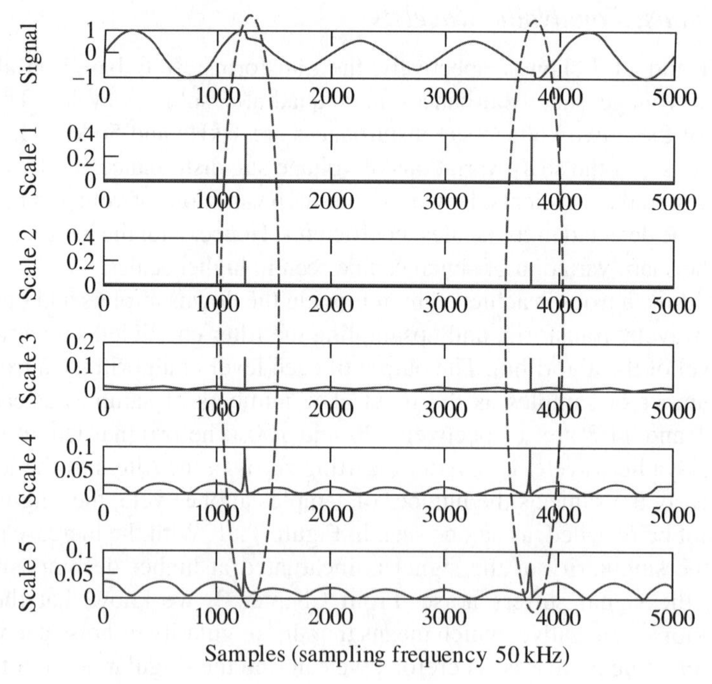

Figure 11.2 Transform results of voltage sag disturbance at sampling frequency of 50 kHz

11.2.2 Detection of disturbance signal based on wavelets with different symmetries

11.2.2.1 Symmetric wavelets and nonsymmetric wavelets

The asymmetry of wavelet bases will make a shift of modulus maxima corresponding to the singular points. This shift makes it difficult to search the modulus maxima curve. Figure 11.3 shows the wavelet transform results of the voltage sag disturbance based on a quadratic spline wavelet and db4 wavelet at five scales. The voltage sag disturbance occurred at samples 1251 and 3650. Tables 11.2 and 11.3 show that the modulus maxima of the symmetric quadratic spline wavelet is coincident with the sharp variation points. However, the modulus maxima of the nonsymmetric db4 wavelet has a deviation from the sharp variation points.

11.2.2.2 Symmetric wavelets and antisymmetric wavelets

Figure 11.4 shows the transform wavelet results of the transient impulse induced by lightning. It shows that the symmetric wavelet bior3.1 has good detection performance for impulse signals, whereas the antisymmetric wavelet rbio3.1 only detects the impulse at low scales. Figure 11.5 displays the wavelet transform results of the voltage sag disturbance induced by fault; it shows that both bior3.1 and rbio3.1 are capable of detecting the voltage sag. Therefore, an antisymmetric wavelet is more suitable for edge detection, and a symmetric wavelet is more suitable for impulse detection.

It is an efficient method to detect power quality disturbance based on the algorithme à trous. A large number of experts and scholars have studied power quality disturbance detection. Readers who are interested in this topic could compare the proposed detection methods and propose new ones.

11.3 Power quality disturbance identification based on wavelet transform

Wavelet transform can provide time, frequency, and amplitude information of signals, and these help to display the signal more comprehensively. Therefore, many researchers have applied wavelet transform to analyze the features of power quality disturbances, which are the classification basis of disturbances.

The general steps of power quality disturbance identification based on wavelet transform are shown in Figure 11.6.

11.3.1 Signal pre-processing wavelet-based threshold de-noising [4]

De-noising methods based on wavelet transform can be categorized into three types according to wavelet types: continuous wavelet transform–based de-noising, discrete wavelet transform–based de-noising, and wavelet packet–based de-noising. According to the threshold, the de-noising methods can be summarized as modulus maxima–based

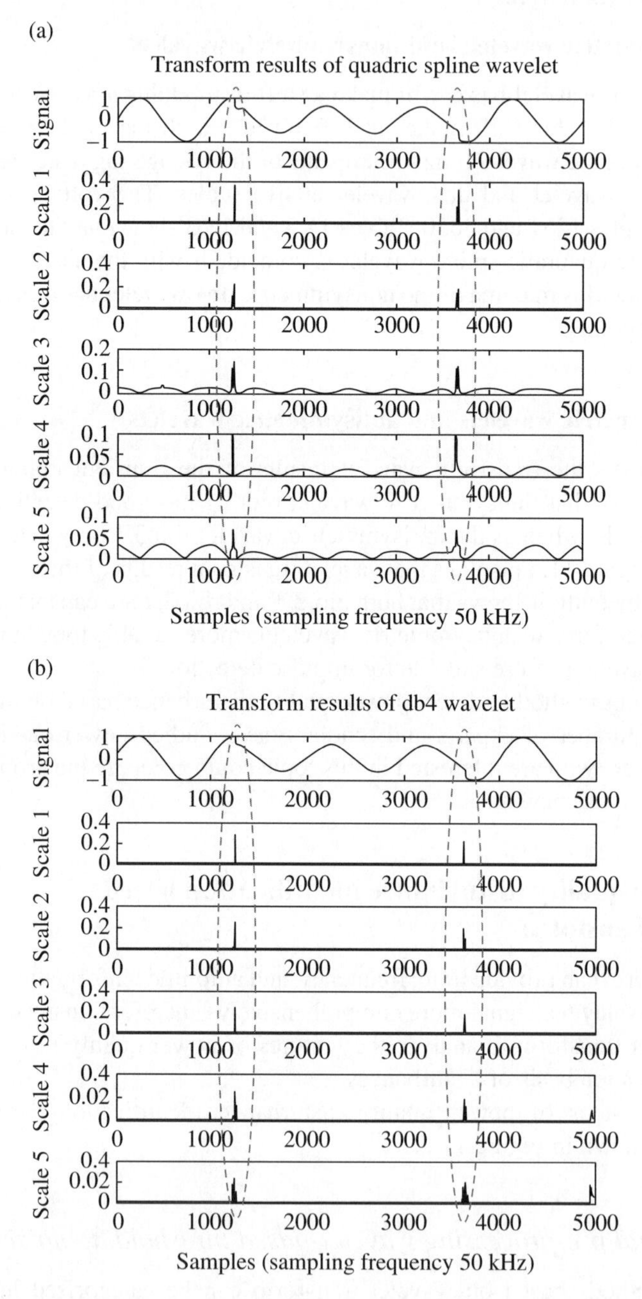

Figure 11.3 Wavelet transform results of voltage sag disturbance

Table 11.2 The location of sharp variations and the corresponding modulus maxima extracted by a quadratic spline wavelet

	Original signal	Scale 1	Scale 2	Scale 3	Scale 4	Scale 5
Location of sharp	1251	1251	1251	1251	1251	1251
variations	3650	3650	3650	3650	3650	3650
Modulus maxima	–	0.3806	0.2112	0.139	0.0964	0.0676
		0.3109	0.1756	0.1195	0.0881	0.0691

Table 11.3 The location of sharp variations and the corresponding modulus maxima extracted by a db4 wavelet

	Original signal	Scale 1	Scale 2	Scale 3	Scale 4	Scale 5
Location of sharp	1251	1249	1249	1250	1251	1254
variations	3650	3648	3648	3649	3650	3653
Modulus maxima	–	0.2741	0.2477	0.2339	0.2338	0.2336
		0.2212	0.1997	0.1888	0.1885	0.1884

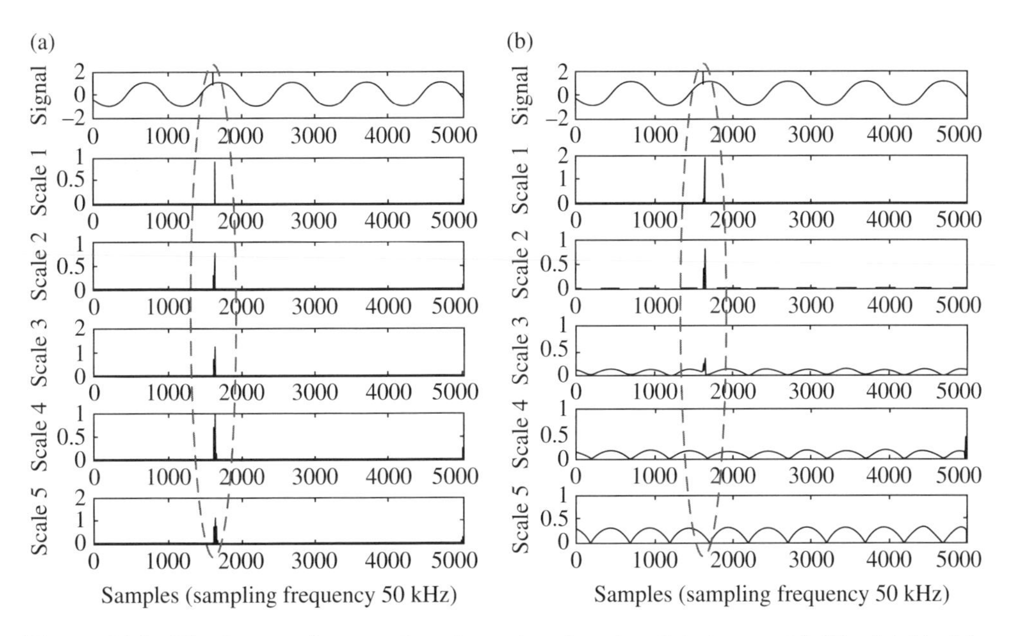

Figure 11.4 Wavelet transform results of transient impulse disturbance: (a) Wavelet bior3.1, (b) wavelet rbio3.1

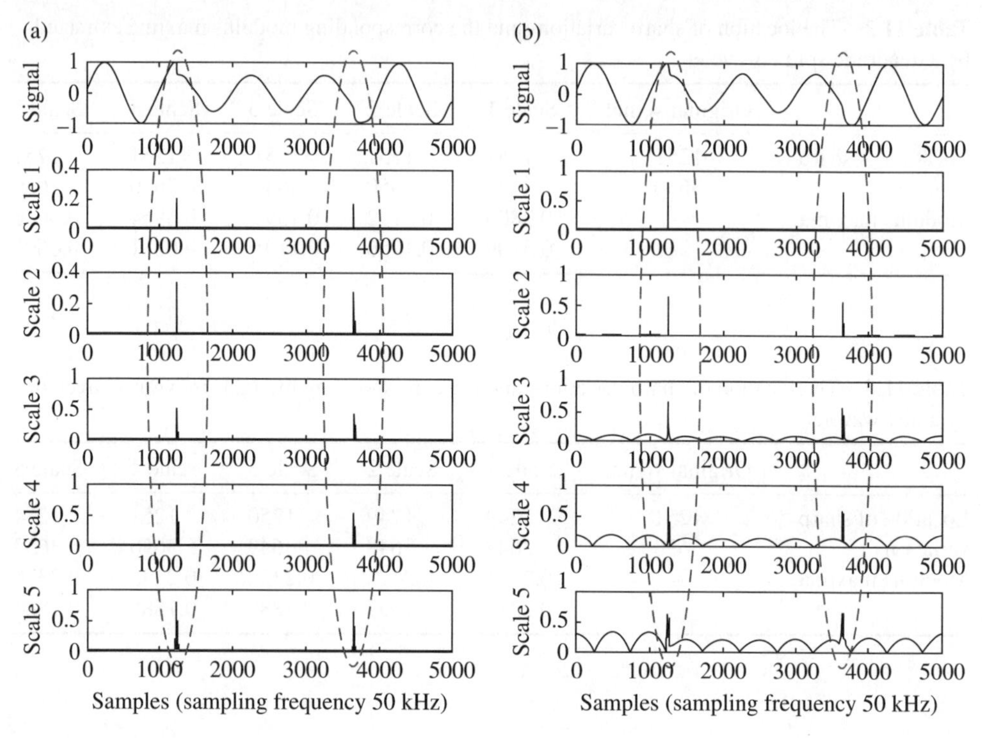

Figure 11.5 Wavelet transform results of voltage sag disturbance: (a) Wavelet bior3.1, (b) wavelet rbio3.1

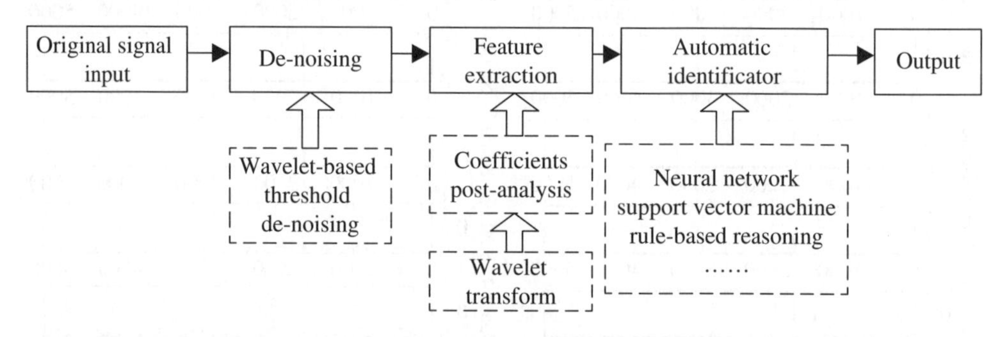

Figure 11.6 The general steps of power quality disturbance identification based on wavelet transform

de-noising, Donoho threshold–based de-noising, and energy threshold–based de-noising. In this subsection, the Donoho threshold–based de-noising method is used for power quality signal de-noising because of its good performance. The Donoho threshold–based de-noising method in general involves three steps:

1. Choose a mother wavelet and a maximum decomposition level. Then, compute the decomposition coefficients at each level.

2. Compute thresholds for each level, and apply the threshold to the coefficients at each level to remove the coefficients below the threshold.
3. Reconstruct the signal with the modified coefficients through the inverse wavelet transform.

Four factors mainly affect de-noising: the threshold type, threshold estimation, wavelet basis selection, and determination of the decomposition level. The threshold type can be a hard threshold, soft threshold, or modified threshold. The soft threshold has a widespread application because it performs better than other threshold types. The commonly used threshold estimation methods include fix threshold, Rigrsure threshold, heuristic threshold, and mini–maxi threshold. One can select different thresholds for different signals and de-noising requirements. The wavelet basis is selected from the consideration of reconstruction ability. In Chapter 5, we know that the wavelet with a linear phase can keep the reconstructed signal undistorted. A B-spline wavelet has a better de-noising effect. Generally speaking, the higher decomposition level could make the signal be separated from noise more easily. Thus, a better de-noising effect could be obtained. However, the higher decomposition level requires a large amount of computation. A proper decomposition level should be determined with consideration of de-noising effect and the amount of computation.

An assessment criterion should be established before assessing an algorithm. Two aspects should be concerned to assess the de-noising algorithm. One is the computation speed and real time; the other is the de-noising effect. Nevertheless, these two aspects are always contradictory. It is necessary to find a compromise between speed and accuracy.

The noise is always white noise in power systems. De-noising is to reduce the white noise and retrieve the desired signal with little loss of detail. Except for the direct observation on the signal after de-noising, we also should find a factor to reflect the similarity between the signal after de-noising and the preset signal without noise, which can impersonally assess the de-noising effect. In traditional assessments, the signal-to-noise ratio is always used to assess the de-noising effect. In fact, for de-noising assessment, not only the overall deviation of the signal but also the local deviation should be considered.

The overall deviation is defined as

$$e_1 = \sqrt{(x - x_1)^2 / N} \tag{11.3}$$

where x is the signal without noise; and x_1 is the signal after de-noising.

The local deviation is defined as

$$e_2 = \max|x - x_1| \Big/ \sqrt{\|(x - x_N)\| / N} \tag{11.4}$$

where x_N is the noisy signal.

The reconstruction factor is defined as

$$e = \gamma_1 e_1 + \gamma_2 e_2 \tag{11.5}$$

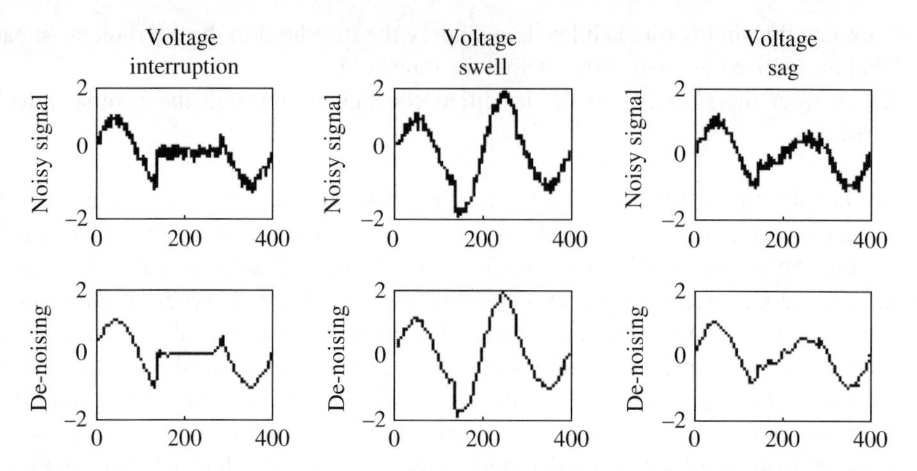

Figure 11.7 De-noising results of voltage interruption, voltage swell, and voltage sag

Table 11.4 Reconstruction factors for de-noising voltage transient disturbances

Disturbance	De-noising method	e_1	e_2	e
Voltage interruption	No de-noising	0.10067	∞	∞
	Wavelet-based de-noising	0.046177	0.13078	0.088478
Voltage swell	No de-noising	0.10046	∞	∞
	Wavelet-based de-noising	0.050372	0.11239	0.079233
Voltage sag	No de-noising	0.10289	∞	∞
	Wavelet-based de-noising	0.047686	0.14176	0.094724

where γ_1 and γ_2 are, respectively, the weighted coefficients of the overall deviation and the local deviation. γ_1 and γ_2 are also called the overall deviation factor and the local deviation factor; $\gamma_1 + \gamma_2 = 1$. We can adjust γ_1 and γ_2 according to the requirements of the overall deviation and the local deviation. Based on the definition, the reconstruction factor is proportional to the deviation between the signal after de-noising and the preset signal without noise. The higher deviation means that the de-noising effect is worse.

Figure 11.7 shows the de-noising results of voltage interruption, voltage swell, and voltage sag. The wavelet transform decomposes the voltage into three levels. The disturbance transient occurs from sample 140 to sample 280. The amplitudes in the transient duration are, respectively, 0 pu for voltage interruption, 1.8 pu for voltage swell, and 0.4 pu for voltage sag. From Table 11.4, we can see that a wavelet-based de-noising method can reduce the noise availability.

11.3.2 Feature extraction

The study objects are the power quality disturbances shown in Appendix A.6. Eight types of single disturbance and two types of multiple disturbances are considered: normal (type A), swell (type B), sag (type C), interruption (type D), spike (type E), oscillatory transients (type F), notch (type G), harmonic (type H), swell and harmonics (type I), and sag and harmonics (type J). An example is used to illustrate how to extract the power quality disturbances by wavelet transform and its postprocessing methods [5].

11.3.2.1 Feature extraction based on DWT

The signals are decomposed by db4 wavelets into four-scale resolution in order to obtain the detail coefficients of each scale. The energy distributions can help to classify disturbance types. The energy of each scale is obtained by Parseval theorem, which can be expressed as follows:

$$E_j = \sum_{k=1}^{N/2^j} \left| D_j(k) \right|^2 \tag{11.6}$$

where E_j is the energy of the jth scale; and $D_j(k)$ represents the detail coefficients of the jth scale.

Hence, the characteristics have been found. Their calculated formula appears in the following equations:

$$\begin{cases} E_1 = E_{50\,\mathrm{Hz}} \\ E_2 = E_{150\,\mathrm{Hz}} + E_{250\,\mathrm{Hz}} + E_{350\,\mathrm{Hz}} \\ E_3 = \sum\limits_{j>7} E_{50\times j\mathrm{Hz}} \end{cases} \tag{11.7}$$

where the first characteristic E_1 stands for the energy E_j that contains the fundamental frequency band; the second characteristic E_2 is the sum of the energy E_j that includes the third harmonic band, the fifth harmonic band, and the seventh harmonic band; and the third characteristic E_3 is the sum of the energy E_j that consists of the rest of the higher frequency bands.

A three-dimensional plot comprising E_1, E_2, and E_3 is developed and shown in Figure 11.8a. The datasets of the plot are the training sets, and the following plots are based on the same datasets. From Figure 11.8a, it is clearly visible that types B, D, F, H, I, and J in the plot can be well recognized from each other. Nevertheless, the other four disturbances, which are not plotted in the figure, cannot be well distinguished by the energy features. To be specific, type A cannot be well distinguished from type E, and types C, D, and G are mixed.

In spite of using energy distributions to classify disturbance types, such factors cannot independently represent each feature for disturbances. Therefore, other information available to enhance classifying PQ types should be found. The wavelet coefficient entropy can reflect the different complex degree of the signals at each scale. Therefore, the wavelet coefficient entropy can be used as another feature for classifying disturbance types. It can be defined as

$$w_j = -\sum_{k=1}^{N/2^j} \mu_j(k) \ln \mu_j(k) \tag{11.8}$$

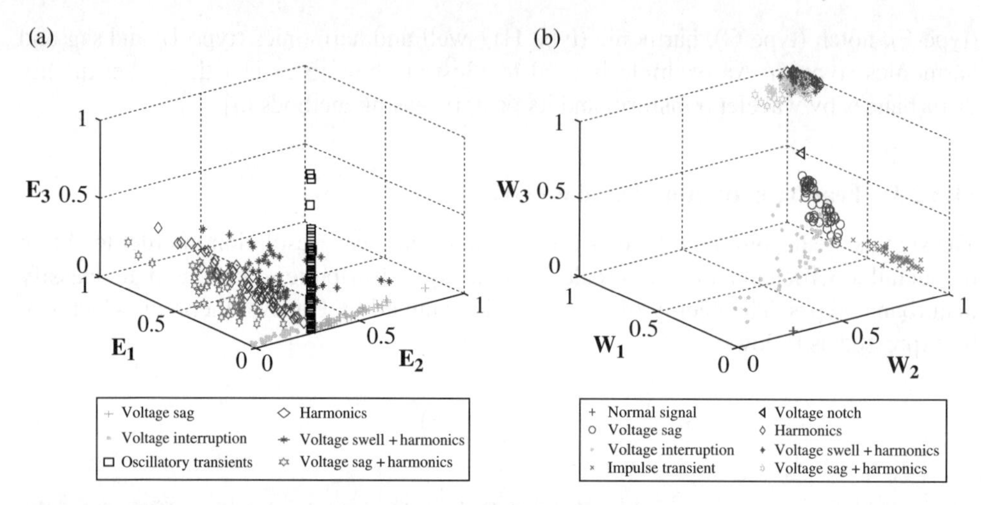

Figure 11.8 Three-dimensional plot of $\mathbf{V}_1$: (a) a three-dimensional plot comprising E_1, E_2, and E_3, (b) a three-dimensional plot comprising W_1, W_2, and W_3

where $\mu_j(k) = D_j(k) \Big/ \sum_{k=1}^{N/2^j} D_j(k)$ and $D_j(k)$ are the detailed coefficients of the signal at the jth scale.

Therefore, three other characteristics have been obtained, as shown in Equation (11.9).

$$\begin{cases} W_1 = w_{50\,\text{Hz}} \\ W_2 = w_{150\,\text{Hz}} + w_{250\,\text{Hz}} + w_{350\,\text{Hz}} \\ W_3 = \sum_{j>7} w_{50\times j\text{Hz}} \end{cases} \tag{11.9}$$

where W_1 is the wavelet coefficient entropy w_j that contains the fundamental frequency band; W_2 is the wavelet coefficient entropy w_j that includes the third harmonic band, the fifth harmonic band, and the seventh harmonic band; and W_3 is the wavelet coefficient entropy w_j that consists of the rest of the higher frequency bands.

The three-dimensional plot comprising W_1, W_2, and W_3 is developed and shown in Figure 11.8b. From Figure 11.8b, it is clearly visible that types A and E can be well identified from each other, and types C, D, and G have separate classification surfaces after adding the features W_1–W_3.

Furthermore, to distinguish variations more clearly containing sags, swells, or interruptions from other disturbances, one takes the root mean square value (RMS) as a measure during the disturbance time at the scale that contains the fundamental frequency component. Therefore, the first feature vector $\mathbf{V}_1$ includes seven parameters:

$$\mathbf{V}_1 = \left[E_1, E_2, E_3, W_1, W_2, W_3, \text{RMS}\right] \tag{11.10}$$

11.3.2.2 Feature extraction based on wavelet packet transform

The second feature vector is extracted by wavelet packet transform (WPT). It is known that the standard deviation of the wavelet packet coefficients of each end node can be taken into account for disturbances classification. The signals are decomposed into four scales by a wavelet packet with a db4 wavelet in order to determine the feature parameters for each event. The standard deviation of the wavelet packet coefficients of each end node is calculated as follows:

$$s_j = \sqrt{\frac{1}{n}\sum_{k=1}^{n}\left[w_j(k) - \overline{w}_j\right]^2}, j = 1,2,\cdots,2^J \tag{11.11}$$

where $w_j(k)$ is the kth wavelet packet coefficient of the jth end node; n is the number of the coefficient; $\overline{w}_j$ means the average of the n coefficients; an J is the number of the wavelet packet decomposition level.

Using the similar rule of E_1–E_3, three characteristics S_1–S_3 can be calculated from the standard deviation of the wavelet packet coefficients of each end node.

$$\begin{cases} S_1 = s_{50\,\mathrm{Hz}} \\ S_2 = s_{150\,\mathrm{Hz}} + s_{250\,\mathrm{Hz}} + s_{350\,\mathrm{Hz}} \\ S_3 = \sum_{j>7} s_{50\times j\mathrm{Hz}} \end{cases} \tag{11.12}$$

where S_1 is the standard deviation of the wavelet packet coefficients s_j that contains the fundamental frequency band; S_2 is the standard deviation of the wavelet packet coefficients s_j that includes the third harmonic band, the fifth harmonic band, and the seventh harmonic band; and S_3 is the standard deviation of the wavelet packet coefficients s_j that consists of the rest of the higher frequency bands.

Similarly, a three-dimensional map comprising S_1, S_2, and S_3 is developed and shown in Figure 11.9a. The figure clearly shows that all types can be classified well except C, E, and G. Specifically, types C and G are confused with D, and type E cannot be well recognized from I.

The standard deviation is easily affected by the noise. Therefore, it influences the classification accuracy. To overcome these difficulties, other characteristics have been found. The energy entropy of each end node through wavelet packet decomposition has been chosen as feature vectors. According to the Parseval theorem, the energy E_j of the jth end node can be expressed by the wavelet packet coefficient $w_j(k)$. The energy entropy can be defined as

$$o_j = -\sum_{k=1}^{n} p_j(k) \ln p_j(k) \tag{11.13}$$

where $p_j(k) = E_j(k) \Big/ \sum_{k=1}^{n} E_j(k)$.

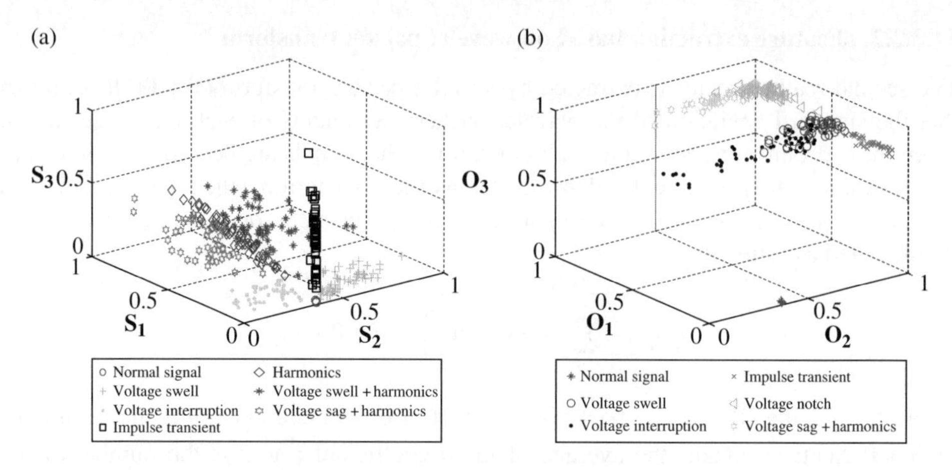

Figure 11.9 Three-dimensional plot of $\mathbf{V}_2$

Among the above, three other characteristics O1–O3 can be calculated by the similar rule of E_1–E_3:

$$\begin{cases} O_1 = o_{50\text{Hz}} \\ O_2 = o_{150\text{Hz}} + o_{250\text{Hz}} + o_{350\text{Hz}} \\ O_3 = \sum_{j>7} o_{50\times j\text{Hz}} \end{cases} \tag{11.14}$$

where O_1 is the energy entropy o_j that contains the fundamental frequency band; O_2 is the energy entropy o_j that includes the third harmonic band, the fifth harmonic band, and the seventh harmonic band; and O_3 is the energy entropy o_j that consists of the rest of the higher frequency bands.

The map of O_1–O_3 appears in Figure 11.9b. This map shows that types E and I has good classification performance between each other, and type G also can be well distinguished from C, but type C still cannot be well recognized from D. This problem can be solved by automatic identification.

Besides, to distinguish variations contain sags, swells, or interruptions from other disturbances more easily, it takes $\bar{E}$ into account here, where $\bar{E}$ is the mean value of the energy during the disturbance time at the scale, which contains the fundamental frequency component.

Therefore, the second feature vector $\mathbf{V}_2$ also includes seven parameters:

$$\mathbf{V}_2 = \left[S_1, S_2, S_3, O_1, O_2, O_3, \bar{E}\right] \tag{11.15}$$

11.3.3 Automatic identification

If we employ artificial intelligence algorithms such as neural networks or support vector machines, the classification results can be obtained easily by inputting directly the feature vectors into the classifier. However, if we employ rule reasoning, we need to set the threshold value according to the extracted feature. Then, the features in different ranges can be categorized.

11.3.3.1 Classification steps by artificial intelligence algorithms

1. Extract the feature vectors from disturbance signal sets. Each disturbance signal set should cover the variation range of the disturbance signals. For example, the amplitudes of sag decrease from 1 pu to 0.1~0.9 pu. Therefore, the amplitudes of the training sets should be selected uniformly from [0.1, 0.9] pu.
2. Input the test sets into the classifier to obtain the results.

11.3.3.2 Classification steps by rule-reasoning algorithms

1. The extracted feature vectors are $\mathbf{V} = [V_1, V_2, \ldots, V_n]$. Divide the features $V_1 \sim V_n$ into different sections. Assume that V_1 is the amplitude feature of disturbances. For $0.1 \le V_1 \le 0.9$, the disturbance is supposed to be voltage sag. For $1.1 \le V_1 \le 1.8$, the disturbance is supposed to be voltage swell. Analogously, the feature vectors $V_2 \sim V_n$ also can be divided.
2. Different feature vectors are fused. The result of global fusion is the ability to overcome the shortage of error recognition to a certain extent and achieve higher classification accuracy.

The two kinds of classifiers have their own advantages. The artificial intelligence algorithms have strong learning ability and are able to recognize the sets that have deviations with the input sets. Their disadvantage is that the algorithms' speed is slow, and they require a large amount of data to learn. The rule-reasoning algorithms have the advantages that they can obtain the results fast, which only require the comparison between the feature value and the threshold value. Their disadvantages are that the extendibility is weak. Once the input sets have any deviation, the classification results are inaccurate. Meanwhile, the threshold values of the rule-reasoning algorithms are difficult to determine.

The selection of the classification methods depends on feature extraction methods. Some feature extraction methods do not have strict physical significance, and so cannot be divided by the threshold value. Hence, these feature extraction methods are more suitable for artificial intelligence algorithms. This is why the artificial intelligence algorithms are so popular.

11.4 Power quality disturbance source location based on wavelet analysis

Shunt capacitors are usually used in distribution networks to provide reactive power compensation for regulating voltage and improving power factors. The capacitor bank-switching transients affect the power quality and may damage the sensitive customer loads. Therefore, in recent years, the power quality issues due to capacitor bank switching have received growing attention. The study of locating disturbance sources due to capacitor switching is of great significance and helpful for solving the power quality problems. As an outstanding signal-processing method, wavelet transform has been applied to many fields relating to power systems. References [6,7] proposed the methods to locate disturbance sources due to capacitor switching based on wavelet transform and wavelet entropy, respectively. In this section, an example from Reference [8] is provided to illustrate the feasibility of wavelet transform in capacitor switching disturbance source location.

The simulation model shown in Figure A.6 in the Appendix is established. Many factors will affect the transient voltage, such as the shunt capacitor location, capacity, load, and distance between the load and the capacitor. The transient voltage caused by the switching capacitor contains abundant information. Moreover, each capacitor has its own transient characteristic. The capacitor can be located with the help of artificial intelligence algorithms [9].

11.4.1 Disturbance features based on wavelet transform

11.4.1.1 Phase to mode transformation

In three-phase power systems, three-phase electrical quantities are usually monitored and analyzed. After phase to mode transformation, only one aerial mode component needs to be analyzed, which greatly decreases the complexity. The aerial mode component in this section is shown in Equation (11.16):

$$U_{\mathrm{m}} = U_{\mathrm{a}} + 2U_{\mathrm{b}} - 3U_{\mathrm{c}} \tag{11.16}$$

Figure 11.10 shows the three-phase voltage and mode voltage signals of busbar 9 when switching capacitor C1. The figure shows that the aerial mode voltage can capture the transient disturbance caused by capacitor switching.

11.4.1.2 The energy feature based on the wavelet frequency band

Figure 11.11a shows the analyzed voltage signal that includes the transient disturbance caused by capacitor switching. By subtracting the fundamental component from the analyzed voltage signal, we obtain the disturbance signal, as shown in Figure 11.11b. The disturbance energy is defined as

$$E = \sum_{k} U_{\mathrm{d}}^{2}(k) \tag{11.17}$$

11.4.1.3 The time feature based on wavelet coefficients

The disturbance energies corresponding to different capacitors switching may have the same order of magnitude in some cases. It is apparent that the energy information is not enough for finding the switched capacitor. Here, we introduce the transient duration to overcome the energy overlap.

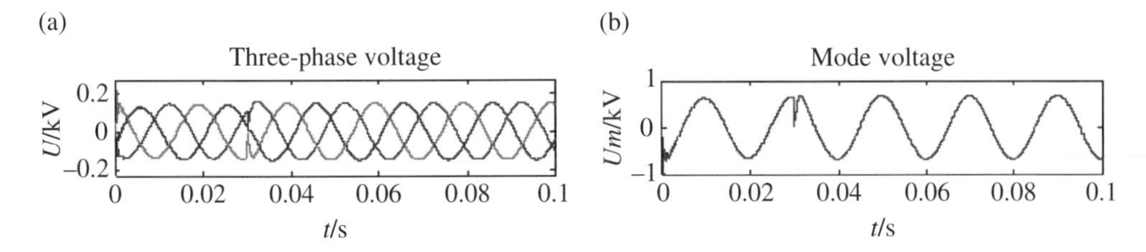

Figure 11.10 Voltage signals of busbar 9 when switching capacitor C1

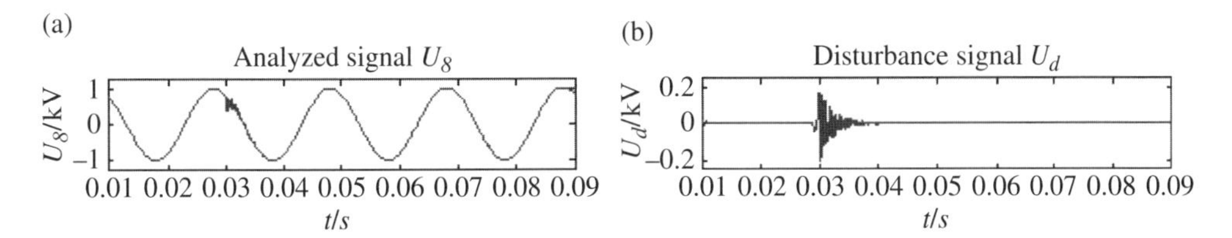

Figure 11.11 Wavelet analysis of the voltage signals of busbar 8 when switching capacitor C1

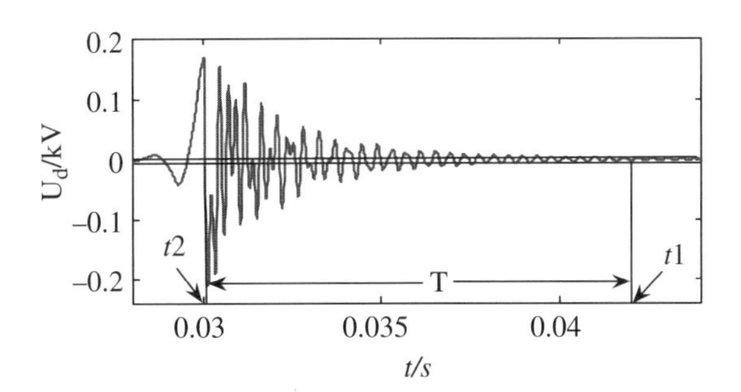

Figure 11.12 Schematic diagram of the transient duration

As shown in Figure 11.12, t_1 is the time when the absolute value of U_d attenuates to no more than 1% of the maximum of U_d. The transient detection time is t_2. The transient duration T is calculated by

$$T = t_1 - t_2 \tag{11.18}$$

11.4.2 Algorithm flow

The disturbance location algorithm employs discrete wavelet transform to extract the disturbance features from the mode voltages of each busbar. The extracted features are disturbance energy and transient duration. Input the extracted features into the trained support vector machine (SVM) to classify the capacitor switching.

The disturbance location flow is illustrated in Figure 11.13. First, three-phase voltage data are imported, and the mode voltage U_m is calculated. Second, the approximation

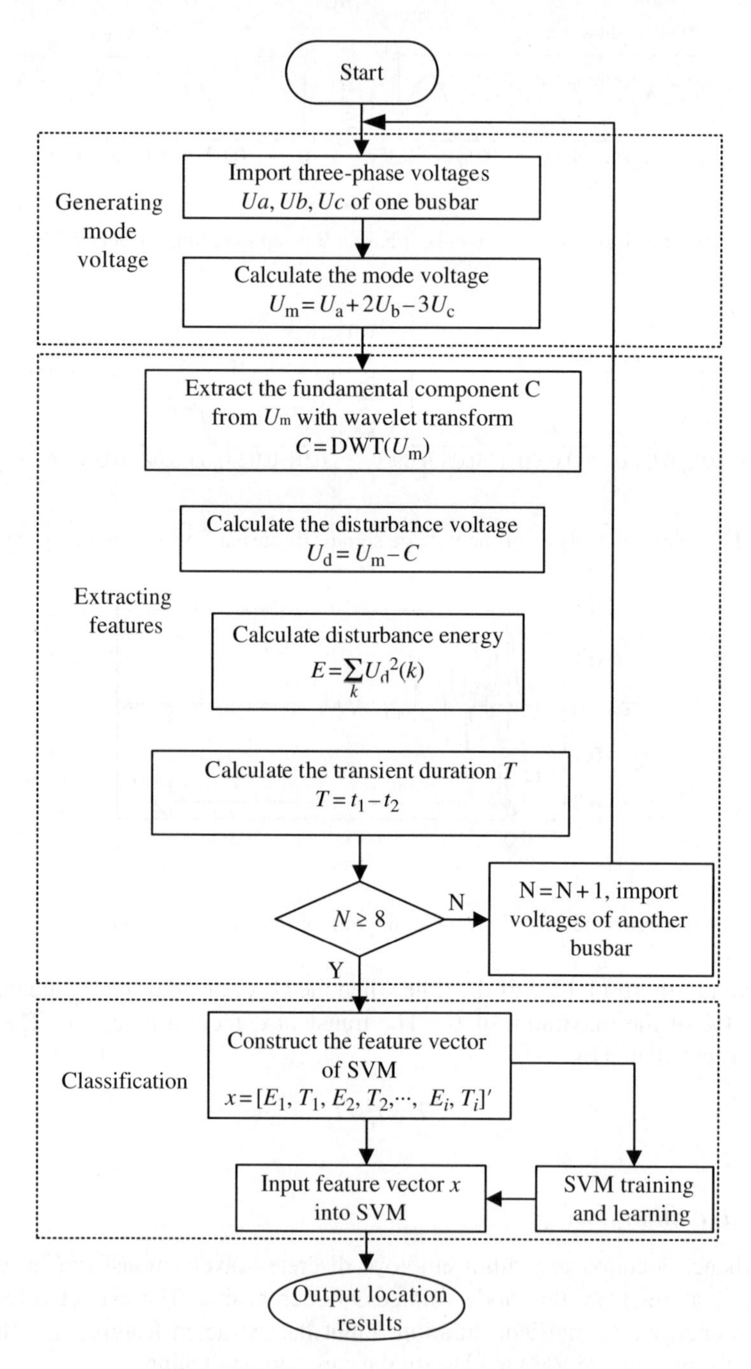

Figure 11.13 Flow chart of the algorithm

coefficient C is obtained using discrete wavelet transform. Then, the disturbance U_d and its energy can be calculated. The transient duration T can also be obtained. Finally, these features are input into an SVM to classify capacitors.

11.4.3 Simulation study

11.4.3.1 Data generation

To provide the required training data and testing data, various simulations are carried out based on the distribution network model in the Appendix. Each capacitor has 200 switching cases, by changing the switching phases in the range of 0–360° with the step of 3.6° in two power–frequency cycles. In total, 600 switching cases are simulated, out of which 240 cases are randomly selected to train the SVMs; the remaining cases are used to test the trained SVMs.

Consider all the possible switching cases to simulate the practical situation. Taking capacitor C1 switching as an example, all the possible capacitor states are shown in Table 11.5. Each possible capacitor state has 50 cases from 0° to 360°.

11.4.3.2 Simulation results of disturbance location due to capacitor switching

The voltages of 18 busbars of the whole distribution network are monitored. The features are extracted from 18 mode voltages in order to obtain the input data of SVM. The location results by SVM are shown in Table 11.6.

Table 11.6 presents some of the switching cases for which SVM is not able to correctly classify the switching capacitor. The misclassified cases are those that occur at very

Table 11.5 Possible state combinations of the capacitors that are switched off

Order	State combinations
1	Both C2 and C3 switched on
2	C2 switched on, C3 switched off
3	C2 switched off, C3 switched on
4	Both C2 and C3 switched off

Table 11.6 The classification results of SVM classifiers

Switched capacitor	Switching cases	Cases with accurate classification	Classification accuracy
C1	120	118	98.33%
C2	120	119	
C3	120	117	

small-voltage magnitudes. However, this misclassification is acceptable. The trained SVM can discriminate 354 cases out of 360 cases accurately; the classification accuracy is 99.17%.

The method introduced above only gives an example to illustrate how to locate the capacitor-switching disturbance source. Besides the energy feature and transient duration feature, many other features can be extracted from the high-frequency transients caused by capacitor switching. Readers can try to figure out which features are better for an individual disturbance source location.

References

[1] John S., Alan C., Power quality. *Power Engineering Journal*, vol. 15, no. 2, pp. 58–64, 2001.

[2] Wang J., Gao J.X., Cao D.X., et al., Dynamic deformation signal extracting model based on a dyadic wavelet transform. *Journal of China University of Mining & Tec hnology*, vol. 36, no. 1, pp. 116–120, 2007.

[3] Li B., Yang D., Yuan G.C., Identification of transformer inrush current based on wavelet transformer. *Power System Technology*, vol. 30, suppl., pp. 384–389, 2006.

[4] Wang L.X., He Z.Y., Zhao J., Zhang H.P., Wavelet transform and mathematical morphology's application in power disturbance signal denosing. *Power System Protection and Control*, vol. 36, no. 24, pp. 30–35, 2008.

[5] Zhang H.P., and He Z.Y., A method for classifying power quality disturbances based on quantum neural network and evidential fusion. *Power and Energy Engineering Conference*, 2009. APPEEC 2009. Asia-Pacific. IEEE, 2009.

[6] Wang J.D., Wang C.S., Capacitor switching disturbance source locating based on wavelet transform. *Electric Power Automation Equipment*, vol. 24, no. 5, pp. 20–23, 2004.

[7] Zhang W., Wang C., Recognition and locating of wavelet entropy based capacitor switching disturbances. *Automation of electric power systems*, vol. 31, no. 7, pp. 71–89, 2007.

[8] Dai M., He Z.Y., Zhao J., Jia Y., A method to locate disturbance source due to capacitor switching in distribution network based on integration of wavelet transform with support vector machine. *Power System Technology*, vol. 34, no. 6, pp. 198–204, 2010.

[9] Abu-Elanien A.E.B., Salama M.M.A.A., Wavelet-ANN Technique for Locating Switched Capacitors in Distribution Systems. *IEEE Transactions on Power Delivery*, vol. 24, no. 1, pp. 400–409, 2009.

12

Wavelet Entropy Definition and Its Application in Detection and Identification of Power Systems' Transient Signals

Albert Einstein said that entropy law is the premier law of all science, which shows the significance of the entropy theory and application. After the tireless efforts of many scholars, entropy has become a concept that is used widely in natural science, engineering science, social science, and the humanities. In order to investigate the uncertainty of information, Claude Shannon from Bell Laboratories, the founder of information science, defined entropy as the average amount of information contained in a source of information according to thermodynamic entropy. We also call it information entropy. The definition of Shannon entropy makes an unprecedented expansion for the application fields of entropy. Information theory has been widely used in many fields after that, and information science has made remarkable achievements.

Information entropy not only solves the problem of measuring information quantitatively, but also provides the basis of the further generalization of information entropy. The basic principle of generalization of information entropy is described as follows. In various fields of natural science, including power systems, and even social science, a large number of random event sets exist with different levels and different categories. For each event set, there will be a set of uncertainties. All these uncertainties can be described uniformly by information entropy. That is to say, all the activity processes that result in the increment or decrease of affirmative, organization, and order can be measured by the changes in information entropy. Thus, the regularity of the event is described quantitatively. This procedure is the generalization of information entropy.

With the development of signal-processing technology, various time–frequency analysis techniques are emerging gradually and widely applied. The frequency domain information

Wavelet Analysis and Transient Signal Processing Applications for Power Systems, First Edition. Zhengyou He.

becomes more and more important. The subband information needs to be made full use of, especially in processing the nonstationary signals. Hence, the extended concepts and definitions of entropy based on frequency information were created. Spectrum entropy is defined in the frequency domain based on the power spectrum. Wavelet entropy is generated by the generalization of information entropy in recent years. The basic idea of wavelet entropy is to process the wavelet transform coefficients as a probability distribution sequences. Thus, the wavelet coefficients at each scale are regarded as the message of a source of signal. The entropy computed by these sequences or coefficients reflects the spare degree of the coefficient matrix (i.e., the order degree of the signal probability distribution). In addition, the distributions in wavelet phase space are different for various signals. Therefore, different entropy measures could be defined to represent different signals based on various principles and processing methods in two-dimensional wavelet phase space. This chapter will introduce several self-defined wavelet entropies, such as wavelet time entropy, wavelet energy entropy, wavelet singular entropy, wavelet time–frequency entropy, and their applications in power systems transient analysis.

12.1 Definitions of six wavelet entropies

12.1.1 Wavelet transform

When a discrete signal $x(k)$ is wavelet transformed, it has high-frequency component coefficients $d_j(k)$ and low-frequency component coefficients $a_j(k)$ at instant k and scale j. The frequency band ranges contained in the signal components $D_j(k)$ and $A_j(k)$ obtained by reconstruction are

$$\begin{cases} D_j(k):\left[2^{-(j+1)}f_s, 2^{-j}f_s\right] \\ A_j(k):\left[0, 2^{-(j+1)}f_s\right] \end{cases} \quad (j=1,2,\cdots,J) \tag{12.1}$$

where f_s is the sampling frequency; and J is the maximal scale. The original signal sequence $x(k)$ can be represented by the sum of all components, namely:

$$x(k)=D_1(k)+A_1(k)=D_1(k)+D_2(k)+A_2(k)=\sum_{j=1}^{J}D_j(k)+A_J(k) \tag{12.2}$$

For identity, replace $A_J(k)$ by $D_{J+1}(k)$, and we obtain

$$x(k)=\sum_{j=1}^{J+1}D_j(k) \tag{12.3}$$

where $D_j(k)$ represents the component of transient signal $x(k)$ at each scale. It is also the multiresolution representations of the signal.

12.1.2 Information entropy and spectrum entropy

12.1.2.1 Information entropy

The uncertainty of any event is associated with its states and probabilities. The set of all possible states is called sample space X $\{x_1, x_2, \ldots, x_N\}$. Each piece of information x_i has a probability $P(x_i) = P_i$, $0 \le P_i \le 1$, $\Sigma P_i = 1$. The self-information of each event x_i is

$$I(x_i) = -\log_\alpha P(x_i) = -\log_\alpha P_i \tag{12.4}$$

As the self-information $I(x_i)$ is a random variable that will change with different information, it is not suitable for measuring information quantity of the whole information source. Therefore, we define the mathematical expectation of the self-information as the entropy of the information source, and the entropy of X is denoted by $H(X)$:

$$H(X) = \mathrm{E}\left[I(X)\right] = \mathrm{E}\left[-\sum_i \log_\alpha P_i\right] = -\sum_i P_i \log_\alpha P_i \tag{12.5}$$

The base a of the algorithm determines the unit of the entropy. When a is 2, e, and 10, the entropy unit is bit, nat, and Hartely, respectively. Customarily, we choose $\alpha = \mathrm{e}$. If all events have the same probabilities, the uncertainty of each event reaches its maximum, and so does the entropy. In other words, the entropy obtains the maximal value when the uncertainty of each event is the same. The entropy of any certain event is zero. Therefore, entropy is the measure of uncertainty.

12.1.2.2 Spectrum entropy

Based on the conception of information entropy and power spectrum, the spectrum entropy is defined in the frequency domain. Letting $X(\omega)$ be the discrete Fourier transform (DFT) of signal $x(k)$, the power spectrum is $S(\omega) = \frac{1}{N}|X(\omega)|^2$. N is the length of signal $x(k)$. Due to the conversion of energy in the transformed procedure from the time domain to frequency domain – namely, $\sum x^2(t)\Delta t = \sum |X(\omega)|^2 \Delta\omega$, $S = \{S_1, S_2, \ldots, S_N\}$ – is a partition of the original signal, so the proportion of the ith power spectrum occupied in the whole spectrum is $H = -\sum_{i=1}^{N} p_i \log p_i$. The corresponding information entropy – namely, power spectrum entropy – is denoted as follows:

$$H = -\sum_{i=1}^{N} p_i \log p_i \tag{12.6}$$

Spectrum entropy is a measure of signal complexity. The narrower the peak of the signal power spectrum is, the smaller the spectrum entropy is. That means the signal is more regular and less complex. The flatter the power spectrum is, the larger the spectrum entropy is. For example, the white noise is an irregular random signal that has a flat power spectrum and large spectrum entropy, which means the signal has high complexity.

12.1.3 Definitions of wavelet entropy measures and their calculation methods

Wavelet entropy theory was put forward by S. Blanco in 1998. He defined wavelet entropy and used it to analyze event-related potentials. Thereafter, many scholars applied wavelet entropy to the analysis of nonstationary signals such as brain waves and to fault detection in the machine field.

When power systems operate from a normal condition to a fault state, the amplitudes and frequency of voltage and current signals will change. The Shannon information entropy will change accordingly. But the simple information entropy seems powerless for the analysis of some singular signals. If we introduce wavelet analysis technique, this problem will be solved. It makes full use of the localization ability of wavelets in both the time domain and frequency domain to combine wavelet analysis with entropy. The combination not only indicates the information representation ability of information entropy but also fuses the information, which can be more useful for analyzing fault signals. Several definitions of wavelet entropy measures are put forward that are similar to those for spectrum entropy. In the following definitions, $E_{jk} = |\mathrm{D}_j(k)|^2$ is the energy spectrum at scale j and instant k. Thus, $E_j = \sum_k E_{jk}$ is the wavelet energy spectrum at scale j. Six kinds of wavelet entropy measures are shown in Figures 12.1, 12.2, 12.3, 12.4, 12.5, and 12.6.

12.1.3.1 Wavelet energy entropy

Definition 12.1 Let $E = E_1, E_2, \cdots E_m$ be wavelet spectra of signal $x(k)$ at m scales. Then, E is a partition of signal energy at the scale domain. According to the characteristic of orthogonal wavelet transform, at a certain time window the total signal energy E is the sum of energy E_j of each component. If $p_j = E_j/E$, then $\sum_j p_j = 1$. We thus define wavelet energy entropy (WEE) as

$$WEE = -\sum_j p_j \log p_j \tag{12.7}$$

From Definition *12.1*, the change law of WEE with time can be obtained as the window slides. The definition in Equation (12.7) reflects the energy distribution of voltage or current in frequency space. Because the wavelet function does not have pulse selection property at either the frequency domain or time domain but instead in a support region, the partition of current or voltage energy at the scale space indicates the energy distribution features in both the time domain and frequency domain.

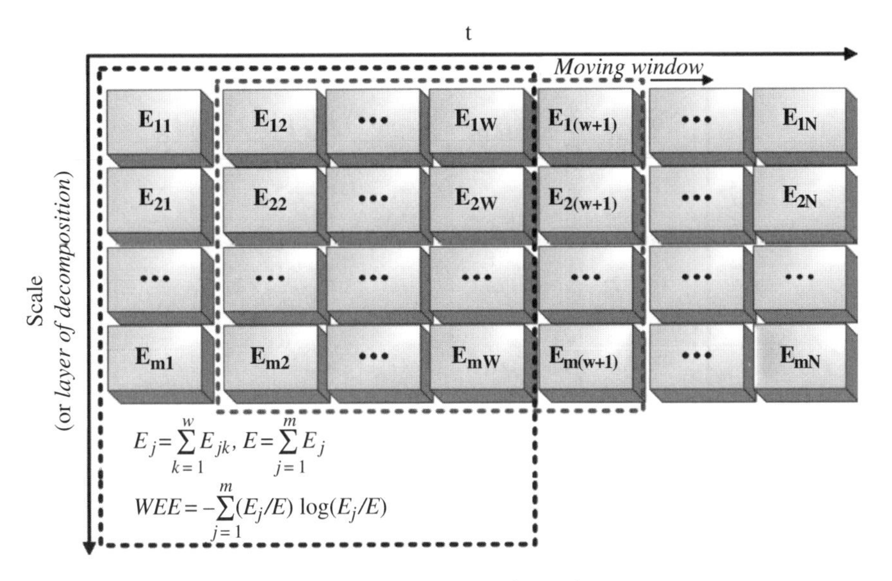

Figure 12.1 Fundamental diagram of wavelet energy entropy

12.1.3.2 Wavelet time entropy

A sliding window $w \in N$ is defined under the wavelet transform result $D_j(k)$. The sliding factor is $\delta \in N$. Then, the sliding window $W(m; w, \delta)$ is described as $\{D_j(k), k = 1 + m\delta, \cdots w + m\delta\}$, $m = 1, 2, \cdots, M$. Divide the sliding window into the following L sections: $W(m; w, \delta) = \bigcup_{l=1}^{L} Z_l$, $\{Z_l = [s_{l-1}, s_l), l = 1, 2, \cdots L\}$, and they do not interest each other. Moreover, $s_0 < s_1 < s_2 < \cdots < s_L$, and $s_0 = \min[W(m; w, \delta)]$, $s_L = \max[W(m; w, \delta)]$.

Definition 12.2 Let $p^m(Z_l)$ represent the probability that wavelet coefficients $D_j(k)$ fall into section Z_l. According to the classic probability theory, $p^m(Z_l)$ is the proportion of the amount of coefficients that fall into Z_l to the total number of coefficients in $W(m; w, \delta)$. Thus, we define the wavelet time entropy (WTE) at scale j as

$$WTE_j(m) = -\sum p^m(Z_l) \log\left(p^m(Z_l)\right), m = 1, 2, \cdots M \tag{12.8}$$

where $M = (N - w)/\delta \in N$. At each scale, WTE is calculated, and the variation curve of WTE also can be drawn. WTE is capable of detecting and locating the change of signals or system parameters, and its calculation burden is low. The wavelet coefficients under the first scale should not be used because the first scale corresponds to the topmost frequency, with the possible introduction of significant measurement error and noise; therefore, this information should not be used. In practical implementation, we usually choose scale $j = 2$ or greater scales.

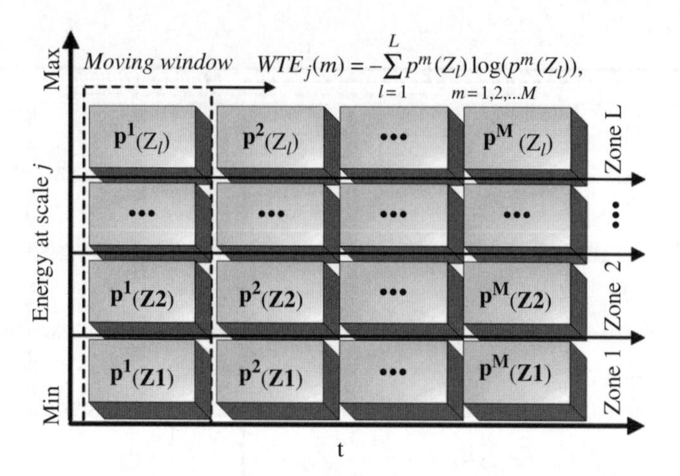

Figure 12.2 Fundamental diagram of wavelet time entropy

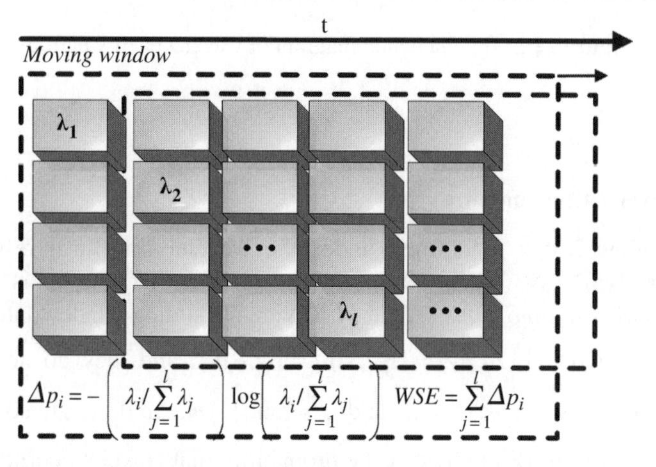

Figure 12.3 Fundamental diagram of wavelet singular entropy

12.1.3.3 Wavelet singular entropy

The wavelet transform results $D_j(k)$ could construct a $m \times n$ matrix $\mathbf{D}_{m \times n}$. According to the singular value decomposition (SVD), for any $m \times n$ matrix $\mathbf{D}$, a $m \times l$ matrix $\mathbf{U}$, a $l \times n$ matrix $\mathbf{V}$, and a $l \times l$ matrix Λ exist, which enable equivalent representation in the SVD form $\mathbf{D}_{m \times n} = \mathbf{U}_{m \times l} \Lambda_{l \times l} \mathbf{V}^T{}_{l \times n}$. The diagonal elements $\lambda_i (i = 1, 2, ...l$ of diagonal matrix Λ are called singular values. These singular values are all nonnegative and arranged in a descending order (i.e., $\lambda_1 \geq \lambda_2 \geq \cdots \lambda_l \geq 0$). Referring to SVD theory, when the signal has no noise or a high signal-to-noise ratio, only a few diagonal singular values are nonzero. The singular values of wavelet decomposition results of signals have the same rule. If there are fewer frequency components, then there are fewer nonzero singular values of the wavelet decomposition.

Definition 12.3 To describe the signal frequency components and their distribution features, we define the l-order wavelet singular entropy (WSE) as follows:

$$WSE_l = \sum_{i=1}^{l} \Delta p_i \tag{12.9}$$

where $\Delta p_i = -\left(\lambda_i / \sum_{j=1}^{l} \lambda_j\right) \log\left(\lambda_i / \sum_{j=1}^{l} \lambda_j\right)$ is the ith-rank increment wavelet singular entropy.

According to this definition, WSE is used to map the correlative wavelet space into independent linearity space, and to indicate the uncertainty of the energy distribution in the time–frequency domain with a high immunity to noise. Due to its way of implementation, WSE is sensitive to the transients produced by the faults, and the fewer modes the transients congregate to, the smaller the WSE is. Therefore, the proposed WSE will be suitable and useful for measuring the uncertainty and complexity of the analyzed signals, and will provide an intuitive and quantitative outcome for the fault diagnosis, which can be utilized to overcome the drawbacks in the previous methodologies.

12.1.3.4 Wavelet time–frequency entropy

Definition 12.4 The discrete wavelet decomposition $D_j(k)$ is in fact a two-dimensional matrix. Thus, we can obtain two vector sequences in columns and rows and define the wavelet time–frequency entropy (WTFE):

$$WTFE(k,j) = \left[WTFE_t(t = kT), WTFE_f\left(a = 2^j\right)\right] \tag{12.10}$$

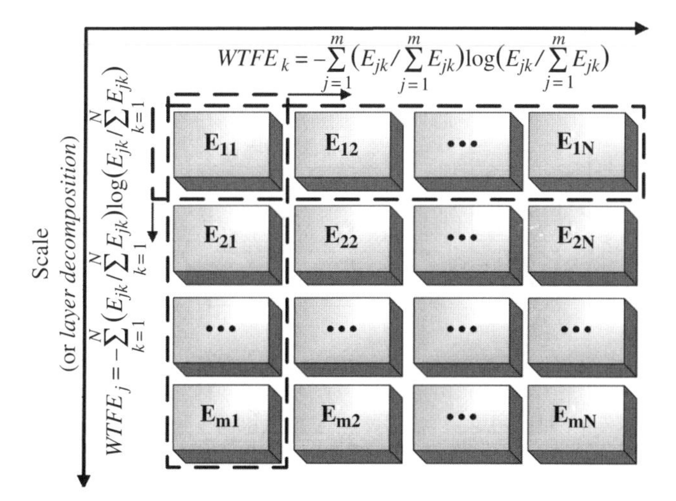

Figure 12.4 Fundamental diagram of wavelet time–frequency entropy

where

$$WTFE_t\left(t=kT\right)=-\sum_j P_{D\left(a=2^j\right)}\ln P_{D\left(a=2^j\right)} \tag{12.11}$$

$$WTFE_f\left(a=2^j\right)=-\sum_k P_{D(t=kT)}\ln P_{D(t=kT)} \tag{12.12}$$

where $PD(a=2^j)=\left|D_j(k)\right|^2/\sum_j\left|D_j(k)\right|^2$, $P_{D(t=kT)}=\left|D_j(k)\right|^2/\sum_k\left|D_j(k)\right|^2$.

The result of WTFE consists of two vectors or sequences. The first vector stretches over all time space, and the second vector stretches over all frequency space. A large entropy value at instant kT indicates there are wavelet coefficients widely distributed in frequency space. On the other hand, a small entropy value indicates that wavelet coefficients congregate at a few frequency points or segments. WTFE is able to measure the signal information features at any given instant and frequency band. Therefore, it can be used to classify different signals and has application potential in the fault detection and diagnosis fields.

12.1.3.5 Wavelet average entropy

The differences among time–frequency distributions of signals are represented by the differences of the energy distribution on a small time–frequency surface within time–frequency space. The homogeneity of energy distribution in each time–frequency area

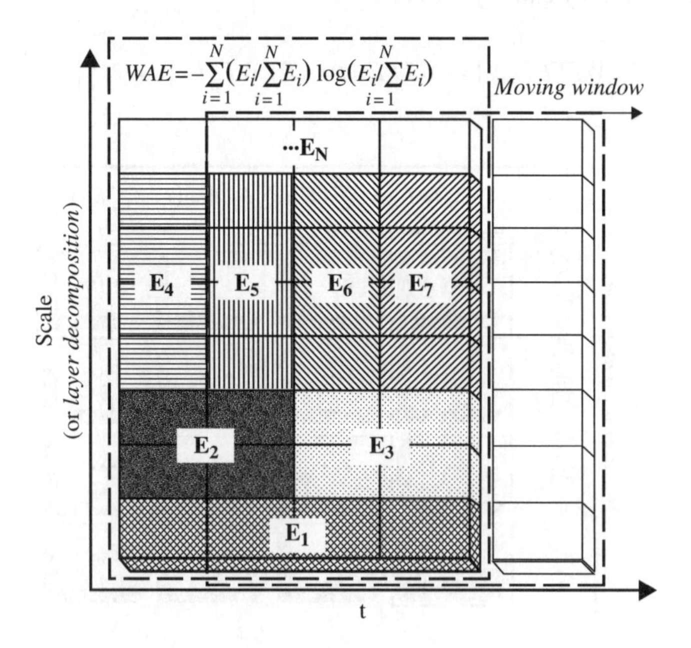

Figure 12.5 Fundamental diagram of wavelet average entropy

reflects the difference of the physical state in the observed system. The homogeneity of time–frequency distribution of sampled current and voltage signals is distinct under normal operation conditions and fault conditions. Information entropy is the measure of probability distribution uniformity. Based on this, we can establish a calculation formula similarly to that for information entropy to describe quantitatively the energy distribution uniformity in the wavelet time–frequency domain. Thus, we define a wavelet entropy – namely, wavelet average entropy (WAE) – which has the property of average.

Definition 12.5 Divide the signal wavelet transform time–frequency plane $(t = kT, a = 2^j)$ into *N* time–frequency windows dilated by scale $[kT - a\Delta_\psi, KT + a\Delta_\psi] \times [\omega^*/a - \Delta_\omega/a, \omega^*/a + \Delta_\omega/a]$, where Δ_ψ and Δ_ω are the time the radii of, respectively, the time window and frequency window. ω^* is the center of frequency window; *T* is the time interval; and *k* is a discrete sequence (0,1, …, *N*). Let $E_i (i = 1, 2, \cdots, N)$ be the energy of each window; the total energy is $E = \sum_{i=1}^{N} E_i$. Normalize the energy using $P_i = E_i/E$, so $\sum_{i=1}^{N} P_i = 1$. Therefore, we define the WAE as

$$WAE = -\sum_{i=1}^{N} P_i \ln P_i \quad (12.13)$$

WAE reflects the average complexity of the whole signal, and it does not change with time or frequency. In some applications, different signals may have similar energy distribution, which makes the feature extraction and classification difficult. According to the basic property of information entropy, the entropy is proportional to energy distribution uniformity. In normal operation conditions, the current and voltage signals contain mainly fundamental components and harmonics. The energy spreads in a few frequency points, which means that the energy distribution is not uniform and the corresponding WAE is small. When a fault occurs on the transmission line, the current and voltage signals contain high-frequency transients, and the frequency spreads widely. The entropy therefore increases.

12.1.3.6 Wavelet distance entropy

Definition 12.6 For discrete wavelet decomposition $D_j(k)$, we can get a vector sequence *D*(*k*) with the variable *k*. According to the definition of correlation distance, we introduce the information calculation method to define the wavelet distance entropy:

$$\text{WDE} = -\sum_{k=l}^{m}\sum_{l=1}^{m} d'_{kl} \ln d'_{kl} \quad (12.14)$$

where $d'_{kl} = d_{kl} / \sum_{k=l}^{N}\sum_{l=1}^{N} d_{kl}$, $d_{kl} = \left\| D(kT) - D(lT) \right\|$, $k, l = 1, 2, \cdots N$.

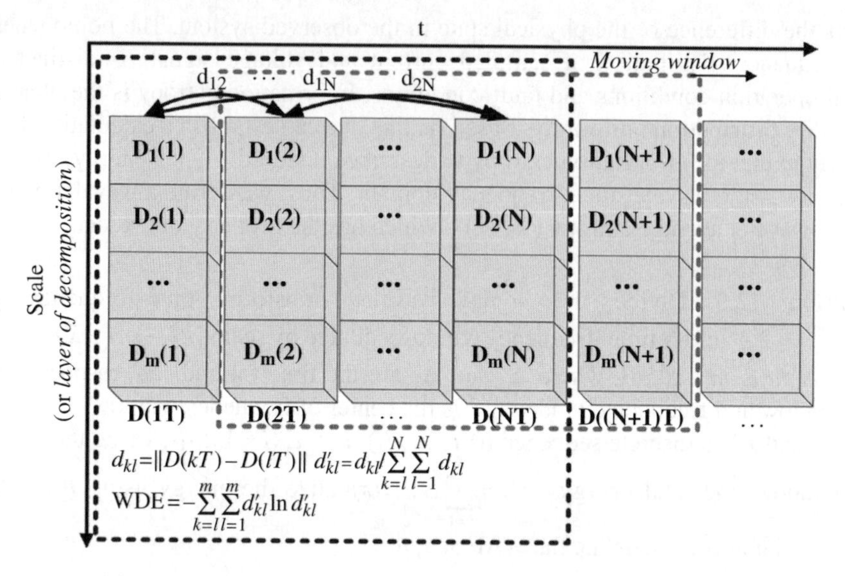

Figure 12.6 Fundamental diagram of wavelet distance entropy

12.2 Detection and identification of power systems' transient signals based on wavelet entropy

Some wavelet entropies defined in Section 12.1 have already been applied to electroencephalograph (EEG) signal analysis in biology and fault diagnosis in the machine field. That is because wavelet entropy presents a particular property of feature extraction of sharp variation signals. Similarly to EEG signals and machine fault signals, the transmission line fault transients are sharp variation signals superimposed by signals with different frequencies. Therefore, wavelet entropy has good application potential in power transients detection and identification.

12.2.1 Wavelet entropy applied to transient signals detection [1]

Simulations of six transients are carried out on the transmission network shown in Figure A.1 in the Appendix. They are breaker switching, capacitor switching, short circuit fault, primary arc fault, lighting interruption, and lighting strike fault. Their outputs appear in Figure A.2.

We start to detect six transients based on wavelet entropy. The window parameter w and δ of WTE are 50 and 1, respectively. The current of phase A and the calculated wavelet entropies are shown in Figure 12.7a–f. Figure 12.7a–f represents the original current of phase A, W_{EE}, W_{TE}, W_{SE}, W_{TFEt}, W_{AE}, and W_{DE}, respectively. We also test the wavelet entropy detection results for two different short-circuit faults; the test results are shown in Figure 12.8.

As seen in Figure 12.7, six wavelet entropies increase significantly at fault instant. Thus, we can identify the fault occurrence and indicate the fault instant. Moreover, by comparison with other transient detection quantities, such as wavelet modulus maxima,

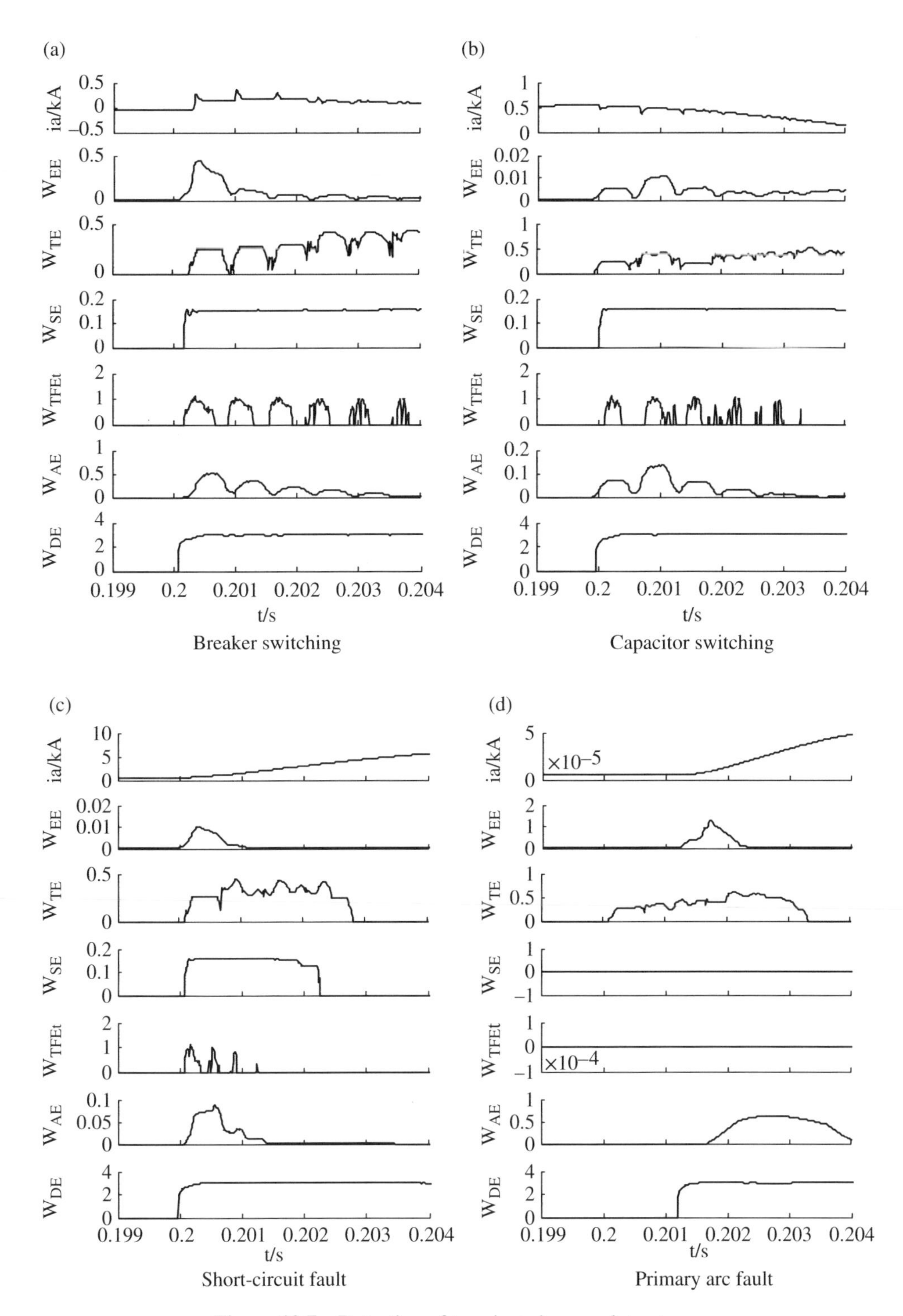

Figure 12.7 Detection of transients by wavelet entropy

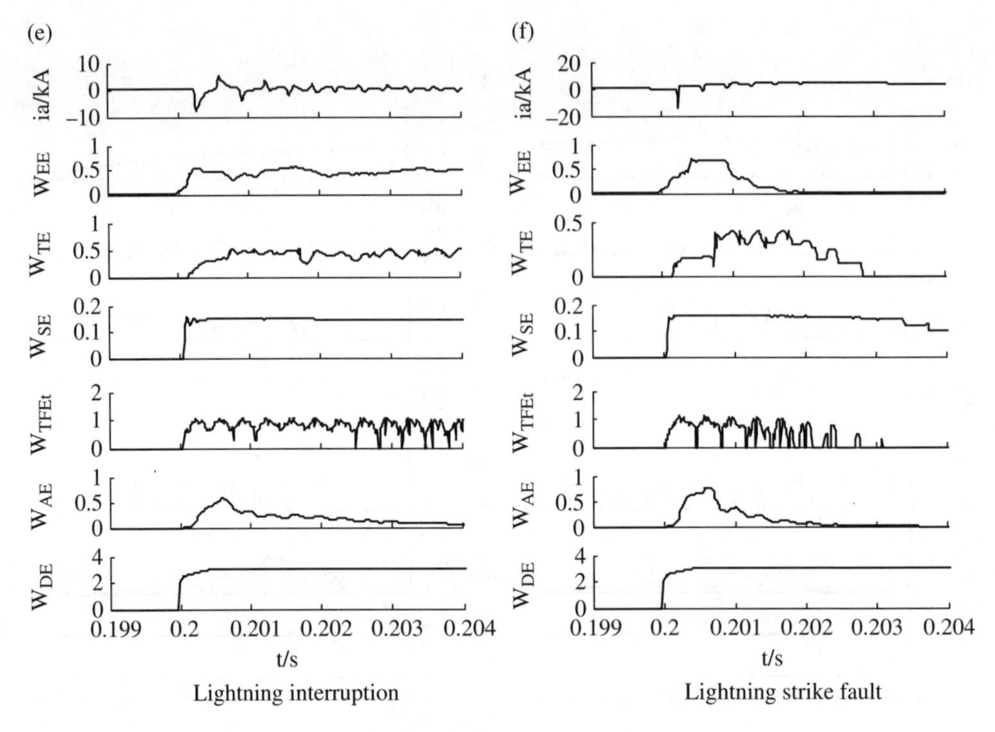

Figure 12.7 (Continued)

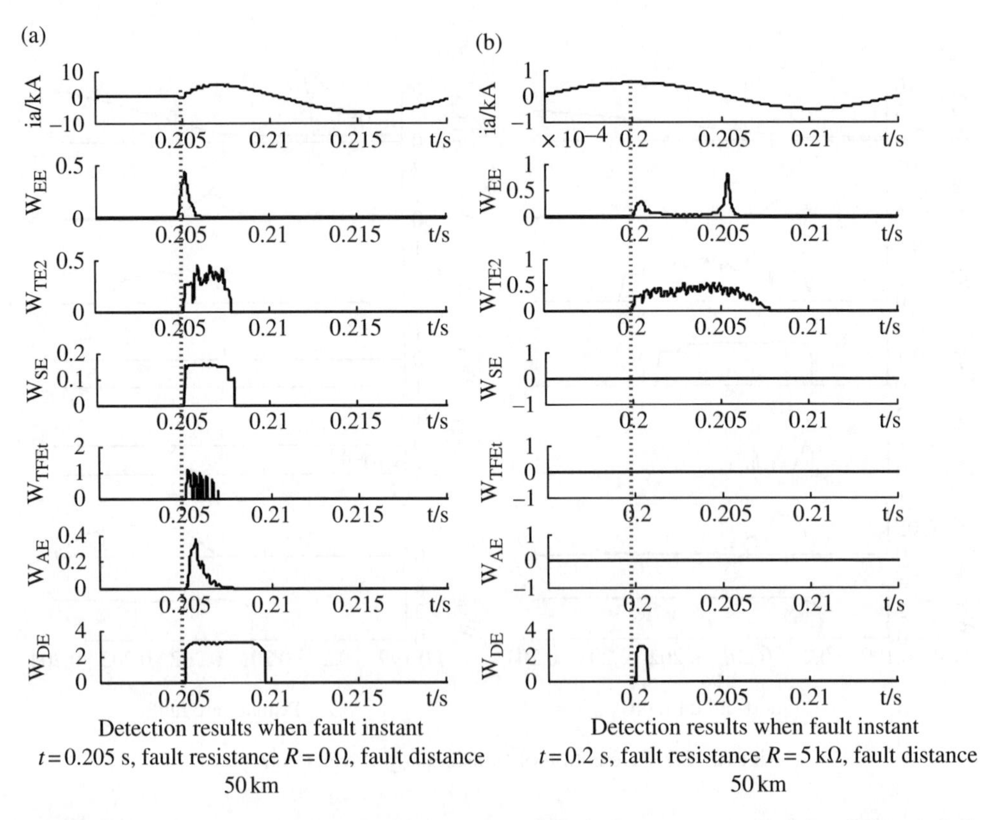

Figure 12.8 Wavelet entropy detection results of fault currents generated by different fault conditions

the wavelet entropies are immune to the amplitude of current and voltage. Some wavelet entropies are even capable of discriminating the permanent faults from temporary faults because they can indicate the fault end instant. Because the transients of primary arc fault or faults with high fault resistance are relatively small, these transients need a long time delay to be detected by W_{AE}. Even they cannot be detected by W_{SE} and W_{TFE}. The real-time capability of W_{EE} and W_{DE} is weak at this moment. Thus, we conclude that W_{TE} has the best detection ability for weak transients. And, for other transients, all six wavelet entropies are capable of detecting the transients in real time.

From the detection effects of six wavelet entropies, it can be seen that W_{SE} and W_{DE} have a faster detecting speed. The amplitude of W_{DE} is greater, so it is helpful for detecting transients accurately and rapidly. However, W_{SE} and W_{DE} are insensitive to weak transients. Other wavelet entropies change rather slowly, but W_{TE} is able to detect the signal sharp variation. Therefore, we should choose proper wavelet entropies or their combination and threshold to identify the transients.

12.2.2 Application of WEE

12.2.2.1 Application of WEE to feature extraction of fault phase selection [2]

Taking the transmission line in the Appendix as an example, simulations are carried out to generate three-phase voltage signals. After multiresolution analysis of three-phase voltage signals using dyadic discrete wavelet, the WEE of each phase is calculated in a certain data window. The window length is selected according to sampling frequency and calculation speed. The calculated WEE under different fault types is shown in Figure 12.9.

It can be seen in Figure 12.9 that when there is a phase A–to-ground fault (AG), phase A has greater W_{EE} than the other two phases. Likewise, for phase B–to-ground faults (BGs) and phase C–to-ground faults (CGs), the faulted phase has greater W_{EE} than the wholesome phases. The W_{EE} of wholesome phases is nearly zero. For phase-to-phase faults or phasev phase-to-ground faults shown in Figure 12.9d–f and Figure 12.9h, the faulted phases have much greater W_{EE} than the wholesome phase. The W_{EE} of the wholesome phase is about zero. When three-phase fault occurred, the W_{EE} of all phases is greater than zero. Based on the analysis above, W_{EE} can be used to identify the fault phases.

12.2.2.2 Application of WEE to transient signals identification [3]

We introduce another application example of identifying transient signals by WEE in this subsection. But first, we extend the definition of WEE.

1. Let E_{jk} be the wavelet energy of wavelet coefficient at scale j and instant k. $E_j = \sum_{k=1}^{N} E_{jk}$ represents the sum of energy of wavelet coefficients at scale j. Letting $p_{jk} = E_{jk}/E_j$, then $\sum_k p_{jk} = 1$. Thus, we define the first wavelet energy entropy as

$$W_{EE1j} = -\sum_k p_{jk} \log p_{jk} \tag{12.15}$$

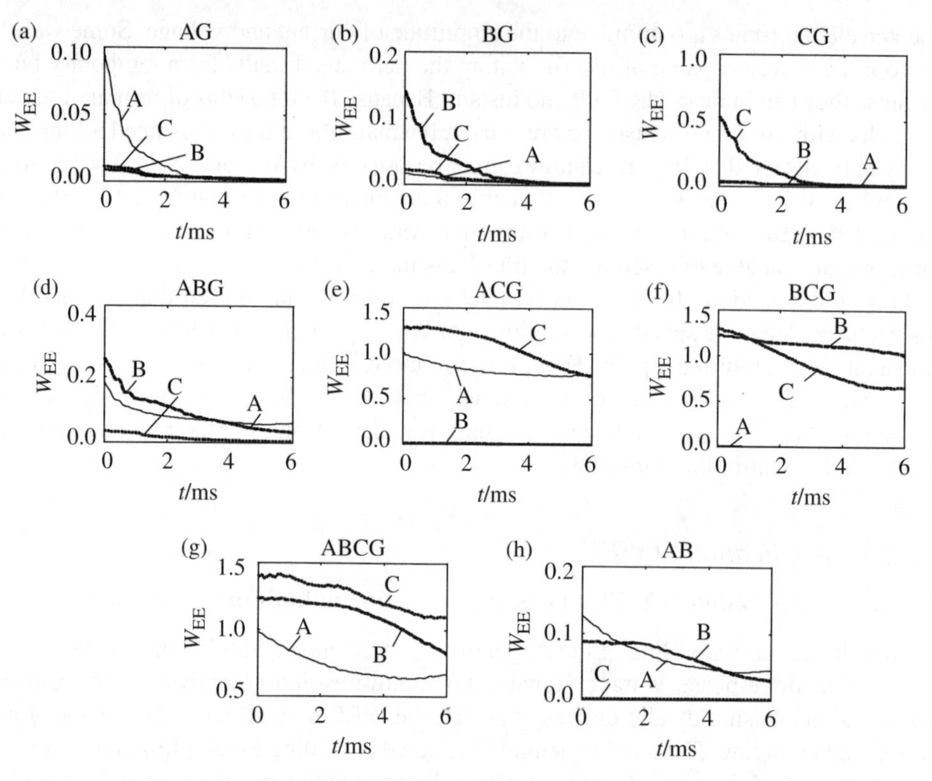

Figure 12.9 WEE of three-phase voltage under different fault types

2. Let $E_{j\min}$ and $E_{j\max}$ be the minimum and maximum of N wavelet energies at scale j. Dividing $(E_{j\min}, E_{j\max})$ into L subintervals, let $\delta_j = (E_{j\max} - E_{j\min})/L$, then the number of energies falling into the ith subinterval is N_{ji}. Letting $p_{ji} = N_{ji}/N$, then $\sum_i p_{ji} = 1$. Thus, we define the second wavelet energy entropy as

$$W_{EE2j} = -\sum_i p_{ji} \log p_{ji} \tag{12.16}$$

The difference of time–frequency distribution of power systems transients can represent the energy distribution differences of different time–frequency sections in the time–frequency plane. The uniformity of energy distribution at each time–frequency section reflects the distinction of transient features. The wavelet energy entropies defined in this chapter are capable of describing the uniformity of energy distribution quantitatively and measure the uncertainty and complexity of transients. Therefore, the defined wavelet energy entropies have good feature extraction and classification ability.

Taking into account five transients – namely, single phase-to-ground fault, breaker switching, capacitor switching, primary arc fault, and lightning – the first wavelet energy entropies and second wavelet energy entropies of five transients are calculated using

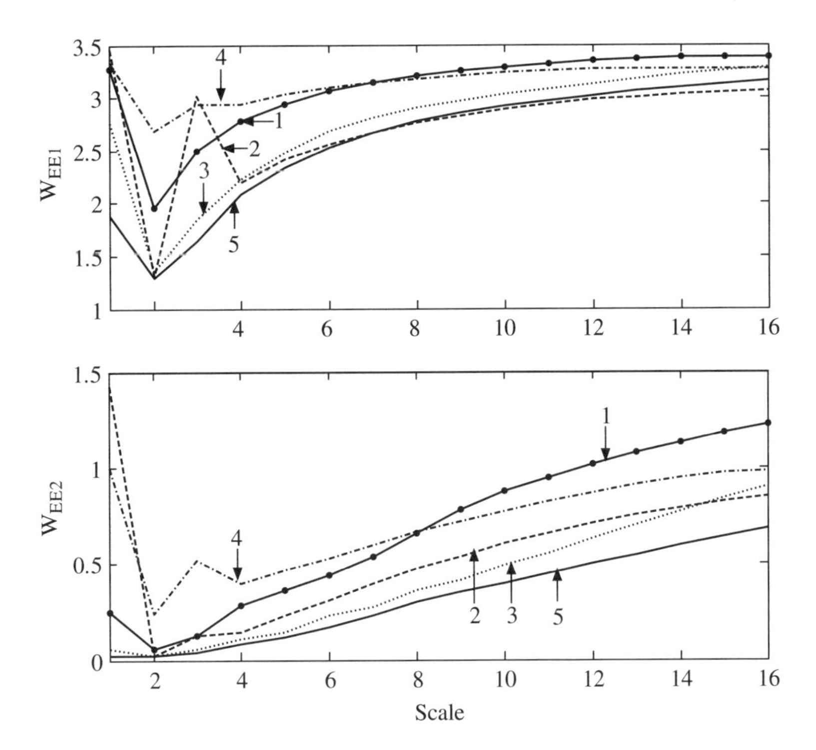

Figure 12.10 Two wavelet energy entropies of five kinds of transients: (1) single phase-to-ground fault, (2) breaker switching, (3) capacitor switching, (4) primary arc fault, (5) lightning

Equations (12.15) and (12.16). Figure 12.10 shows the calculated wavelet energy entropies. The first wavelet energy entropy is similar to the second wavelet energy entropy. As shown in Figure 12.10, both wavelet energy entropies have minima at scale 2. The wavelet energy entropies of single phase-to-ground fault is similar to that of primary arc fault. These two wavelet energy entropies have a crossover point in the neighborhood of scale 8, which indicates that these two transients have high similarity and are hard to distinguish from each other. The transient signal of breaker switching has two local minima in the neighborhood of scale 2 and scale 3. In the aspect of amplitudes of wavelet energy entropy, the amplitudes of wavelet energy entropies corresponding to arc fault, single phase-to-ground fault, breaker switching, capacitor switching, and lightning are in descending order from scale 2 to scale 8. When the scale is greater than 8, the amplitude order still has a certain rule, although the order changes to some degree. For example, the wavelet energy entropy of lightning is minimal, whereas the wavelet energy entropy of capacitor switching is the second-most minimal.

In order to investigate the universal law of distribution curves and values of five transient signals, each transient is simulated under different conditions to obtain 20 cases. Calculate the first and second wavelet energy entropies for each case. The average values of two wavelet energy entropies in 20 cases are shown in Figure 12.11. It can be seen that the wavelet energy entropies represent the similar distribution rule.

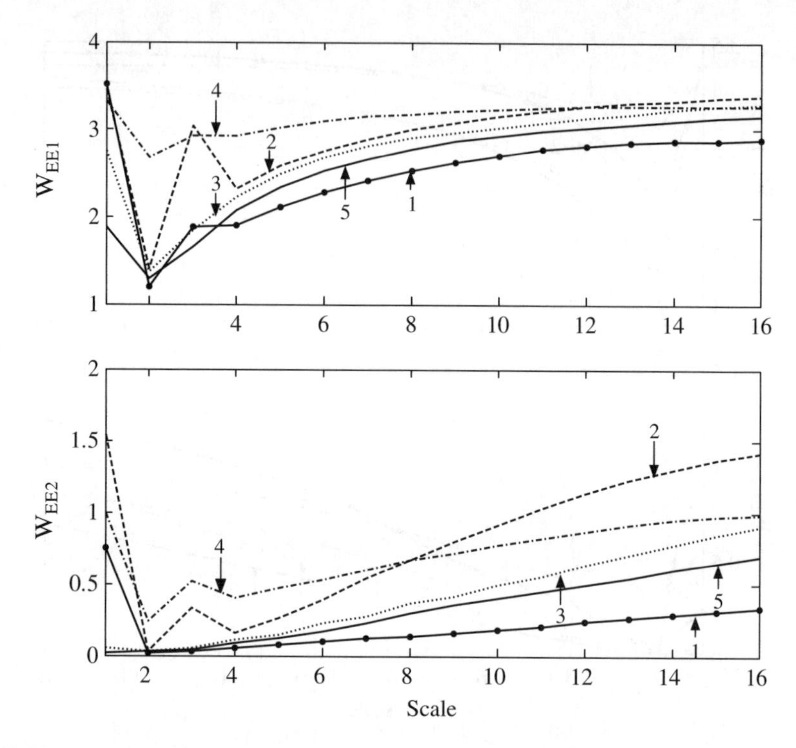

Figure 12.11 The average value of two wavelet energy entropies of five kinds of transients under different conditions

Meanwhile, the analysis of wavelet energy entropies distribution for each transient signal indicates that the first wavelet energy entropy of lightning nearly coincides with the second wavelet energy entropy regarding changing the amplitude, the rise time and fall time of the waveform, or the lightning strike inception time. For breaker switching, the switching time does not affect the shape of the wavelet energy entropies curves. The two wavelet energy entropies of capacitor switching are nearly immune to the transient inception time and amplitude. For single phase-to-ground fault, different fault resistances change the amplitudes of two wavelet energy entropies. However, the distribution rule still exists.

Hence, one can see that wavelet energy entropy has better performance in extracting features than the wavelet energy statistic for breaker switching, capacitor switching, and lightning strikes. These features are immune to fault conditions and are easy to identify. However, it is difficult to identify single phase-to-ground faults and arc faults by wavelet energy entropy. Fortunately, these two transients are faults that should be cleared by protection.

12.2.3 Application of wavelet time–frequency entropy [4]

Wavelet time–frequency entropy measures consist of two vectors. One vector has the time ergodicity, and the other vector has frequency ergodicity. In this section, wavelet time–frequency entropy with frequency ergodicity is mainly applied to identify six transient

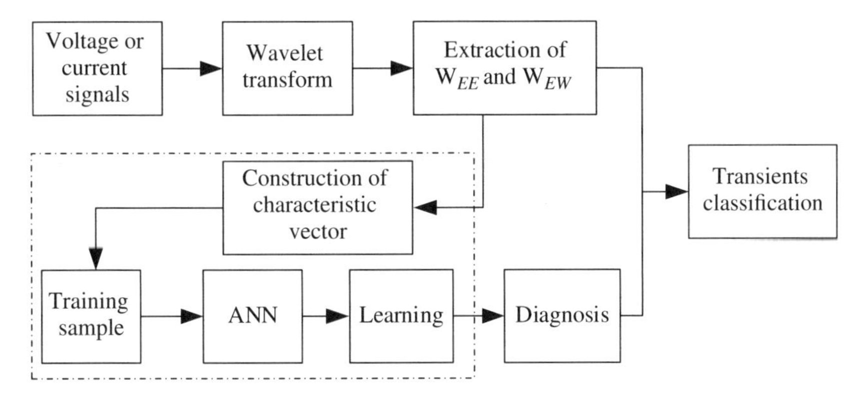

Figure 12.12 Identifying process of transient signals

signals: breaker switching, capacitor switching, single phase-to-ground fault, arc fault, lightning interruption, and lightning strike. Figure 12.12 shows the flow chart for identifying transient signals by wavelet time–frequency entropy and wavelet entropy weight. The feature extraction process based on wavelets is described in detail. The neural network in the dotted box is not described here.

12.2.3.1 Definition of wavelet entropy weight

The weight represents the importance of each character in the decision process and reflects the reliability of each character in representing the original signal feature. Therefore, the process of combining the wavelet entropy and weight is discussed in this subsection. The wavelet entropy weight is defined as follows.

Definition 12.7

1. *Standardize the wavelet coefficients $D_j(k)$ to obtain the dimensionless knowledge data.* The purpose is to eliminate the unit and measurement differences among parameters, and to build the standard submatrix for all the parameters. There are many dimension-eliminating methods, and the following equation is adopted in this book:

$$q_{j,k} = \frac{\min_j \left|D_j(k) - D_0(k)\right| + \lambda \max_j \left|D_j(k) - D_0(k)\right|}{\left|D_j(k) - D_0(k)\right| + \lambda \max_j \left|D_j(k) - D_0(k)\right|} \quad k = 1, 2, \cdots, N \tag{12.17}$$

$D_0 = \{D_0(1), D_0(2), \ldots, D_0(N)\}$ are the wavelet coefficients of normal steady-state signals. $D_j = \{D_j(1), D_j(2), \ldots, D_j(N)\}$ are the wavelet coefficients of each type of transients to be identified. Constant λ is called the identification coefficient and is usually chosen as $\lambda = 0.5$ in engineering applications.

2. *The entropy at each scale (frequency band) is*

$$En_j = -\sum_{k=1}^{n} \frac{q_{j,k}}{\sum_{k=1}^{n} q_{j,k}} \ln \frac{q_{j,k}}{\sum_{k=1}^{n} q_{j,k}} \quad j = 1,2,\cdots,M \tag{12.18}$$

Its value reflects the difference between each type of transient and steady-state signal at this frequency band. Larger values are more effective in reflecting the differences; thus, they are more important. The maximum entropy is $En_{\max} = \ln N$.

3. *Standardize En_j*. The relative importance of the entropy at frequency band j is $e_j = En_j / En_{\max}$.
4. *Define the W_{EW} as*

$$W_{EW\,j} = \frac{1 - e_j}{M - \sum_{j}^{M} e_j} \tag{12.19}$$

12.2.3.2 Feature extraction

The zero mode current $i_0 = \frac{1}{3}(i_a + i_b + i_c)$ of six transients is obtained from the EMTDC (Manitoba Hydro) simulation, and their W_{TFEf} and W_{EW} at scales 1–16 are calculated according to Equations (12.11) and (12.19). The waves of W_{TFEf} and W_{EW} in half cycle are shown in Figure 12.13 with a sampling frequency of 100 kHz.

From the trends of W_{TFEf} and W_{EW} in Figure 12.13, six transients can be divided into three groups: group A includes breaker switching and capacitor switching; group B includes primary arc fault; and group C includes short-circuit fault, lightning disturbance, and lightning strike fault. In order to analyze the transient feature content in W_{TFEf} and W_{EW}, more simulations are performed under various conditions. W_{TFEf} and W_{EW} are calculated at a sampling frequency of 100 kHz and scale 16. The waves of W_{TFEf} and W_{EW} are shown in Figures 12.14, 12.15, 12.16, 12.17, 12.18, and 12.19. There are 10 curves in each figure, which stand for 10 conditions.

Some conclusions are obtained from Figures 12.14, 12.15, 12.16, 12.17, 12.18, and 12.19: (i) under different conditions, W_{TFEf} of breaker switching and capacitor switching varies within scales between 5.8 and 6.4, and the W_{EW} of breaker switching and capacitor switching is nearly the same; (ii) W_{TFEf} and W_{EW} are almost the same when a lightning strike occurs at different times or with different lightning currents; and (iii) the distributions of other W_{TFEf} and W_{EW} are different, but the trends are nearly consistent and are concentrated in a small range. From this analysis, it could appear that different conditions have little effect on the distribution differences of W_{TFEf} and W_{EW}, so using them to analyze transient signals is feasible.

Figure 12.20 shows the mean distribution of W_{TFEf} and W_{EW} under different conditions that can distinguish most transients obviously. W_{TFEf} and W_{EW} of breaker switching and

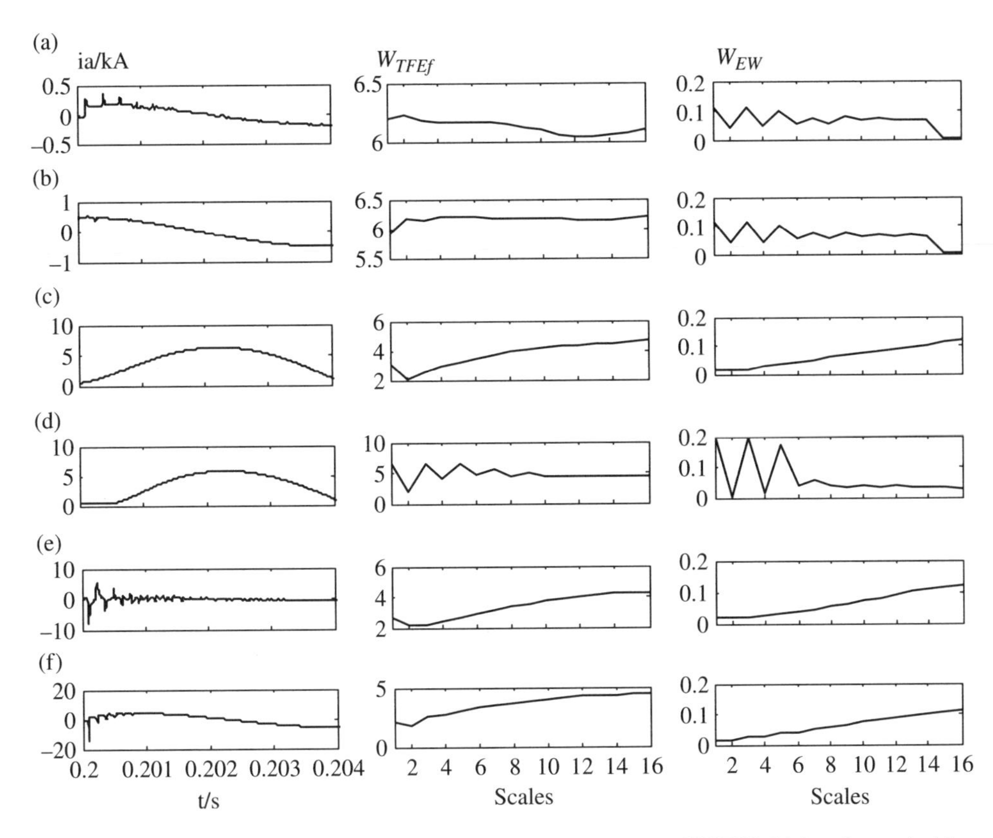

Figure 12.13 Six transient signals and their multiscale W_{TFEf} and WEW: (a) breaker switching, (b) capacitor switching, (c) single phase-to-ground fault, (d) primary arc fault, (e) lightning interrupt, (f) lightning strike

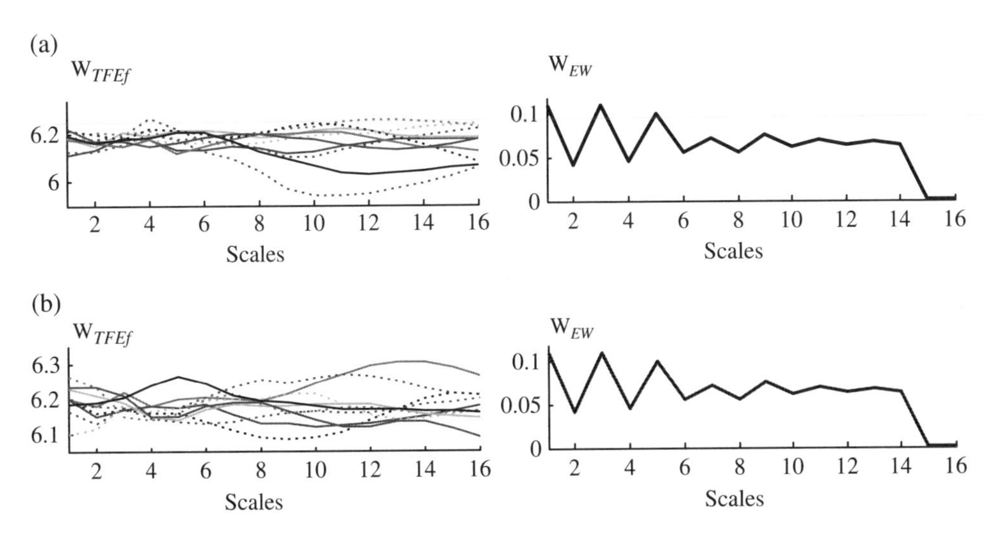

Figure 12.14 W_{TFEf} and W_{EW} of breaker switching under different conditions: (a) different switch closing time of breaker (t = 0.201, 0.202, 0.203, 0.204, 0.205, 0.206, 0.207, 0.208, 0.209, 0.21 s), (b) switching closing of an unideal breaker with different closing resistances (R = 0, 5, 10, 20, 30, 40, 50, 60, 80, 100 Ω)

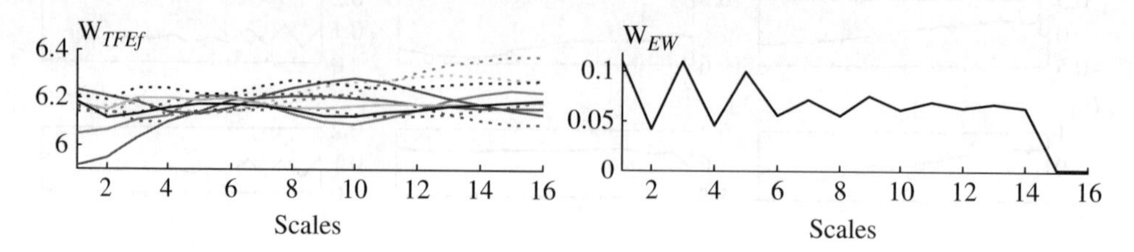

Figure 12.15 W_{TFEf} and W_{EW} of capacitor switching at different times

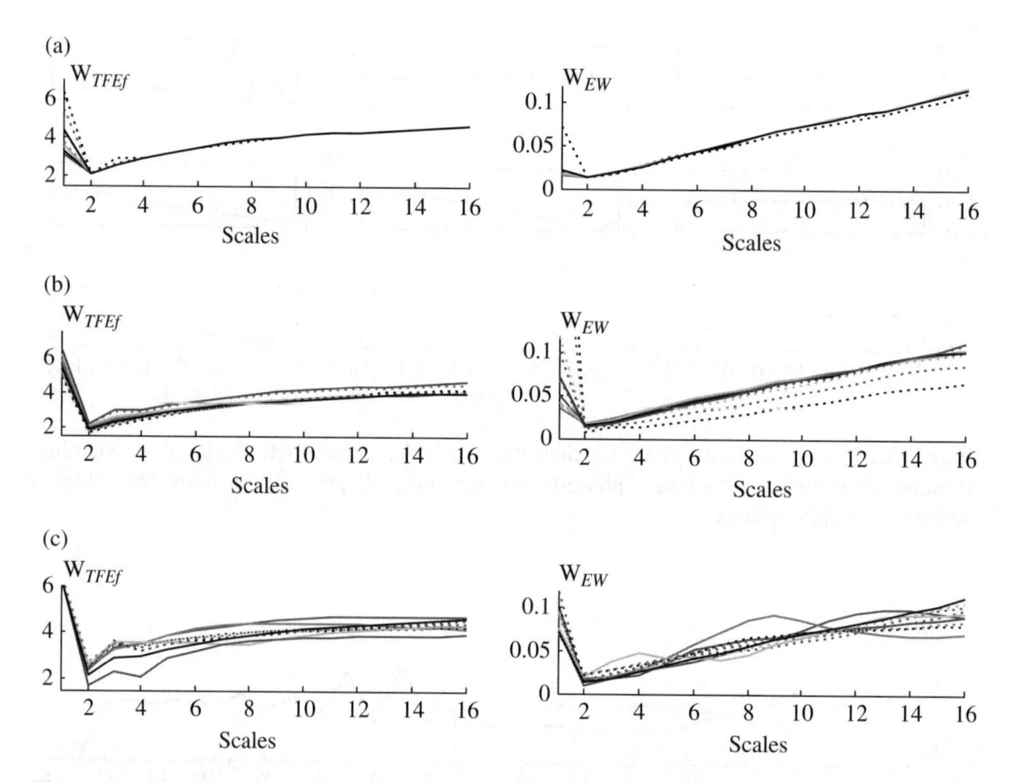

Figure 12.16 W_{TFEf} and W_{EW} of single phase-to-ground fault under different conditions: (a) different fault times (t = 0.201, 0.202, 0.203, 0.204, 0.205, 0.206, 0.207, 0.208, 0.209, 0.21 s), (b) different fault resistances (R = 0, 5, 50, 100, 200, 300, 400, 500, 1000, 2000 Ω), (c) different fault locations (d = 5, 10, 20, 30, 50, 55, 60, 70, 80, 90 km)

capacitor switching are the same, whereas other transients have their own characteristics. Consequently, group C mentioned above can be divided into three subtypes: short-circuit fault (C_1), lightning disturbance (C_2), and lightning strike fault (C_3). Therefore, all six transients could be classified into five kinds: switching operation (includes breaker switching and capacitor switching), short-circuit fault, primary arc fault, lightning disturbance, and fault lightning strike.

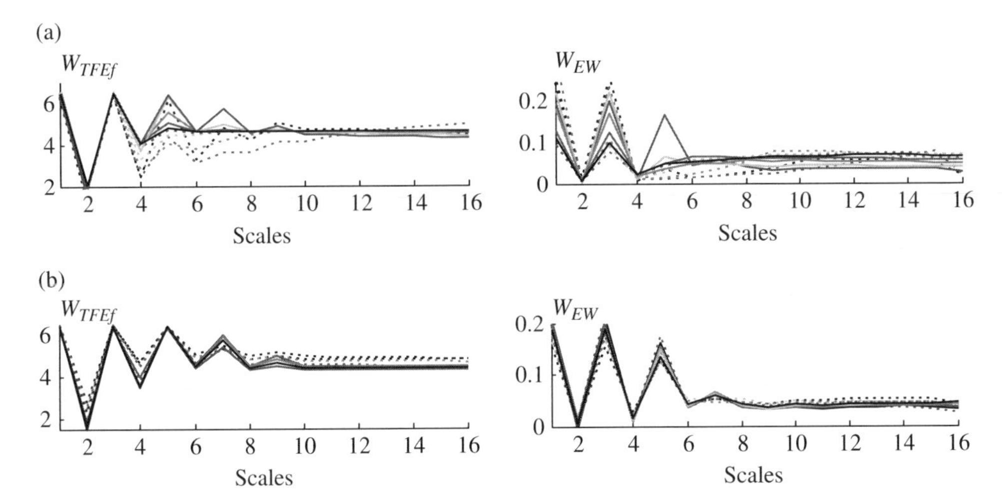

Figure 12.17 W_{TFEf} and W_{EW} of primary arc fault under different conditions: (a) different fault times ($t = 0.201$, 0.202, 0.203, 0.204, 0.205, 0.206, 0.207, 0.208, 0.209, 0.21 s), (b) different fault locations ($d = 5$, 10, 20, 30, 50, 55, 60, 70, 80, 90 km)

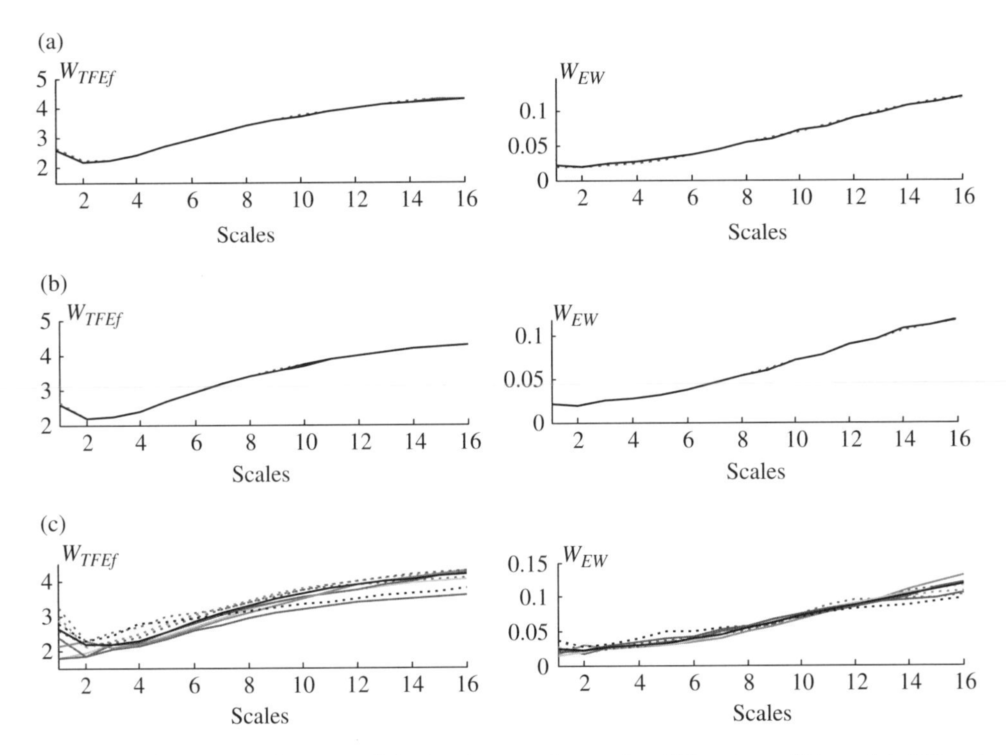

Figure 12.18 W_{TFEf} and W_{EW} of lightning interruption under different conditions: (a) different lightning currents ($I_0 = 5,10,12,15,18$ kA; $\tau_1/\tau_2 = 1.2/50$ μs, 4/100 μs), (b) different lightning strike times ($t = 0.201$, 0.202, 0.203, 0.204, 0.205, 0.206, 0.207, 0.208, 0.209, 0.21 s), (c) different lightning strike locations ($d = 5$, 10, 20, 30, 50, 55, 60, 70, 80, 90 km)

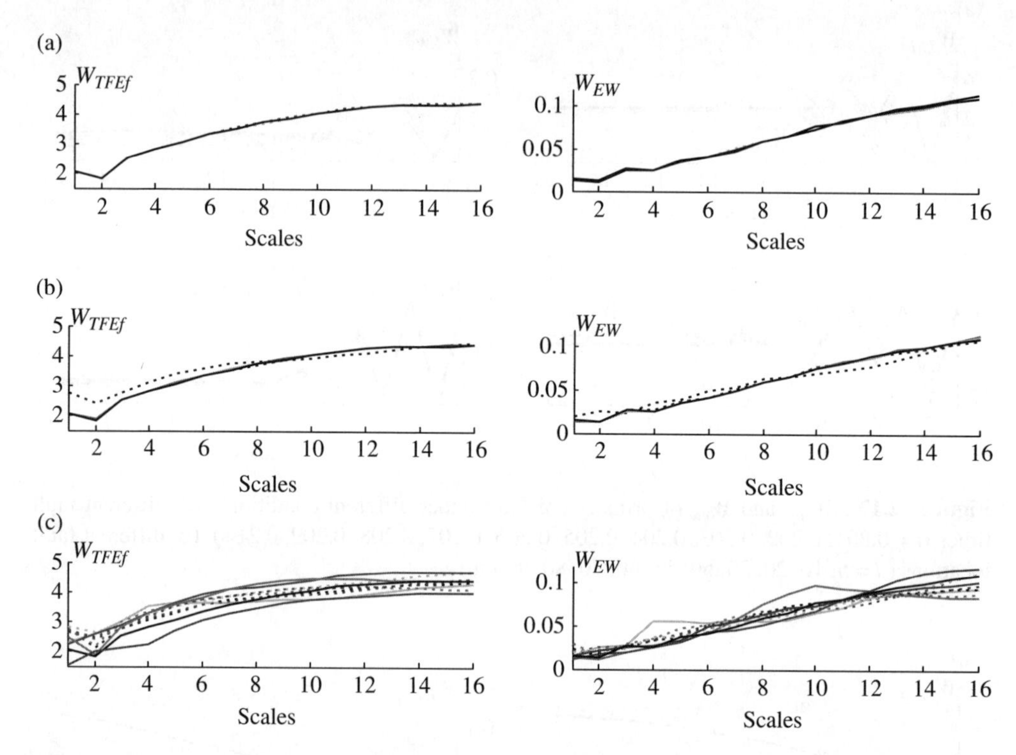

Figure 12.19 W_{TFEf} and W_{EW} of lightning strike under different conditions: (a) different lightning currents (I_0 = 30, 45, 50, 75, 100 kA; τ_1/τ_2 = 1.2/50 μs, 4/100 μs), (b) different lightning times (t = 0.201, 0.202, 0.203, 0.204, 0.205, 0.206, 0.207, 0.208, 0.209, 0.21 s), (c) different lightning strike locations (d = 5, 10, 20, 30, 50, 55, 60, 70, 80, 90 km)

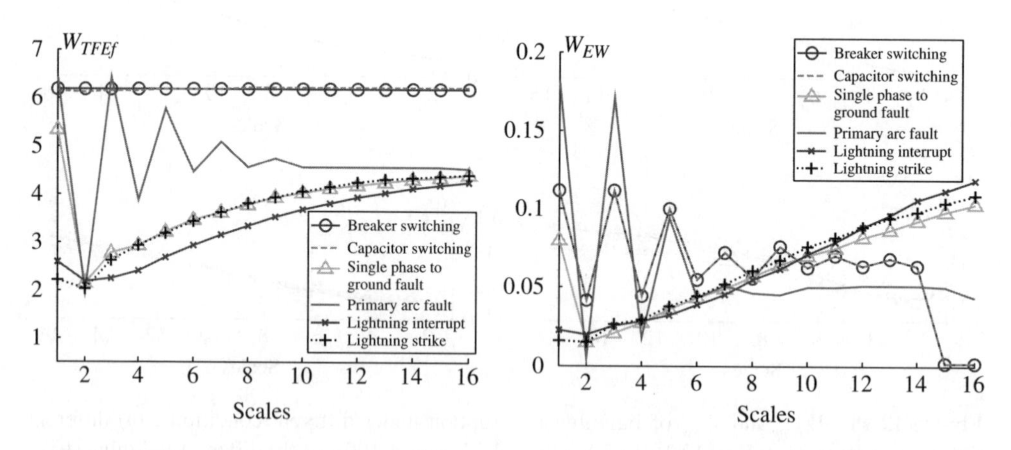

Figure 12.20 The average W_{TFEf} and W_{EW} of five transient signals under various conditions

Based on qualitative analysis of multiple scales, transients can be successfully identified directly by eye inspection but still cannot be identified by computer (or other intelligent machines) precisely. Precise classification can be achieved by neural networks or other artificial intelligence methods.

12.2.4 Application of wavelet singular entropy [5]

According to the definition given in this chapter, WSE is used to map the correlative wavelet space into independent linearity space, and to indicate the uncertainty of the energy distribution in the time–frequency domain with a high immunity to noise. Due to its way of implementation, WSE is sensitive to the transients produced by the faults, and the fewer modes the transients congregate to, the smaller the WSE is. Therefore, WSE will be suitable and useful for measuring the uncertainty and complexity of the analyzed signals, and will provide an intuitive and quantitative outcome for fault diagnosis that can be utilized to overcome the drawbacks in the previous methodologies.

12.2.4.1 Application of WSE in fault classification

The WSE value is different between the faulty phase and sound phase, and the WSE values of the faulty phase are much greater than those of the sound phase. However, the uncertainty in system and fault conditions would influence the WSE value as well as the classification results. In order to remove the effect of uncertainty and obtain reliable fault classification under various conditions, three indicators (m_a, m_b, and m_c) are introduced here to indicate the WSE value of each phase.

$$m_a = \sum_{k=4}^{8}\left(\frac{WSE_k(a)}{WSE_k(b)} + \frac{WSE_k(a)}{WSE_k(c)}\right) \tag{12.20}$$

$$m_b = \sum_{k=4}^{8}\left(\frac{WSE_k(b)}{WSE_k(a)} + \frac{WSE_k(b)}{WSE_k(c)}\right) \tag{12.21}$$

$$m_c = \sum_{k=4}^{8}\left(\frac{WSE_k(c)}{WSE_k(a)} + \frac{WSE_k(c)}{WSE_k(b)}\right) \tag{12.22}$$

where a, b, and c indicate phases A, B, and C, respectively; and WSE_k denotes the k-order WSE of the phase signal during the first half cycle after fault inception. The WSE ratio, such as $\frac{WSE_k(a)}{WSE_k(b)}$, indicates the relative differences between the faulty phase and sound phase, which is used to highlight the differences of faulty phases. The summation of WSE ratios is used to accumulate the differences of faulty phases under a different-order WSE. By virtue of this treatment, one can obtain more accurate and reliable fault classification. The flow chart of WSE-based fault classification is shown in Figure 12.21.

In Figure 12.21, the values of threshold ε_1, ε_2 are determined according to the system situation, generally in virtue of m_a, m_b, and m_c. When a three-phase-to-ground fault occurs, ε_1 and ε_2 are set to be about the minimum and maximum, respectively, of m_a, m_b, and m_c. ε is set to be no less than ε_2. In this book, we set $\varepsilon_1 = 9$, $\varepsilon_2 = 11$, and $\varepsilon = 1.5$.

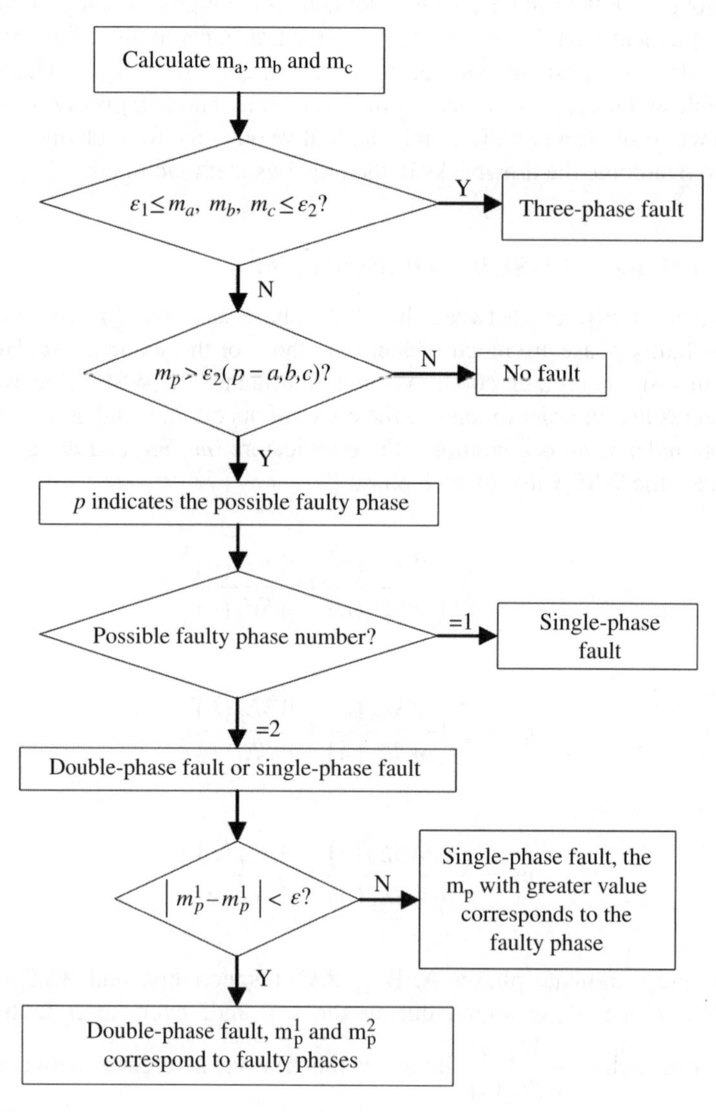

Figure 12.21 Flowchart of the WSE-based fault classification algorithm

12.2.4.2 Fault classification tests in simulation based on WSE

The system shown in Figure A.4 is used for simulation tests. Also, the white-noise model has been included in this simulation, and the signal-to-noise ratio (SNR) is set to 40 according to the actual system. The current as well as the voltage transient of each phase, which is measured in one end, are analyzed in the case of fault classification. Their performances are similar to each other. Therefore, we only illuminate the test results of voltage transient to verify the algorithm. The sampling frequency is set to be 20 kHz, the db4 mother wavelet and four-scaled wavelet transform are chosen, and we take the 200-sample-long sequence (i.e., half-cycle data after fault inception) as the input of WSE.

The simulation results under different fault types, fault times, fault resistances, and fault locations are shown in Figures 12.22, 12.23, 12.24, and 12.25.

The summary of all fault classification tests is shown in Table 12.1. It can be seen that the faults can be classified correctly, benefiting from the proposed WSE-based algorithm, which proves the reliability of the WSE-based algorithm and the validity of the threshold

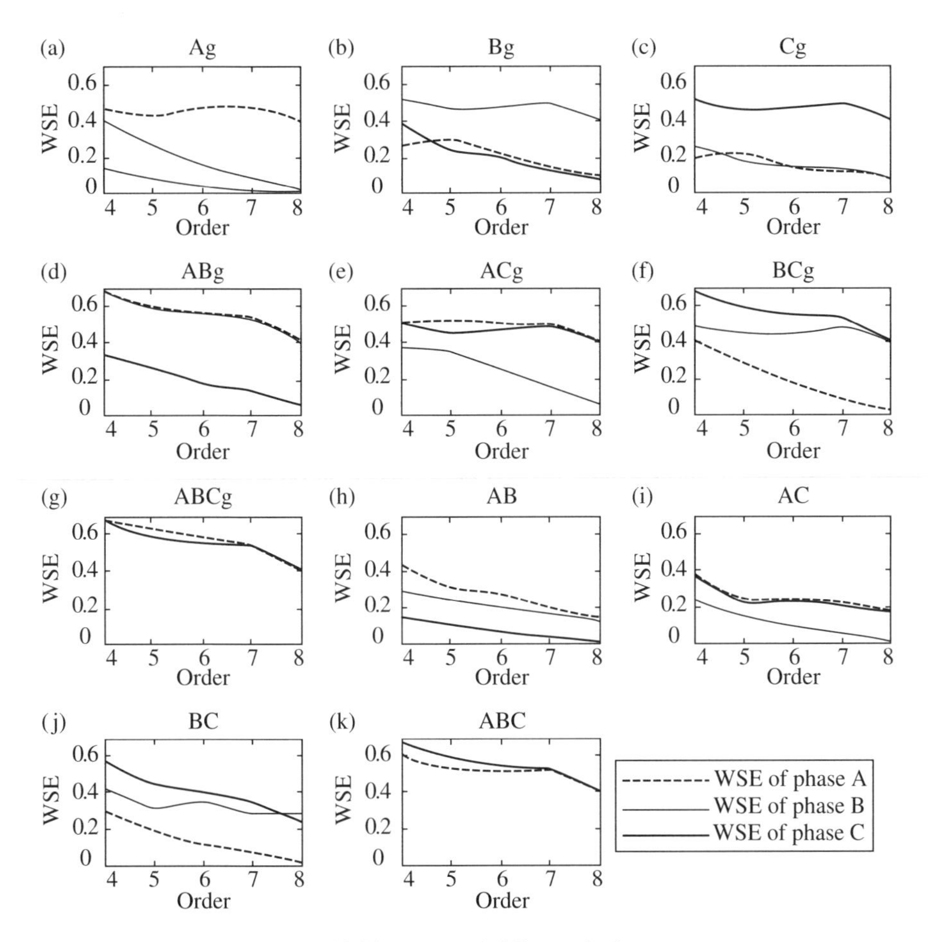

Figure 12.22 WSE of different fault types

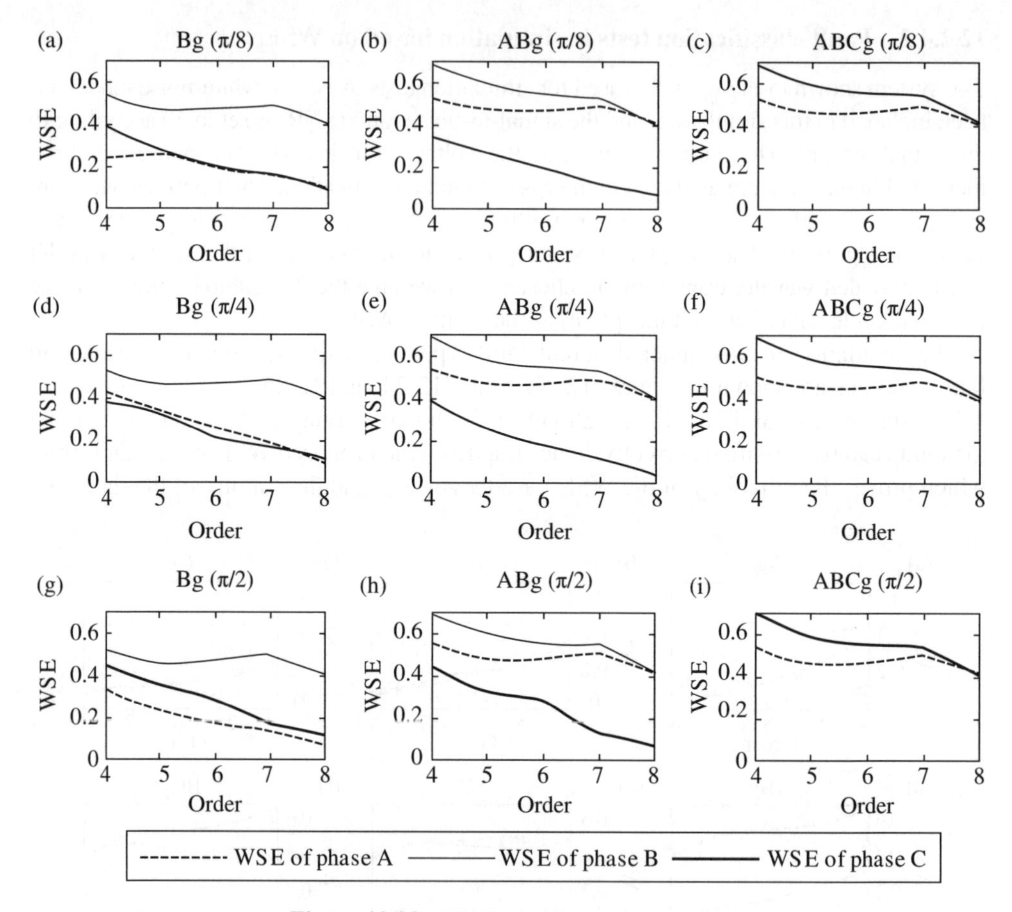

Figure 12.23 WSE of different fault times

selection in the application. This proposed fault classification algorithm is provided with high accuracy and is immune to different conditions, such as fault types, fault times, fault resistances, and fault locations. Because the proposed algorithm only requires the first half cycle of the postfault signal, it has good real-time performance.

12.3 Fault diagnosis based on wavelet entropy fusing model

Due to the diversity, uncertainty and complexity of faults, it will be difficult to diagnose faults accurately with single wavelet entropy. If various wavelet entropies can extract features from different perspectives, the extracted features will assistant each other and provide more reasonable and comprehensive results. Therefore, information fusion technology is introduced to fuse the fault features extracted by various wavelet entropies. The fused fault features are capable of reflecting the fault more reliably. This section will put forward an information fusion method of wavelet entropy, with evidence, and its application in fault discrimination.

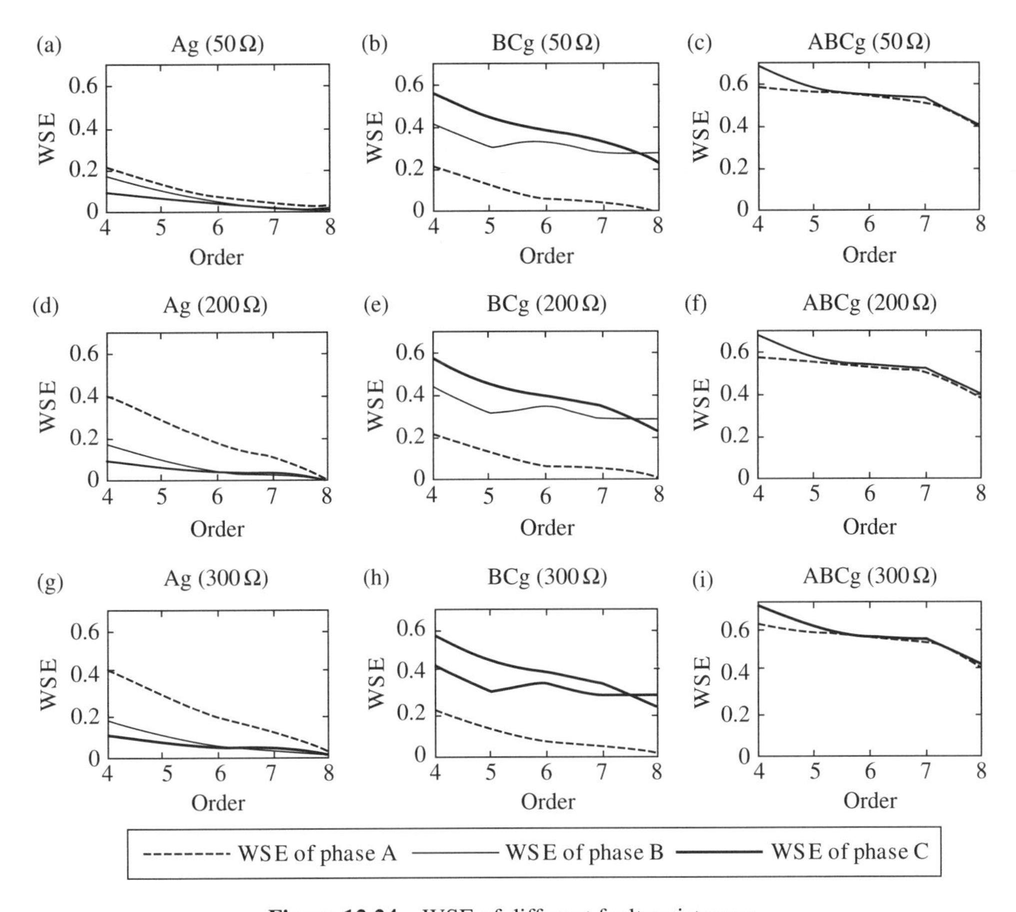

Figure 12.24 WSE of different fault resistances

12.3.1 Fusing model of different wavelet entropies based on Dempster–Shafer (DS) evidence theory [6]

12.3.1.1 Basic theory of DS evidence fusion

Now that we want to fuse the features, a fusion algorithm is needed. As an important method in the information fusion field, DS evidence fusion theory provides an evidence fusion algorithm to combine various forms of evidence into new evidence. This new evidence collates the information of various evidence and focuses on the common support points of various evidence.

Definition 12.8 Let F be the framework of discernment, where F contains p types of faults. Let set function m: $2^F \rightarrow [0,1]$ (2^F is the power set of F) meet $m(\Phi) = 0$ and $\sum_{F_k \subseteq F} m(F_k) = 1, (k = 1, 2, \ldots, p)$, where F_k denotes fault of type k, and m denotes the basic probability assignment on discernment framework F. $\forall F_k \in 2^F$, $m(F_k)$ is called the basic probability number of F_k, and it represents the uncertainty or supporting rate of a type k fault.

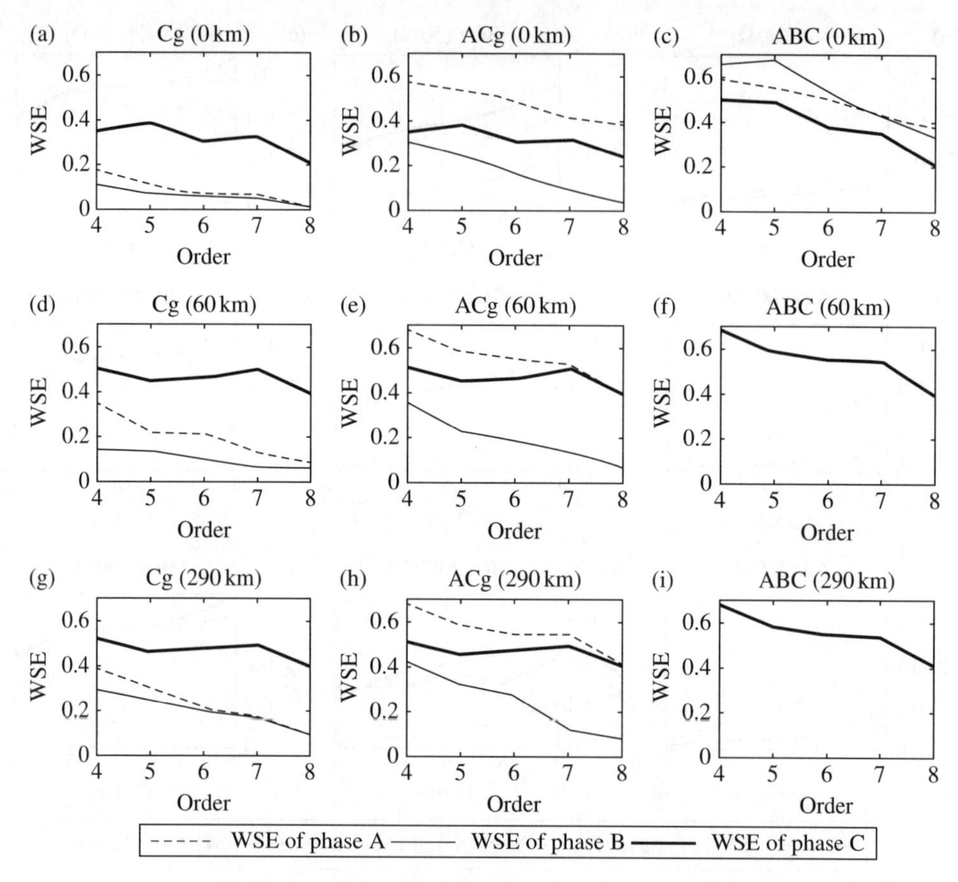

Figure 12.25 WSE of different fault locations

Table 12.1 Summary of results obtained in fault classification (t_A is the fault inception angle of phase A; R is the fault resistance; and d is the fault location)

Simulation variable	Variable details	Other conditions	Number of samples	Failed fault classification
Fault type	11 types of fault repeat for five times	$t_A = 0°$, $R = 0\ \Omega$, $d = 50$ km	55	0
Fault inception angle	From 0° to 360° step by 5°	$R = 0\ \Omega$, $d = 50$ km	72	0
Fault resistance	From 0 Ω to 300 Ω step by 5 Ω	$t_A = 0°$, $d = 50$ km	61	0
Fault location	From 0 km to 300 km step by 5 km	$t_A = 0°$, $R = 0\ \Omega$	61	0

Definition 12.9 Let $m_1, m_2, \cdots, m_q$ be the basic probability on the discernment framework F; the orthogonal sum of the multiple distribution function $m = m_1 \oplus m_2 \oplus \cdots \oplus m_q$ is expressed in Equation (12.23):

$$\begin{cases} m(\Phi) = 0 \\ m(F_k) - K \sum_{\cap F_i = F_k} \prod_{1 \le j \le q} m_j(F_i), F_k \ne \Phi \\ \quad (k, i = 1, 2, \ldots, p) \end{cases} \tag{12.23}$$

where

$$K^{-1} = 1 - \sum_{\cap F_i = \Phi} \prod_{1 \le j \le q} m_j(F_i) \tag{12.24}$$

Equations (12.23) and (12.24) are the core of DS evidence theory, and they can combine several individual forms of evidence. K is often interpreted as a measure of conflict between the different sources. The larger K is, the more the sources are conflicting, and the less sense their combination has. Finally, the orthogonal sum does not exist when $K = \infty$, and in that case, the sources are said to be totally or flatly contradictory, and it is no longer possible to combine them. Therefore, we give a threshold to K and consider that the combination has no sense when K is larger than the threshold; in that case, we need to find new sources of evidence or amend the DS evidence theory itself.

12.3.1.2 Fusing model of different wavelet entropies

1. *Pattern identification framework*
   The pattern identification framework of DS evidence theory is a pattern set that contains the basic elements of all the possible judgment results or decision conclusions. In power systems fault investigation, the final decision is pointing out the fault state of power systems. Therefore, all possible fault states make up the identification framework of fault investigation.
   Assuming that there are p fault states in all, these p fault states make up the identification framework. Let this framework be F:

$$F = \{F_k \,|\, k = 1, 2, \ldots, p\} \tag{12.25}$$

   where F_k represents the type k fault state.
2. *Constructing the basic probability assignment based on a norm's weighted average*
   As mentioned before, the basic probability assignment m represents the supporting degree of fault state F_k according to the evidence. For any possible fault states, $m(F_k)$ measures the uncertainty or supporting degree of a type k fault.

Basic probability assignment is the foundation of DS evidence theory, and different assignment functions will bring different fusing models. To simplify the fusing model, this book uses the wavelet entropies to form the assignment function. If X is a $p \times q$ matrix formed by q types of wavelet entropies corresponding to p types of faults, then the basic probability corresponding to an i-type fault and j-type entropy is defined as follows:

$$m_j(F_i) = \frac{(1-\mu_j)x_{ij}^{\ 2}}{\|x_j\|_2^2} \tag{12.26}$$

where $i = 1,2,\ldots,p$; $j = 1,2,\ldots,q$; μ_j is the uncertainty of the j-type evidence and is deduced by statistical or historical data; and $\|x_j\|_2^2$ represents the secondary square norm of j type entropy and can describe the distance between entropies. Equation (12.26) constructs the basic probability assignment-based weighted average norm according to the wavelet entropies' characteristics.

3. *Establishing evidence a fusing model*
   Taking WSE, WEE, WTFE, and WDE as the evidence, we can obtain four groups of basic probability assignments.

   WSE evidence 1: $m_1(F_1)$, $m_1(F_2)$, ……, $m_1(F_p)$, $m_1(F)$;
   WEE evidence 2: $m_2(F_1)$, $m_2(F_2)$, ……, $m_2(F_p)$, $m_2(F)$;
   WTFE evidence 3: $m_3(F_1)$, $m_3(F_2)$, ……, $m_3(F_p)$, $m_3(F)$;
   WDE evidence 4: $m_4(F_1)$, $m_4(F_2)$, ……, $m_4(F_p)$, $m_4(F)$.

   The task of evidence fusion is to combine different wavelet entropies into the final evidence. Then, the combined probability assignment is obtained by fusing the basic probability assignments according to Equations (12.23) and (12.24).

12.3.1.3 Decision making based on basic probability

Fault decision making based on DS evidence theory usually meets the following terms: the actual fault type possesses the largest basic probability, and the probability difference between the actual fault type and other fault types must be greater than a threshold.

In fault identification, assume that there are two possible fault states $F_i, F_j \subset F$, and these two fault states have the largest basic probability compared to other fault states. That is to say, $m(F_i) = \max[m(F_k)]$, $m(F_j) = \max[m(F_k)]$, $F_k \subset F$, $k = 1,2,\ldots,p$, $k \neq i$, and $k \neq j$. F_i is the decision-making result if

$$\begin{cases} m(F_k) > \varepsilon_1 \\ m(F_i) - m(F_j) > \varepsilon_2 \end{cases} \tag{12.27}$$

where ε_1 and ε_2 are the presetting threshold values: ε_1 is set to make sure that the basic probability of F_i is large enough to make the decision, and ε_2 is set to make sure that the fault decision has enough discrimination.

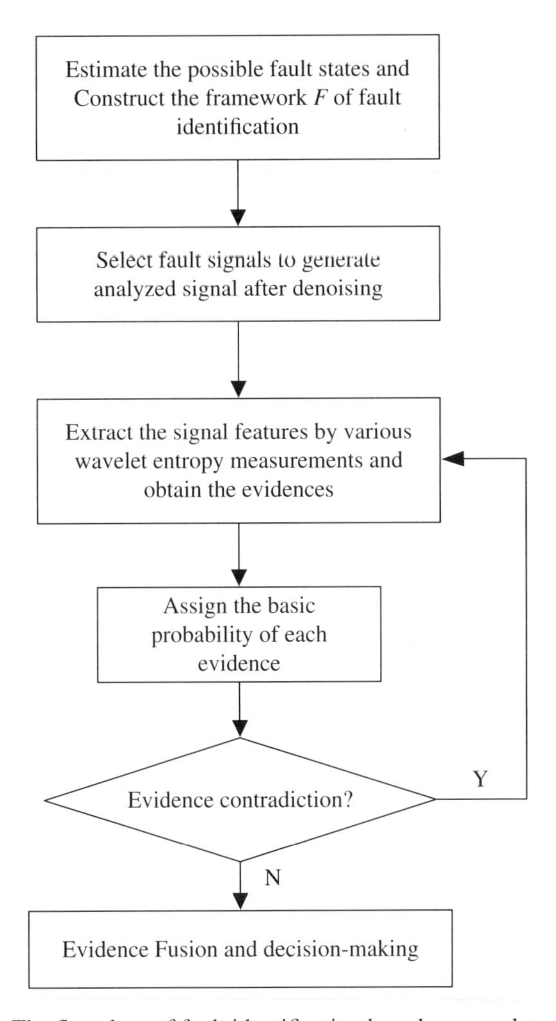

Figure 12.26 The flowchart of fault identification based on wavelet entropies fusion

In conclusion, the flowchart of fault identification appears in Figure 12.26.

First, various wavelet entropies are used to extract characteristics, and the evidence for fusing is formed. Then, the weighted average norm is employed to assign the basic probability of each evidence. Finally, evidence is fused by the DS theory, and the decision is made for fault identification.

12.3.2 Fault classification based on a fusing model of wavelet entropy [6]

The simulated 500 kV transmission system is shown in the Appendix. Take the identification of single phase-to-ground fault in power transmission lines as an example. During a single phase-to-ground fault, the fault phase will have more abundant transients than the nonfault phase due to the superimposition of fault components. The time–frequency distribution in the fault phase will be much more complex than that in the nonfault phase. So the wavelet entropies between the fault phase and nonfault phase are different. Taking the voltage signal as the information source, the WSE, WEE, WTFE, and WDE are calculated as the features, and they are named as the evidence m_1, m_2, m_3, and m_4. For direct comparison, we assume that the fuzzy uncertainties μ of four wavelet entropies are zero. Then, by means of the basic probability assignment in Equation (12.26) and the evidence combination principles in Equations (12.23) to (12.24), as well as the decision-making conditions in Equation (12.27), the fault phase can be identified. The decision-making thresholds ε_1 and ε_2 are selected as 0.6 and 0.45 according to the historical data and experience analysis.

1. The identification results of single phase-to-ground fault by means of the fusion of two arbitrary wavelet entropies are shown in Table 12.2.
2. The identification results of single phase-to-ground fault based on the fusion of three arbitrary wavelet entropies are shown in Table 12.3.
3. The identification results of single phase-to-ground fault based on the fusion of four arbitrary wavelet entropies are shown in Table 12.4.

The simulation results in Tables 12.2, 12.3, and 12.4 show that the fusion of multiple wavelet entropies could improve the supporting rate for a single phase-to-ground fault. From the comparison of these three tables, we know that the more fused evidence there

Table 12.2 The results of fault phase identification based on the fusion of two arbitrary wavelet entropies

Fault type	Fusion type	Phase A	Phase B	Phase C	Decision making
Ag	WSE m_1	0.62624	0.19226	0.18150	Uncertain
	WEE m_2	0.38377	0.28129	0.33494	Uncertain
	Combination m	0.67660	0.15225	0.17115	Ag
Bg	WTFE m_3	0.15238	0.56182	0.28580	Uncertain
	WDE m_4	0.26319	0.69416	0.04265	Uncertain
	Combination m	0.09068	0.88176	0.02756	Bg
Cg	WSE m_1	0.23080	0.04389	0.72531	Uncertain
	WTFE m_3	0.24720	0.14514	0.60766	Cg
	Combination m	0.11316	0.01264	0.87420	Cg
Ag	WSE m_1	0.62624	0.19226	0.18150	Uncertain
	WDE m_4	0.58721	0.18673	0.22606	Uncertain
	Combination m	0.82699	0.08073	0.09227	Ag
Bg	WEE m_2	0.26458	0.53286	0.20256	Uncertain
	WDE m_4	0.26319	0.69416	0.04265	Uncertain
	Combination m	0.15538	0.82535	0.01927	Bg
Cg	WEE m_2	0.24065	0.27641	0.48294	Uncertain
	WTFE m_3	0.24720	0.14514	0.60766	Uncertain
	Combination m	0.15134	0.10206	0.74659	Cg

Table 12.3 The results of fault phase identification based on the fusion of three arbitrary wavelet entropies

Fault type	Fusion type	Phase A	Phase B	Phase C	Decision making
Ag	WSE m_1	0.62624	0.19226	0.18150	Uncertain
	WEE m_2	0.38377	0.28129	0.33494	Uncertain
	WTFE m_3	0.57348	0.26243	0.16409	Uncertain
	Combination m	0.85081	0.08761	0.06157	Ag
Bg	WSE m_1	0.18763	0.61248	0.19989	Uncertain
	WEE m_2	0.26458	0.53286	0.20256	Uncertain
	WDE m_4	0.26319	0.69416	0.04265	Uncertain
	Combination m	0.05413	0.93871	0.00715	Bg
Cg	WSE m_1	0.23080	0.04389	0.72531	Uncertain
	WTFE m_3	0.24720	0.14514	0.60766	Uncertain
	WDE m_4	0.23564	0.23963	0.52473	Uncertain
	Combination m	0.05459	0.00619	0.93920	Cg
Ag	WEE m_2	0.38377	0.28129	0.33494	Uncertain
	WTFE m_3	0.57348	0.26243	0.16409	Uncertain
	WDE m_4	0.58721	0.18673	0.22606	Uncertain
	Combination m	0.83140	0.08867	0.07992	Ag
Bg	WSE m_1	0.18763	0.61248	0.19989	Uncertain
	WEE m_2	0.26458	0.53286	0.20256	Uncertain
	WTFE m_3	0.15238	0.56182	0.28580	Uncertain
	Combination m	0.03735	0.9055	0.05714	Bg
Cg	WSE m_1	0.23080	0.04389	0.72531	Uncertain
	WEE m_2	0.24065	0.27641	0.48294	Uncertain
	WDE m_4	0.23564	0.23963	0.52473	Uncertain
	Combination m	0.06550	0.01455	0.91994	Cg

Table 12.4 The results of fault phase identification based on the fusion of four arbitrary wavelet entropies

Fault type	Fusion type	Phase A	Phase B	Phase C	Decision making
Ag	WSE m_1	0.62624	0.19226	0.18150	Uncertain
	WEE m_2	0.38377	0.28129	0.33494	Uncertain
	WTFE m_3	0.57348	0.26243	0.16409	Uncertain
	WDE m_4	0.58721	0.18673	0.22606	Uncertain
	Combination m	0.94286	0.03087	0.02627	Ag
Bg	WSE m_1	0.18763	0.61248	0.19989	Uncertain
	WEE m_2	0.26458	0.53286	0.20256	Uncertain
	WTFE m_3	0.15238	0.56182	0.28580	Uncertain
	WDE m_4	0.26319	0.69416	0.04265	Uncertain
	Combination m	0.01534	0.98085	0.00380	Bg
Cg	WSE m_1	0.23080	0.04389	0.72531	Uncertain
	WEE m_2	0.24065	0.27641	0.48294	Uncertain
	WTFE m_3	0.24720	0.14514	0.60766	Cg
	WDE m_4	0.23564	0.23963	0.52473	Uncertain
	Combination m	0.02804	0.00365	0.96829	Cg

is, the higher the supporting rate for the fault is. In terms of phase B–to-ground faults, after the fusion of two wavelet entropies' evidence, the supporting rate for fault type improves from 0.7 to 0.82. After the fusion of three wavelet entropies' evidence, the supporting rate for fault type improves to more than 0.9. After the fusion of four wavelet entropies' evidence, the supporting rate for fault type improves to more than 0.98. The evidence fusion could improve the confidence level greatly. In terms of phase A–to-ground fault, due to the effect of uncertain system factors such as fault instant and fault line conditions, the supporting rates for various fault types based on WEE m_2 are 0.38377, 0.28129, and 0.33494. Therefore, it is difficult to distinguish the fault type. However, after the fusion of multiple wavelet entropies' evidence, the supporting rate for phase A–to-ground fault is more than 0.67, 0.83, and 0.94, which is capable of distinguishing the fault type well.

It is apparent that, after the fusion of wavelet entropies' evidence, such evidence can enhance the supporting rate for the fault phase while decreasing the supporting rate for the nonfault phase. This result improves the identification accuracy.

12.3.3 Fault line identification for an ineffectively grounded distribution system [6]

In an ineffectively grounded distribution system, the fault transient signals are always dynamic, unstable, and nonstationary. Therefore, it is not easy to extract the transient features of fault signals accurately. The extracted fault features may be insufficient for fault line identification, especially for weak faults. Because wavelet entropies are capable of making statistical analysis to transient signals, they can capture the fault features sufficiently. The evidence fusion methods based on wavelet entropies provide reliable and feasible solutions for fault line identification because they make full use of various fault transients from different perspectives.

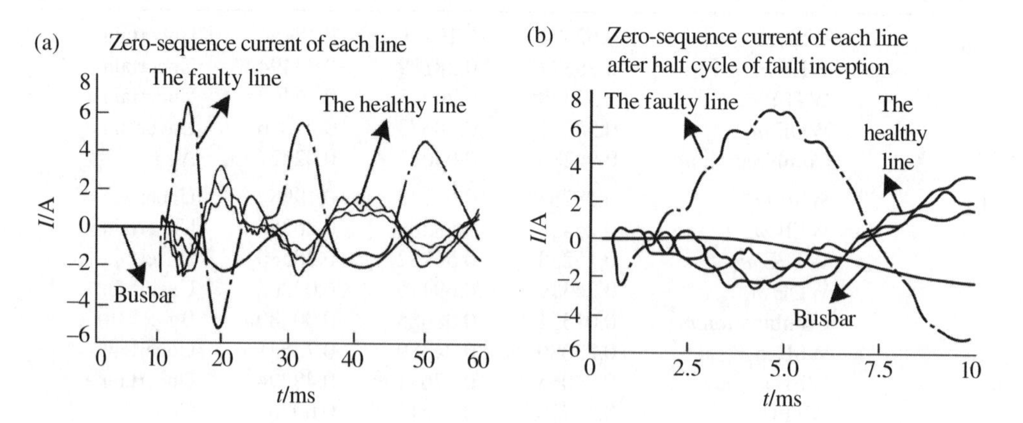

Figure 12.27 Zero sequence of feeders and bus

Table 12.5 Results of fusion of wavelet entropies' evidence for fault line identification

	Line L_1	Line L_2	Line L_3	Line L_4	Fuzzy uncertainty μ
WSE m_1	0.12948	0.10535	0.08934	0.57583	0.10
WEE m_2	0.12006	0.22031	0.20283	0.30679	0.15
WTFE m_3	0.11921	0.09652	0.15503	0.48923	0.14
WDE m_4	0.27751	0.18267	0.14743	0.21239	0.18
Combination m	0.02562	0.02039	0.02063	0.91453	0.01883

The simulations are carried out on the distribution system shown in the Appendix by removing the feeder L_5 and changing the lengths of feeders L_1, L_2, L_3, and L_4 to 10 km, 15 km, 25 km, and 20 km, respectively. The fault feeder is L_4. Take the zero-sequence currents as the information source. From the curves in Figure 12.27, we can see that the zero-sequence current of the fault line is obviously larger than that of the nonfault line. Wavelet entropies within the half cycle after the fault inception are calculated and used as evidence. Using the statistic experience, we assume that the fuzzy uncertainties of WSE, WEE, WTFE, and WDE are, respectively, 0.1, 0.15, 0.14, and 0.18. Thus, wavelet entropies are fused according to Equations (12.23), (12.24), (12.25), and (12.26), and the diagnostic results are shown in Table 12.5.

Table 12.5 demonstrates that the fault line will be identified falsely if we take the single wavelet entropy as the evidence. For example, the fault-supporting rates of each line are, respectively, $m_4(L_1) = 0.27751$, $m_4(L_2) = 0.18267$, $m_4(L_3) = 0.14743$, and $m_4(L_4) = 0.21239$, with the decision-making thresholds $\varepsilon_1 = 0.6$ and $\varepsilon_2 = 0.45$. The fault-supporting reliability of WDE of line L_1 is the greatest. Then, line L_1 is identified as the fault line rather than line L_4. The reason for wrong identification is mainly that the fault signals are disturbed or WDE makes inaccurate statistics for fault features in the time–frequency domain due to the effects of the uncertainties. Therefore, the single wavelet entropy may result in wrong fault line identification. However, after the fusion of four wavelet entropies, the fault-supporting rate based on combined evidence improve up to 0.91453, whereas the fault-supporting rates based on each piece of single wavelet entropy evidence are $m_1(L_4) = 0.57583$, $m_2(L_4) = 0.30679$, $m_3(L_4) = 0.48923$, and $m_4(L_4) = 0.21239$. The combined evidence highlights the supporting rate of the fault line and weakens the supporting rate of the nonfault line. The fuzzy uncertainty of evidence is also weakened. Then, line L_4 is easy to identify as the fault line according to Equation (12.27).

The simulation results in Table 12.5 indicate that the fusion of four wavelet entropies improves the supporting rate for fault lines and avoids the misdiagnosis with single entropy evidence.

The study results of fault type identification for high-voltage transmission lines and fault line identification for noneffectively grounded distribution systems show that the information fusion technique based on wavelet entropies evidence has good prospects in fault diagnosis in power systems. In actual application, the fault-recording data are used to calculate the wavelet entropies. The wavelet entropies are taken as evidence that is fused to identify the fault accurately and reliably.

References

[1] Cheng X.Q., A Study of electric power transient signal detection and classification based on wavelet entropy measure. *Southwest Jiaotong University*, 2003.

[2] Zhang B., He Z.Y., Qian Q.Q., A Faulty phase selector based on wavelet energy entropy and fuzzy logic. *Power System Technology*, vol. **30**, no. 15, pp. 30–35, 2006.

[3] He Z.Y., Chen X.Q., A study of electric power system transient signals identification method based on multi-scales energy statistic and wavelet energy entropy. *Proceedings of the CSEE*, vol. **26**, no. 10, pp. 33–39, 2006.

[4] Chen X.Q., He Z.Y., Recoginition of power quality disturbance signals based on wavelet and wavelet entropy weight. *Electric Power Science and Engineering*, no. 1, 2006.

[5] He Z.Y., Fu L., Mai R.K. et al., Study on wavelet singular entropy and its application to faulty phase selection in hv transmission lines. *Proceedings of the CSEE*, vol. **27**, no. 1, pp. 31–36, 2007.

[6] Fu L., He Z.Y., Mai R.K. et al., Information fusion method of entropy evidences and its application to fault diagnosis in power system. *Proceedings of the CSEE*, vol. **28**, no. 13, pp. 64–69, 2008.

Appendix A

Simulation Models

A.1 500 kV transmission line simulation model

Figure A.1 shows the 500 kV transmission line simulation model. I use the frequency-dependent distributed model of the transmission line to accurately include the influences of the fault-induced transients. The parameters of the line are as follows.

The positive-sequence impedance is $Z_1 = 0.02473 + j0.2872\ \Omega/\text{km}$. The zero-sequence impedance is $Z_0 = 0.2245 + j1.02\ \Omega/\text{km}$. The positive-sequence shunt capacitance is $C_1 = 0.0132\,\mu\text{F/km}$. The zero-sequence shunt capacitance is $C_0 = 0.0090\,\mu\text{F/km}$. The busbar stray capacitance is $1\,\mu\text{F}$.

The breaker switching, capacitor switching, single phase to ground fault, first arc, lightning interruption, and lightning-induced fault are simulated in Figure A.1. The corresponding transient signals are shown in Figure A.2.

1. *Breaker switching*: In Figure A.1, switch 1, switch 2, and switch 4 are closed, and switch 3 is open in the beginning. At t = 0.2s, switch 3 is closed, and the breaker-switching transients are generated.
2. *Capacitor switching*: The phase-to-ground capacitor banks near power source A are switched on or off to generate the capacitor-switching transients.
3. *Single phase to ground fault.*
4. *First arc.*

The arc model used in this book is based on the energy balance of the arc column. This model describes the arc in air by a differential equation of the arc conductance, as shown in Equation (A.1).

$$\frac{dg_P}{dt} = \frac{1}{T_P}\left(G_P - g_P\right) \tag{A.1}$$

Wavelet Analysis and Transient Signal Processing Applications for Power Systems, First Edition. Zhengyou He.

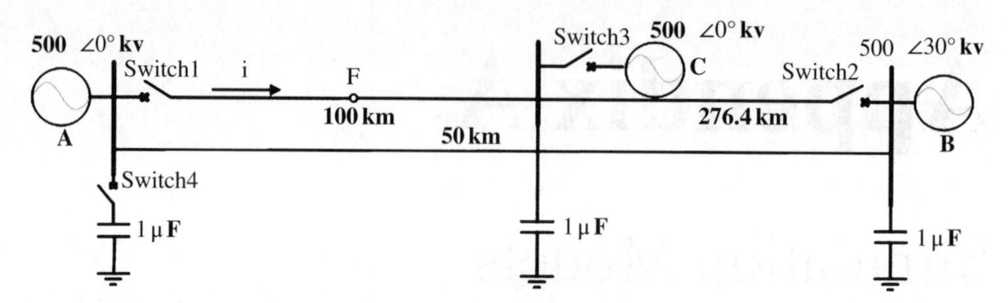

Figure A.1 500 kV transmission line simulation model

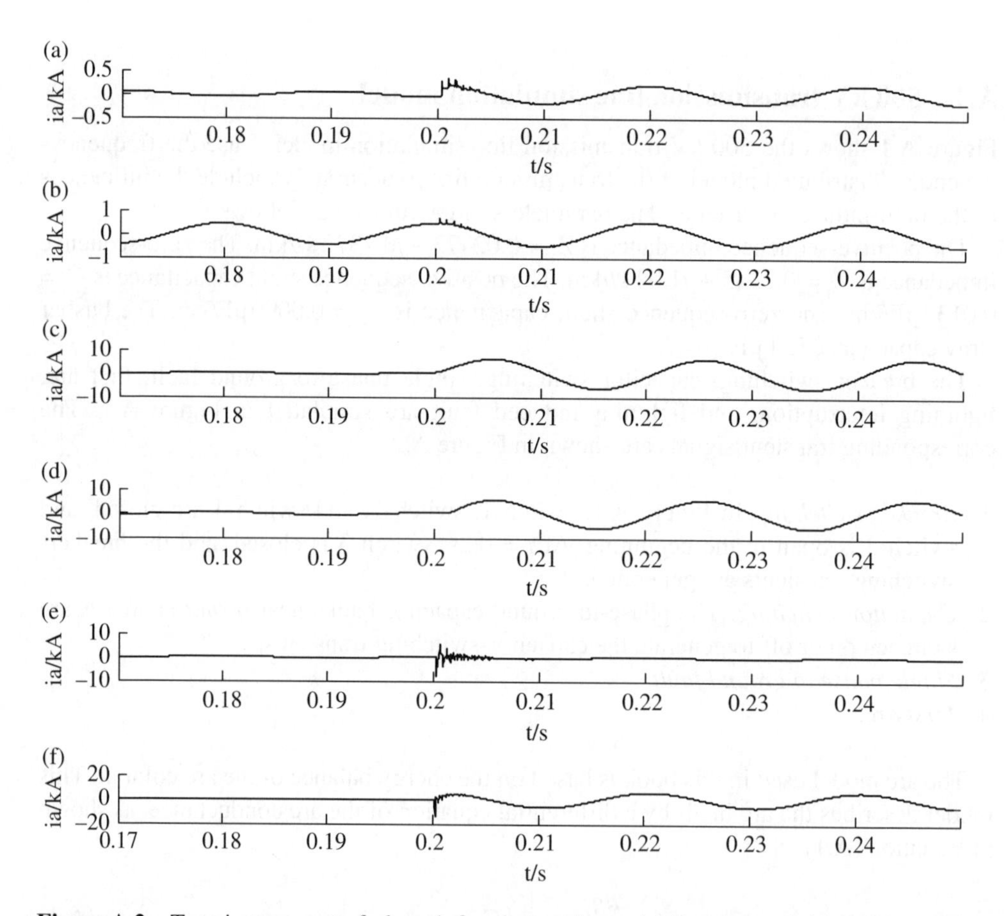

Figure A.2 Transient current of phase A for six transients: (a) Breaker switching, (b) capacitor switching, (c) single phase to ground fault, (d) first arc, (e) lightning disturbance, and (f) fault induced by lightning

where g_p is the time-varying arc conductance; G_p is the stationary arc conductance, which is defined as $G_p = \frac{|i|}{V_P L_P}$; T_P is the arc time constant, which is defined as $T_P = \frac{\alpha I_P}{L_P}$; and α is the coefficient, which is about 2.85×10^{-5} for the great current arc, $I_P = 1.4\,\text{kA}\sim14\,\text{kA}$ and $V_P = 15\,\text{V/cm}$. The arc length L_p is 10 m in general. The arc in Equation (A.1) is solved in the Laplace domain, as shown in Equation (A.2):

$$g_p(s) = \frac{G_p(s)}{T_P + 1} = \frac{G}{T_p + 1} \cdot |i(s)| \tag{A.2}$$

Lightning stroke includes nonfault lightning and faulty lightning. A current source controlled by lightning current is added at the fault point F to simulate the nonfault lightning stroke. The faulty lightning stroke is simulated by generating a short-circuit fault after the lightning occurs. A standard biexponential equation simulates the lightning currents; the equation is described as

$$\begin{aligned} i &= I_0\left(e^{-\alpha t} - e^{-\beta t}\right) \\ \alpha &= 1/\tau_2 \\ \beta &= 1/\tau_1 \end{aligned} \tag{A.3}$$

where, I_0 is the amplitude of lightning currents; and τ_1 and τ_2 are, respectively, the front time and time to half value.

A.2 The simplified 500 kV two-terminal power system: example 1

Figure A.3 shows a 500 kV power system that is simplified from Figure A.1. The bus-bar stray capacitance is 0.1 μF. Other parameters are the same as the parameters in Figure A.1.

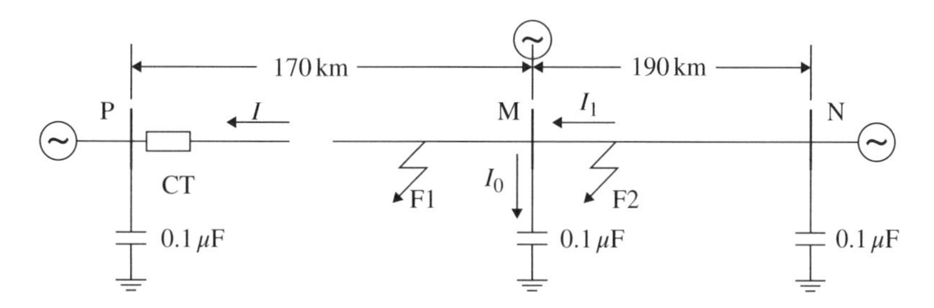

Figure A.3 The simplified power system: example 1

A.3 The simplified 500 kV two-terminal power system: example 2

Figure A.4 shows a 500 kV two-terminal power system with a line length of 300 km. The parameters of the line are as follows: the positive-sequence parameters are $r_1 = 0.035\ \Omega/\text{km}$, $x_1 = 0.424\ \Omega/\text{km}$, and $b_1 = 2.726 \times 10^{-6}$ S/km; and the zero-sequence parameters are $r_0 = 0.3\ \Omega/\text{km}$ and $x_0 = 1.143\ \Omega/\text{km}$.

A.4 Neutral ineffectively grounded power system

A radial power distribution network is described in Figure A.5. It is a typical 10 kV distribution system whose bus is fed from a 110 kV 50 Hz network via a Δy_n transformer. Five feeders emanate from the bus. The lengths of lines L_1, L_2, L_3, L_4, and L_5 are, respectively, 6 km, 12 km, 9 km, 17 km, and 20 km.

The line parameters are as follows: the positive-sequence parameters are $R_1 = 0.45\ \Omega/\text{km}$, $L_1 = 1.1714$ mH/km, and $C_1 = 0.061\ \mu\text{F/km}$; and the zero-sequence parameters are $R_0 = 0.70\ \Omega/\text{km}$, $L_0 = 3.9065$ mH/km, and $C_0 = 0.038\ \mu\text{F/km}$.

The Peterson coil, which provides 108% overcompensation of the ground fault current, is used. The parameters of the Peterson coil are $R_L = 6.777\ \Omega$ and $L = 0.2623$ H.

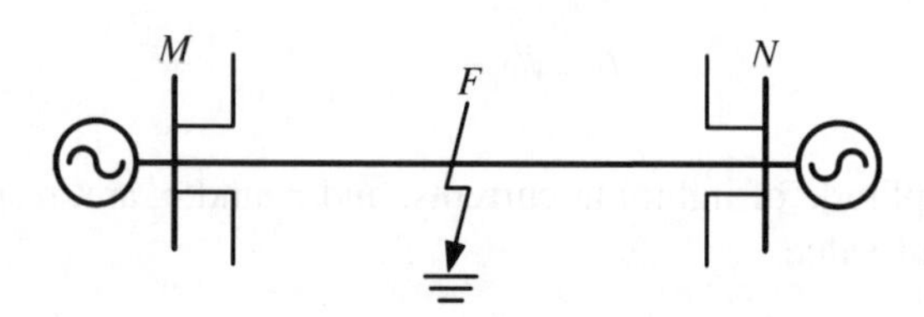

Figure A.4 The simplified power system: example 2

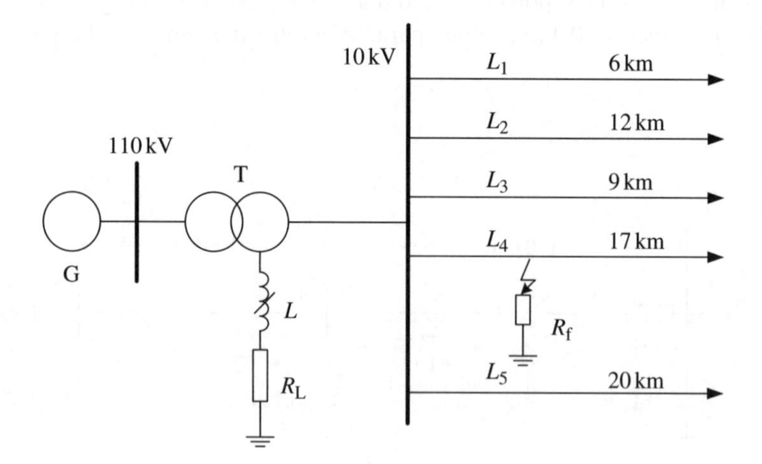

Figure A.5 Neutral ineffectively grounded power system

A.5 Distribution network with capacitor banks

Figure A.6 shows a distribution network model with capacitor banks, in which three feeders connect with busbar 2, busbar 3, and busbar 4. Busbar 1 is the public busbar. Three capacitor banks C1, C2, and C3 are installed at busbar 9, busbar 1, and busbar 16, respectively, in the network.

The parameters of the system model are shown in Tables A.1, A.2, and A.3.

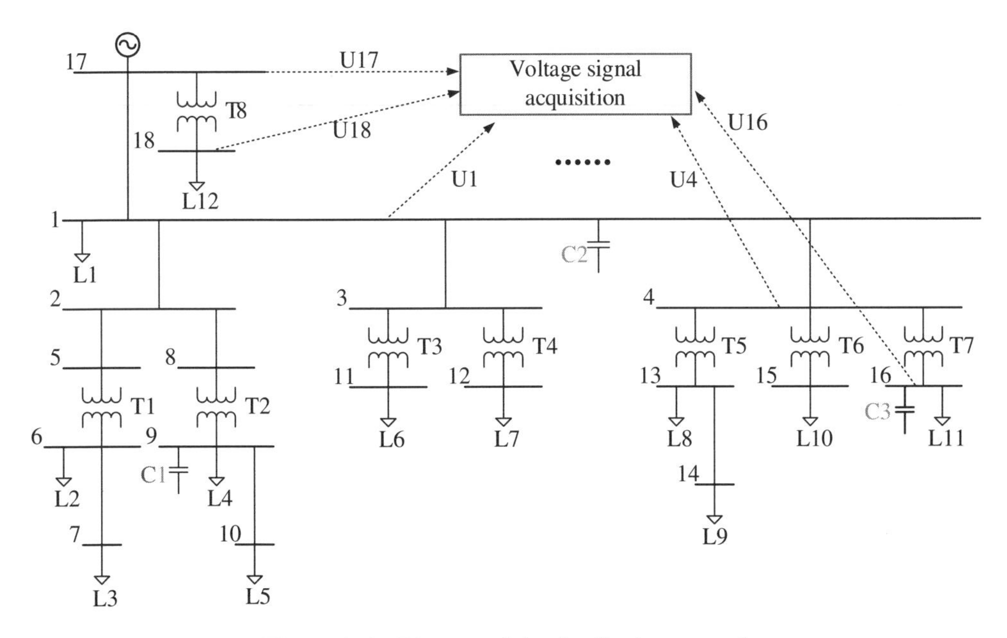

Figure A.6 Diagram of the distribution network

Table A.1 Parameters of the power source

Capacity, MVA	Voltage class, kV	Frequency, Hz
20	10.0	50.0

Table A.2 The load data

Load	Voltage, kV	Active power, MW	Load	Voltage, kV	Active power, MW
L1	10	4	L7	3.01	1.237
L2	0.35	0.578	L8	0.35	0.703
L3	0.35	0.168	L9	0.35	0.62
L4	0.35	0.831	L10	1.74	0.353
L5	0.35	0.084	L11	1.74	2.65
L6	0.35	0.084	L12	0.35	0.478

Table A.3 Transformer data

Transformer	Capacity, MVA	Ratio, kV	Transformer	Capacity, MVA	Ratio, kV
T1	1.5	10/0.35	T5	1.5	10/0.35
T2	1.5	10/0.35	T6	1.5	10/1.237
T3	1.25	10/0.35	T7	3.75	10/1.237
T4	1.725	10/3.01	T8	1.5	10/0.35

Table A.4 Models of power quality disturbances

Types	Models	Parameters
Normal signal	$v_1(t) = \sin(\omega_o t)$	$\omega_o = 2\pi f$
Voltage swell	$v_2(t) = \sin(\omega_o t) \times \{1 + \alpha[u(t-t_1) - u(t-t_2)]\}$	$0.1 \le \alpha \le 0.9$ $T \le t_2 - t_1 \le 6T$
Voltage sag	$v_3(t) = \sin(\omega_o t) \times \{1 - \alpha[u(t-t_1) - u(t-t_2)]\}$	$0.1 \le \alpha \le 0.9$ $T \le t_2 - t_1 \le 6T$
Voltage interruption	$v_4(t) = \sin(\omega_o t) \times \{1 - \alpha[u(t-t_1) - u(t-t_2)]\}$	$0.9 < \alpha \le 1$ $T \le t_2 - t_1 \le 6T$
Impulse transient	$v_5(t) = \sin(\omega_o t) + \alpha[u(t-t_1) - u(t-t_2)]e^{(-t+t_1)}$	$1 \le \alpha \le 2$ $\frac{T}{20} \le t_2 - t_1 \le \frac{T}{10}$
Oscillation transient	$v_6(t) = \sin\omega_o t + \alpha e^{-c(t-t_1)} \times [u(t-t_1) - u(t-t_2)]\sin(\beta\omega_o t)$	$0.1 \le \alpha \le 0.8$ $T \le t_2 - t_1 \le 3T$ $10 \le \beta \le 30$
Voltage notch	$v_7(t) = \sin(\omega_o t) - \sum_{i=0}^{10} \alpha(u(t_2 + i \cdot T) - u(t_1 + i \cdot T))$	$0.1 \le \alpha \le 0.4$ $\frac{T}{20} \le t_2 - t_1 \le \frac{T}{2}$
Harmonics	$v_8(t) = \sin(\omega_o t) + \alpha_3 \sin(3\omega_o t) + \alpha_5 \sin(5\omega_o t) + \alpha_7 \sin(7\omega_o t)$	$0.05 \le \alpha_i \le 0.3$
Voltage swell and harmonics	$v_9(t) = v_2(t) + v_8(t) - \sin(\omega_o t)$	—
Voltage sag and harmonics	$v_{10}(t) = v_3(t) + v_8(t) - \sin(\omega_o t)$	—

A.6 Power quality disturbances

Seven types of single disturbance and two types of multiple disturbances are considered, along with the normal signal: normal (type A), swell (type B), sag (type C), interruption (type D), spike (type E), oscillatory transients (type F), notch (type G), harmonic (type H), swell and harmonics (type I), and sag and harmonics (type J). See Table A.4 for the simulation models.

Author Index

Wavelet Analysis and Transient Signal Processing Applications for Power Systems, First Edition. Zhengyou He.

Subject Index